Lecture Notes in Mathematics　　2057

Editors:
J.-M. Morel, Cachan
B. Teissier, Paris

For further volumes:
http://www.springer.com/series/304

Andrzej Cegielski

Iterative Methods
for Fixed Point Problems
in Hilbert Spaces

 Springer

Andrzej Cegielski
University of Zielona Góra
Faculty of Mathematics,
  Computer Science and Econometrics
Zielona Góra, Poland

ISBN 978-3-642-30900-7   ISBN 978-3-642-30901-4 (eBook)
DOI 10.1007/978-3-642-30901-4
Springer Heidelberg New York Dordrecht London

Lecture Notes in Mathematics ISSN print edition: 0075-8434
                        ISSN electronic edition: 1617-9692

Library of Congress Control Number: 2012945521

Mathematics Subject Classification (2010): 47-02, 49-02, 65-02, 90-02, 47H09, 47J25, 37C25,
                        65F10, 65K15, 90C25

© Springer-Verlag Berlin Heidelberg 2012
This work is subject to copyright. All rights are reserved by the Publisher, whether the whole or part of
the material is concerned, specifically the rights of translation, reprinting, reuse of illustrations, recitation,
broadcasting, reproduction on microfilms or in any other physical way, and transmission or information
storage and retrieval, electronic adaptation, computer software, or by similar or dissimilar methodology
now known or hereafter developed. Exempted from this legal reservation are brief excerpts in connection
with reviews or scholarly analysis or material supplied specifically for the purpose of being entered
and executed on a computer system, for exclusive use by the purchaser of the work. Duplication of
this publication or parts thereof is permitted only under the provisions of the Copyright Law of the
Publisher's location, in its current version, and permission for use must always be obtained from Springer.
Permissions for use may be obtained through RightsLink at the Copyright Clearance Center. Violations
are liable to prosecution under the respective Copyright Law.
The use of general descriptive names, registered names, trademarks, service marks, etc. in this publication
does not imply, even in the absence of a specific statement, that such names are exempt from the relevant
protective laws and regulations and therefore free for general use.
While the advice and information in this book are believed to be true and accurate at the date of
publication, neither the authors nor the editors nor the publisher can accept any legal responsibility for
any errors or omissions that may be made. The publisher makes no warranty, express or implied, with
respect to the material contained herein.

Printed on acid-free paper

Springer is part of Springer Science+Business Media (www.springer.com)

To my family:
  Elżbieta, Joanna, Gustaw,
  Szymon and Katarzyna

Motto:
If you want to find the source, you have to go up, against the current

[John Paul II]

Preface

In this monograph we deal with iteration methods for finding fixed points (if they exist) of nonexpansive operators defined on a Hilbert space, i.e., operators T having the property

$$d(Tx, Ty) \leq cd(x, y) \text{ for all } x, y \in X \tag{1}$$

and for some constant $c \in [0, 1]$. The origin of these methods dates back to 1920, when Stefan Banach (1892–1945) formulated his famous contraction mapping principle. Banach proved that if $T : X \to X$ is a contraction (an operator satisfying (1) with $c < 1$) defined on a complete metric space, then T has a unique fixed point x^*, i.e., a point for which $Tx^* = x^*$. Furthermore, for any $x \in X$ a sequence $\{T^k x\}_{k=0}^{\infty}$ converges geometrically to x^*. Many practical problems can be reduced to finding a fixed point of a nonexpansive operator or a common fixed point of a family of nonexpansive operators. A simple example is a system of linear equations $Ax = b$. Any solution of this system can be identified with a common fixed point of (nonexpansive) operators of orthogonal projections onto hyperplanes corresponding to particular equations of the system.

Under some additional conditions a nonexpansive operator has a fixed point (see, e.g., the Browder–Göhde–Kirk fixed point theorem), but the Banach theorem does not guarantee the convergence of the sequence $\{T^k x\}_{k=0}^{\infty}$. Therefore, it is of great interest to develop methods for finding fixed points of nonexpansive operators. The first iterative method for solving a linear system was proposed in 1937 by a Polish mathematician, Stefan Kaczmarz (1890–1945), in a very short paper (3 pages) "Angenäherte Auflösung von Systemen linearer Gleichungen" ("Approximate solution of systems of linear equations") published in *Bulletin International de l'Académie Polonaise des Sciences et des Lettres*. A year later, an Italian mathematician, Gianfranco Cimmino (1908–1989), proposed another iterative method for a linear system. He published his result in a paper "Calcolo approssimato per le soluzioni dei sistemi di equazioni lineari" ("Approximate computation of the solutions of linear systems") in a *La Ricerca Scientifica*. As opposed to earlier methods for solving systems of linear equations, both methods are motivated rather by geometrical operations than algebraic ones. The main operation in both methods is a cyclic (in the Kaczmarz method) or simultaneous (in the Cimmino method) orthogonal projection onto hyperplanes described by particular equations of the linear system. The results of Kaczmarz and Cimmino were disregarded for several decades except few references. A great interest in these results started in the 1970s, when it suddenly turned out that the Kaczmarz and Cimmmino methods can be efficiently applied in computed tomography (CT), because the mathematical model of CT can be reduced to a solution of large systems of linear equations with sparse and unstructured matrices, and the methods behave well for such systems. At the end of the twentieth century the interest in the Kaczmarz and Cimmino results was still increasing. Both results have become fundamental to modern iterative methods for fixed point problems for nonexpansive operators.

The third result which influenced the development of this area concerns alternative projections onto two subspaces of a Hilbert space. The result was published in 1950 by John von Neumann in a paper "Functional Operators—Vol. II. The Geometry of Orthogonal Spaces" in *Annals of Mathematical Studies*. Von Neumann proved that a sequence generated by his method converges to the projection of the starting point onto the intersection of the subspaces. The results of Kaczmarz, Cimmino, and von Neumann have been generalized several times in the last decades. Today it is known that the convergence holds for an essentially wider class of operators than orthogonal projections onto hyperplanes, e.g., for nonexpansive operators or for quasi-nonexpansive operators, satisfying some additional conditions. All three methods belong today to classical iterative methods for finding fixed points of nonexpansive operators defined on a Hilbert space. These methods served as the basis for several methods, e.g., for: the Landweber method, projected Landweber method, Douglas–Rachford method, sequential projection methods, methods of cyclic and simultaneous subgradient projections, Dos Santos method of extrapolated simultaneous subgradient projections, reflection-projection method, surrogate projection method, and many others. Irrespective of their theoretical value, they have found application in many areas of mathematics, physics, and technology. The most spectacular application of the methods is an intensity-modulated radiation therapy (IMRT).

Iterative methods for finding fixed points of nonexpansive operators in Hilbert spaces have been described in many publications. In this monograph we try to present the methods in a consolidated way. We introduce some classes of operators, give their properties, define iterative methods generated by operators from these classes, and present general convergence theorems. On this basis we present the conditions under which particular methods converge. A large part of the results presented in this monograph can be found in various forms in the literature. We tried, however, to show that the convergence of a big class of iteration methods follows from general properties of some classes of operators and from some general convergence theorems, in particular from Opial's theorem or from its modifications. This theorem was presented in 1967 by a Polish mathematician, Zdzisław Opial (1930–1974), in a paper "Weak convergence of the sequence of successive approximations for nonexpansive mappings" published in the *Bulletin of the American Mathematical Society*.

In this monograph we work with operators defined on a real Hilbert space, although a part of the results presented herein holds for wider classes of spaces. The monograph is divided into five chapters. In Chap. 1, we recall basic definitions and facts from linear algebra, functional analysis, and convex analysis which we apply in the further part of the monograph. In Chap. 2, we introduce some classes of algorithmic operators (i.e., operators which generate some algorithms or iterative methods): nonexpansive operators, quasi-nonexpansive operators, relaxed quasi-nonexpansive operators, cutter operators, firmly nonexpansive operators, metric projection, relaxed firmly nonexpansive operators, averaged operators, strongly nonexpansive operators, and generalized relaxations of algorithmic operators. Then we present the properties of these classes, in particular the relationships among these classes and their closedness with respect to some algebraic operations on the operators from these classes. In Chap. 3, we analyze the convergence properties of the sequences generated by the operators introduced in Chap. 2. Opial's theorem, its generalizations, and modifications for sequences generated by a subclass of the class of quasi-nonexpansive operators play the key role here. In Chap. 4, we apply the properties of classes of operators presented in Chap. 2 to constructions of operators used in many iterative methods for fixed point problems. In this chapter we also give the properties of the following: the alternating projection, simultaneous projection, cyclic projection, Landweber operator, projected Landweber operator, and some generalizations and extrapolations of these operators. In Chap. 5, we apply the results presented in the previous chapters in order to show the convergence of sequences generated by many iterative methods for fixed point problems, some of which are known in the literature, but several are new. The notions and facts presented in the book are illustrated with 61 figures. Each chapter is followed by several exercises.

Many persons have looked through successive versions of the monograph. I am deeply grateful for their helpful remarks. In particular, I would like to express my thanks to Prof. Simeon Reich from the Technion (Israel), Prof. Yair Censor from the University of Haifa (Israel), Prof. Diethard Pallaschke from the University of Karlsruhe (Germany), and Prof. Heinz Bauschke from the University of British Columbia in Okanagan (Canada). Their valuable remarks have contributed to substantial improvements in successive versions of the monograph and have consolidated me in my aim to give the monograph its final shape. I am also very grateful to my colleagues from the University of Zielona Góra, who looked through the final version of the monograph and also made some useful remarks: Prof. Michał Kisielewicz, Prof. Krzysztof Przesławski, and Prof. Jerzy Motyl. I would like to express my thanks to my Ph.D. student, Rafał Zalas, for his help in the preparation of figures which illustrate the notions and facts presented herein and to Danuta Michalak for the technical composition of the monograph. Last but not least, I would like to express my deep gratitude to my wife, Elżbieta, for her understanding and assistance during the preparation of the monograph.

Zielona Góra, Poland   Andrzej Cegielski
April 2012

Contents

Chapter 1
Introduction

1.1 Background

In this section we present the notation, definitions and basic facts of convex subsets and convex functions defined on a Hilbert space, convergence and differentiation properties, properties of matrices, etc., which will be used in further parts of the book.

1.1.1 Hilbert Space

A linear space $\mathcal{H}$ with an inner product $\langle \cdot, \cdot \rangle$, which is complete with respect to the norm $\|\cdot\| := \sqrt{\langle \cdot, \cdot \rangle}$ induced by this inner product is called a *Hilbert space*.

In what follows, we consider a real Hilbert space $\mathcal{H}$, and denote by X a nonempty closed convex subset of $\mathcal{H}$ and by I a finite subset $\{1, 2, \ldots m\} \subseteq \mathbb{N}$, where $m \in \mathbb{N}$.

1.1.1.1 Properties of the Inner Product

The following equalities hold for all $x, y \in \mathcal{H}$:

$$\|x + y\|^2 = \|x\|^2 + \|y\|^2 + 2\langle x, y \rangle, \tag{1.1}$$

$$\|x - y\|^2 = \|x\|^2 + \|y\|^2 - 2\langle x, y \rangle \tag{1.2}$$

and

$$\langle x + y, x - y \rangle = \|x\|^2 - \|y\|^2. \tag{1.3}$$

Cauchy–Schwarz inequality. The following inequality holds for all $x, y \in \mathcal{H}$:

$$-\|x\| \, \|y\| \leq \langle x, y \rangle \leq \|x\| \, \|y\|. \tag{1.4}$$

A. Cegielski, *Iterative Methods for Fixed Point Problems in Hilbert Spaces*,
Lecture Notes in Mathematics 2057, DOI 10.1007/978-3-642-30901-4_1,
© Springer-Verlag Berlin Heidelberg 2012

Furthermore,

(i) $\langle x, y \rangle = \|x\| \|y\|$ if and only if x and y are *positive linearly dependent vectors*, i.e., $\alpha x = \beta y$ for some $\alpha, \beta \geq 0$,

(ii) $\langle x, y \rangle = -\|x\| \|y\|$ if and only if x and y are *negative linearly dependent vectors*, i.e., $\alpha x = -\beta y$ for some $\alpha, \beta \geq 0$.

Moreover, one can take $\alpha = \|y\|$ and $\beta = \|x\|$ in (i) and in (ii).

Triangle inequality. The following inequality holds for all $x, y \in \mathcal{H}$:

$$\big| \|x\| - \|y\| \big| \leq \|x + y\| \leq \|x\| + \|y\|. \tag{1.5}$$

Furthermore,

(i) $\|x + y\| = \|x\| + \|y\|$ if and only if x and y are positive linearly dependent,

(ii) $\|x + y\| = \big| \|x\| - \|y\| \big|$ if and only if x and y are negative linearly dependent.

Property (i) is known as the *strict convexity of the norm* and can also be formulated equivalently in the form:

(iii) For any $\lambda \in (0, 1)$ there holds

$$\|x\| = \|y\| = \|(1 - \lambda)x + \lambda y\| \implies x = y. \tag{1.6}$$

Convexity of the norm. Let $w = (\omega_1, \ldots, \omega_m) \in \mathbb{R}_+^m$ (all coordinates of w are nonnegative). The following inequality holds for all $x_i \in \mathcal{H}$, $i = 1, 2, \ldots, m$,

$$\left\| \sum_{i=1}^m \omega_i x_i \right\| \leq \sum_{i=1}^m \omega_i \|x_i\|. \tag{1.7}$$

If, moreover, $\omega_i > 0$, $i = 1, 2, \ldots, m$, then the equality holds in (1.7) if and only if all x_i, $i = 1, 2, \ldots, m$, are pairwise positive linearly dependent.

Let $w = (\omega_1, \ldots, \omega_m) \in \mathbb{R}^m$. The following equalities hold for all $x_i \in \mathcal{H}$, $i = 1, 2, \ldots, m$,

$$\left\| \sum_{i=1}^m \omega_i x_i \right\|^2 = \sum_{i,j=1}^m \omega_i \omega_j \langle x_i, x_j \rangle = \sum_{i=1}^m \omega_i^2 \|x_i\|^2 + \sum_{i,j=1, i \neq j}^m 2\omega_i \omega_j \langle x_i, x_j \rangle.$$

Parallelogram law. The following equality holds for all $x, y \in \mathcal{H}$:

$$\|x + y\|^2 + \|x - y\|^2 = 2(\|x\|^2 + \|y\|^2). \tag{1.8}$$

Theorem 1.1.1. *A Banach space with a norm $\|\cdot\|$ is a Hilbert space with an inner product defined by the equality*

$$\langle x, y \rangle = \frac{1}{2}(\|x\|^2 + \|y\|^2 - \|x - y\|^2),$$

if and only if the parallelogram law holds.

1.1.1.2 Examples of Hilbert Spaces

Example 1.1.2. The *Euclidean space* $\mathbb{R}^n := \{x = (\xi_1, \xi_2, \ldots, \xi_n) : \xi_j \in \mathbb{R}, j = 1, 2, \ldots, n\}$. An element $x = (\xi_1, \xi_2, \ldots, \xi_n)$ of $\mathbb{R}^n$ can be identified with a matrix

of type $n \times 1$ (column vector), i.e., we write $x = \begin{bmatrix} \xi_1 \\ \xi_2 \\ \vdots \\ \xi_n \end{bmatrix}$ or $x = [\xi_1, \xi_2, \ldots, \xi_n]^\mathsf{T}$.

The Euclidean space is equipped with the *standard inner product* defined by

$$\langle x, y \rangle := x^\mathsf{T} y = \sum_{j=1}^{n} \xi_j \eta_j,$$

where $x = (\xi_1, \ldots, \xi_n)$, $y = (\eta_1, \ldots, \eta_n)$, and with the norm $\|x\| := (x^\mathsf{T} x)^{\frac{1}{2}}$.

Example 1.1.3. The space $\mathbb{R}^m$ with the inner product induced by a positive definite matrix $G = [g_{ij}]_{m \times m}$, defined by

$$\langle w, v \rangle_G := w^\mathsf{T} G v = \sum_{i=1}^{m} \sum_{j=1}^{m} g_{ij} \omega_i v_j, \tag{1.9}$$

where $w = (\omega_1, \ldots, \omega_m)$, $v = (v_1, \ldots, v_m)$, and with the norm $\|u\|_G := (u^\mathsf{T} G u)^{\frac{1}{2}}$.

Example 1.1.4. The space l_2 of real sequences $x = (\xi_1, \xi_2, \ldots)$ which are square-summable, i.e.,

$$l_2 := \{x : \xi_k \in \mathbb{R}, k = 1, 2, \ldots \text{ and } \sum_{k=1}^{\infty} \xi_k^2 < \infty\},$$

with the inner product $\langle x, y \rangle := \sum_{k=1}^{\infty} \xi_k \eta_k$, and with the norm $\|x\| := (\sum_{k=1}^{\infty} \xi_k^2)^{\frac{1}{2}}$.

Example 1.1.5. $L_2([a, b])$, where $a, b \in [-\infty, +\infty]$, $a < b$—the space of (equivalence classes of) Lebesgue measurable functions $f : \mathbb{R} \to \mathbb{R}$ which are square-integrable, i.e.,

$$L_2(a, b) := \{f : \int_a^b f^2(x) dx < \infty\}$$

with the inner product

$$\langle f, g \rangle := \int_a^b f(x)g(x)dx,$$

and with the norm

$$\|f\| := (\int_a^b f^2(x)dx)^{\frac{1}{2}}.$$

Example 1.1.6. Let $\mathcal{H}_i$ be a Hilbert space with an inner product $\langle \cdot, \cdot \rangle_{\mathcal{H}_i}$ and with the norm $\|\cdot\|_{\mathcal{H}_i}$ induced by this inner product, $i \in I := \{1, 2, \ldots, m\}$. Let $\mathcal{H} = \mathcal{H}_1 \times \ldots \times \mathcal{H}_m$. The function $\langle \cdot, \cdot \rangle : \mathcal{H} \times \mathcal{H} \to \mathbb{R}$ defined by

$$\langle x, y \rangle_w := \sum_{i=1}^m \omega_i \langle x_i, y_i \rangle_{\mathcal{H}_i}, \qquad (1.10)$$

where $x = (x_1, \ldots, x_m)$, $y = (y_1, \ldots, y_m) \in \mathcal{H}$, $w = (\omega_1, \ldots, \omega_m) \in \mathbb{R}_{++}^m$ (all coordinates of w are positive), is an inner product in $\mathcal{H}$, and the function $\|\cdot\| : \mathcal{H} \to \mathbb{R}$ defined by the equality

$$\|x\|_w := (\sum_{i=1}^m \omega_i \|x_i\|_{\mathcal{H}_i}^2)^{\frac{1}{2}}$$

is the norm in $\mathcal{H}$ induced by this inner product. The space $\mathcal{H}$ with the inner product $\langle \cdot, \cdot \rangle_w$ defined by (1.10) is a Hilbert space and is called a *product Hilbert space*.

1.1.1.3 Weak Convergence in Hilbert Spaces

We say that a sequence $\{x^k\}_{k=0}^\infty$ of elements of a Hilbert space $\mathcal{H}$ *converges weakly* to $x \in \mathcal{H}$ if for any $y \in \mathcal{H}$ the sequence $\{\langle y, x^k \rangle\}_{k=0}^\infty$ converges to $\langle y, x \rangle$. We call the point x a *weak limit* of the sequence $\{x^k\}_{k=0}^\infty$ and write $x^k \rightharpoonup x$. If a subsequence $\{x^{n_k}\}_{k=0}^\infty \subseteq \{x^k\}_{k=0}^\infty$ converges weakly to x, then x is called a *weak cluster point* of the sequence $\{x^k\}_{k=0}^\infty$. We say that $\{x^k\}_{k=0}^\infty$ *converges* (*strongly*) to x if $\lim_{k \to \infty} \|x^k - x\| = 0$.

Weakly convergent sequences have the following properties:

1. A weakly convergent sequence $\{x^k\}_{k=0}^\infty \subseteq \mathcal{H}$ has exactly one weak limit.
2. A weakly convergent sequence $\{x^k\}_{k=0}^\infty \subseteq \mathcal{H}$ is bounded.
3. A bounded sequence $\{x^k\}_{k=0}^\infty \subseteq \mathcal{H}$ includes a weakly convergent subsequence.
4. If a sequence $\{x^k\}_{k=0}^\infty \subseteq \mathcal{H}$ is bounded and has exactly one weak cluster point $x \in \mathcal{H}$, then $x^k \rightharpoonup x$.
5. If a sequence $\{x^k\}_{k=0}^\infty$ converges to $x \in \mathcal{H}$, then it converges weakly to $x \in \mathcal{H}$.
6. A weakly convergent sequence $\{x^k\}_{k=0}^\infty$ of a finite dimensional Hilbert space $\mathcal{H}$ is convergent.

Note that a bounded sequence $\{x^k\}_{k=0}^{\infty}$ of a Hilbert space $\mathcal{H}$ needs not to include a convergent subsequence.

1.1.2 Notations and Basic Facts

1.1.2.1 Euclidean Space

Elements of the Euclidean space $\mathbb{R}^n$ can be represented as vectors with real coordinates, e.g., $x = (\xi_1, \ldots, \xi_n)$, $y = (\eta_1, \ldots, \eta_n)$, $z = (\zeta_1, \zeta_2, \ldots, \zeta_n)$ or as $n \times 1$ matrices (column vectors), e.g., $x = [\xi_1, \ldots, \xi_n]^\top$, $y = [\eta_1, \ldots, \eta_n]^\top$, $z = [\zeta_1, \zeta_2, \ldots, \zeta_n]^\top$, where ξ_j, η_j, ζ_i are *coordinates* of x, y and z respectively, $j = 1, 2, \ldots, n$. We usually denote the coordinates of a vector in the Euclidean space with Greek letters. Let $x, y \in \mathbb{R}^n$. Then, $x \geq 0$ means that all coordinates of the vector x are nonnegative and $x > 0$ means that all coordinates of the vector x are positive (the symbols $x \leq 0$ and $x < 0$ are defined similarly). Furthermore, $x \leq y$ denotes that $y - x \geq 0$ and $x \geq y$ denotes that $x - y \geq 0$. We use the following notation: $\max\{x, y\} := (\max\{\xi_1, \eta_1\}, \ldots, \max\{\xi_n, \eta_n\})$; $\min\{x, y\} := (\min\{\xi_1, \eta_1\}, \ldots, \min\{\xi_n, \eta_n\})$; $x_+ := \max\{x, 0\}$; $x_- := \max\{-x, 0\}$; $x = x_+ - x_-$; $\mathbb{R}_+^n := \{x \in \mathbb{R}^n : x \geq 0\}$ denotes the *nonnegative orthant*; $\mathbb{R}_-^m := \{x \in \mathbb{R}^n : x \leq 0\}$ denotes the *nonpositive orthant*; $\mathbb{R}_{++}^m := \{x \in \mathbb{R}^n : x > 0\}$ denotes the *positive orthant*; $e_j := (0, \ldots, 0, 1, 0, \ldots, 0) \in \mathbb{R}^m$ denotes the jth *unit vector*, i.e., a vector, for which the jth coordinate equals 1 and the others are zeros; e denotes a vector with all coordinates equal to 1, i.e., $e = (1, \ldots, 1)$; $\Delta_m := \{u \in \mathbb{R}^m : u \geq 0, e^\top u = 1\}$ denotes the *standard simplex*; $\mathbb{N}$ denotes the set of positive integers; For a finite subset $J \subseteq \mathbb{N}$, the symbol $|J|$ denotes the number of elements of J.

1.1.2.2 Subsets of a Hilbert Space

Let $V \subseteq \mathcal{H}$. The subset

$$V^\perp := \{y \in \mathcal{H} : \langle y, x \rangle = 0 \text{ for all } x \in V\}$$

denotes the *subspace orthogonal to* V. Let $V \subseteq \mathcal{H}$ be a closed subspace. Then $\mathcal{H} = V \oplus V^\perp$, i.e., for any $x \in \mathcal{H}$, there are unique $v \in V$ and $w \in V^\perp$ such that $x = v + w$. The subset

$$B(x, \rho) := \{y \in \mathcal{H} : \|y - x\| \leq \rho\}$$

denotes a *ball* with a centre $x \in \mathcal{H}$ and radius $\rho > 0$. C' denotes the complement of a subset $C \subseteq \mathcal{H}$, i.e., $C' = \mathcal{H} \backslash C$. The subset

$$C + D := \{z \in \mathcal{H} : z = x + y, x \in C, y \in D\}$$

denotes the *Minkowski sum* of $C, D \subseteq \mathcal{H}$. In particular,

$$C + a := C + \{a\} = \{x + a : x \in C\},$$

where $C \subseteq \mathcal{H}$ and $a \in \mathcal{H}$. The subsets int C, cl C and bd C denote the *interior*, the *closure* and the *boundary* of a subset $C \subseteq \mathcal{H}$, respectively. The subset

$$S(x, \rho) := \text{bd } B(x, \rho) = \{y \in \mathcal{H} : \|y - x\| = \rho\}$$

denotes a *sphere* with a centre $x \in \mathcal{H}$ and radius $\rho > 0$. Lin S and Lin a denote a linear subspace generated by the subset $S \subseteq \mathcal{H}$ and by the vector $a \in \mathcal{H}$, respectively. The subset

$$[a, b] := \{z \in \mathcal{H} : z = (1 - \lambda)a + \lambda b, \lambda \in [0, 1]\}$$

denotes a *segment* with endpoints $a, b \in \mathcal{H}$. The subset

$$\operatorname*{Argmin}_{x \in C} f(x) := \{z \in C : f(z) \leq f(x) \text{ for all } x \in C\},$$

where $C \subseteq \mathcal{H}$ and $f : C \to \mathbb{R}$ is called a *subset of minimizers* of a function f. An element of $\operatorname{Argmin}_{x \in C} f(x)$ is called a *minimizer* of f and is denoted by $\operatorname{argmin}_{x \in C} f(x)$. The subset

$$H(a, \beta) := \{x \in \mathcal{H} : \langle a, x \rangle = \beta\},$$

where $a \in \mathcal{H}$, $a \neq 0$ and $\beta \in \mathbb{R}$ is called a *hyperplane* in a Hilbert space $\mathcal{H}$. The hyperplane $H(a, \beta)$ is the boundary of two *half-spaces*

$$H_+(a, \beta) := \{x \in \mathcal{H} : \langle a, x \rangle \geq \beta\}$$

and

$$H_-(a, \beta) := \{x \in \mathcal{H} : \langle a, x \rangle \leq \beta\}.$$

A subset $K \subseteq \mathcal{H}$ is called an *affine subspace* if for all $x, y \in K$ and for any $\lambda \in \mathbb{R}$ we have $(1 - \lambda)x + \lambda y \in K$. The number

$$\sphericalangle(a, b) := \arccos \frac{\langle a, b \rangle}{\|a\| \cdot \|b\|}$$

denotes an *angle* between nonzero vectors $a, b \in \mathcal{H}$.

1.1.2.3 Functions

Let $X \subseteq \mathcal{H}$, $f_i : X \to \mathbb{R}$, $i \in I$. The function $f := \max_{i \in I} f_i$ is defined by

$$f(x) := \max\{f_i(x) : i \in I\}$$

In particular, $f_+ := \max\{f, 0\}$, $f_- := \max\{-f, 0\}$.

Let $f : X \to \mathbb{R}$. The subset

$$S(f, \alpha) = \{x \in X : f(x) \leq \alpha\}$$

is called a *sublevel set* of f at a level $\alpha \in \mathbb{R}$. The subset

$$\text{epi } f := \{(x, \rho) \in X \times \mathbb{R} : \rho \geq f(x)\} \subseteq X \times \mathbb{R}$$

is called an *epigraph* of f. We say that f is *lower semi-continuous* at $x \in X$ if $\liminf_{y \to x} f(y) \geq f(x)$. We say that f is *weakly lower semi-continuous* at $x \in X$ if $\liminf_{y \to x} f(y) \geq f(x)$. We say that a function $f : \mathcal{H} \to \mathbb{R}$ is *coercive* if $\lim_{\|x\| \to \infty} f(x) = +\infty$. We say that a function $f : X \to \mathbb{R}$ attains the *global minimum* at a point $x^* \in X$ if $f(x^*) \leq f(x)$ for all $x \in X$. We say that a function $f : X \to \mathbb{R}$ attains a *local minimum* at a point $x^* \in X$ if there exists $\delta > 0$ such that $f(x^*) \leq f(x)$ for all $x \in X \cap B(x^*, \delta)$.

Theorem 1.1.7. *A function $f : X \to \mathbb{R}$ is (weakly) lower semi-continuous if and only if epi f is a (weakly) closed subset of $X \times \mathbb{R}$.*

Theorem 1.1.8. *If a function $f : X \to \mathbb{R}$ is weakly lower semi-continuous on a weakly closed and bounded subset $X \subseteq \mathcal{H}$, then f is bounded from below and attains its minimum.*

Theorem 1.1.9. *If a function $f : \mathcal{H} \to \mathbb{R}$ is weakly lower semi-continuous and coercive and $X \subseteq \mathcal{H}$ is weakly closed, then f attains the global minimum on X.*

Definition 1.1.10. If a limit $\lim_{t \downarrow 0} \dfrac{f(x + ts) - f(x)}{t}$ exists, then we call it a *directional derivative* of a function $f : \mathcal{H} \to \mathbb{R}$ at a point $x \in \mathcal{H}$ and in a direction $s \in \mathcal{H}$ and we denote it by $f'(x, s)$.

If a function $f : \mathcal{H} \to \mathbb{R}$ attains minimum at $x \in \mathcal{H}$ and $f'(x, s)$ exists for some $s \in \mathcal{H}$, then $f'(x, s) \geq 0$.

Definition 1.1.11. We say that a function $f : \mathcal{H} \to \mathbb{R}$ is *Fréchet-differentiable* or, shortly, *differentiable* at $x \in \mathcal{H}$ if there exists $y \in \mathcal{H}$ such that

$$f(x + h) = f(x) + \langle y, h \rangle + o(\|h\|),$$

where $\lim_{t \to 0} \dfrac{o(t)}{t} = 0$. The element y is called a *derivative* (or *differential*) of f at x and is denoted by $f'(x)$.

The derivative of a differentiable function is uniquely determined.

Definition 1.1.12. We say that a function $f : \mathcal{H} \to \mathbb{R}$ is *Gâteaux-differentiable* at a point $x \in \mathcal{H}$ if it has directional derivatives $f'(x, s)$ for all $s \in \mathcal{H}$ and

$$f'(x, s) = \langle g, s \rangle$$

holds for some $g \in \mathcal{H}$. The element g is called a *Gâteaux-derivative* or *Gâteaux-differential* of f at x and is denoted by $Df(x)$.

The Gâteaux-differential of a Gâteaux-differentiable function is uniquely determined.

Theorem 1.1.13. *If a function $f : \mathcal{H} \to \mathbb{R}$ is Fréchet-differentiable at $x \in \mathcal{H}$, then it is Gâteaux-differentiable at x and $f'(x) = Df(x)$.*

Theorem 1.1.14. *If a function $f : \mathcal{H} \to \mathbb{R}$ is Gâteaux-differentiable in a neighborhood of x and the Gâteaux-differential Df is continuous at x, then f is Fréchet-differentiable at x.*

If $\mathcal{H}$ is a Euclidean space ($\mathcal{H} = \mathbb{R}^n$ and $\langle \cdot, \cdot \rangle$ is the standard inner product), then the derivative $Df(x)$ is called a *gradient* of f at the point x and is denoted by $\nabla f(x)$. There holds the equality $\nabla f(x) = (\frac{\partial f(x)}{\partial \xi_1}, \ldots \frac{\partial f(x)}{\partial \xi_n})$, where $\frac{\partial f(x)}{\partial \xi_j}$ denotes a *partial derivative* of f at the point $x = (\xi_1, \ldots, \xi_n)$ with respect to ξ_j. Let $x \in \mathbb{R}^n$ and assume that $f : \mathbb{R}^n \to \mathbb{R}$ has partial derivatives of the second order $\frac{\partial^2 f}{\partial \xi_i \partial \xi_j}(x)$ for all $i, j = 1, 2, \ldots, n$. Then the matrix

$$\nabla^2 f(x) := \begin{bmatrix} \frac{\partial^2 f}{\partial \xi_1^2}(x) & \cdots & \frac{\partial^2 f}{\partial \xi_1 \partial \xi_n}(x) \\ \cdots & \cdots & \cdots \\ \frac{\partial^2 f}{\partial \xi_n \partial \xi_1}(x) & \cdots & \frac{\partial^2 f}{\partial \xi_n^2}(x) \end{bmatrix}$$

is called the *Hessian* of f at ξ.

Theorem 1.1.15 (Schwarz). *Let $f : \mathbb{R}^n \to \mathbb{R}$. If $\frac{\partial^2 f}{\partial \xi_i \partial \xi_j}$ are continuous at a point $x \in \mathbb{R}^n$, $i, j = 1, 2, \ldots, n$, then the Hessian is a symmetric matrix.*

1.1.2.4   Operators

Let $\mathcal{H}, \mathcal{H}_1, \mathcal{H}_2, \mathcal{H}_3$ be Hilbert spaces. The operator $\mathrm{Id} : \mathcal{H} \to \mathcal{H}$ denotes the *identity*, i.e., $\mathrm{Id}\, x = x$ for all $x \in \mathcal{H}$.

Definition 1.1.16. Let $A : \mathcal{H}_1 \to \mathcal{H}_2$ be a bounded linear operator. The number

$$\|A\| := \sup_{x : 0 < \|x\| \leq 1} \frac{\|Ax\|_{\mathcal{H}_2}}{\|x\|_{\mathcal{H}_1}},$$

where $\|\cdot\|_{\mathcal{H}_1}$ and $\|\cdot\|_{\mathcal{H}_2}$ denote the norms in $\mathcal{H}_1$ and $\mathcal{H}_2$, respectively, is called a *norm* of A in $L(\mathcal{H}_1, \mathcal{H}_2)$.

Definition 1.1.17. Let $A : \mathcal{H}_1 \to \mathcal{H}_2$ be a bounded linear operator. An operator $A^* : \mathcal{H}_2 \to \mathcal{H}_1$ with the property

$$\langle Ax, y \rangle_{\mathcal{H}_2} = \langle x, A^*y \rangle_{\mathcal{H}_1}$$

for all $x \in \mathcal{H}_1$ and $y \in \mathcal{H}_2$, where $\langle \cdot, \cdot \rangle_{\mathcal{H}_1}$ and $\langle \cdot, \cdot \rangle_{\mathcal{H}_2}$ are inner products in $\mathcal{H}_1$ and $\mathcal{H}_2$, respectively, is called an *adjoint operator*.

For every A an adjoint operator A^* exists and is uniquely determined. Futhermore, A^* is a bounded linear operator.

Definition 1.1.18. We say that a bounded linear operator $A : \mathcal{H} \to \mathcal{H}$ is *unitary* if $A^*A = \mathrm{Id}$ holds.

Definition 1.1.19. We say that a bounded linear operator $A : \mathcal{H} \to \mathcal{H}$ is *self-adjoint*, if $A^* = A$.

Definition 1.1.20. We say that a self-adjoint operator $A : \mathcal{H} \to \mathcal{H}$ is *nonnegative* if $\langle Ax, x \rangle \geq m \|x\|^2$ for some constant $m \geq 0$ and for all $x \in \mathcal{H}$. If $m > 0$, then we say that A is *positive* (or *elliptic*).

Theorem 1.1.21. *For any bounded linear operator $A : \mathcal{H}_1 \to \mathcal{H}_2$ it holds*

$$\|A\|^2 = \|A^*\|^2 = \|A^*A\| = \|AA^*\|.$$

Theorem 1.1.22. *If a bounded linear operator $A : \mathcal{H} \to \mathcal{H}$ is self-adjoint, then $\|A\| = \sup_{\|x\| \leq 1} \langle Ax, x \rangle$.*

Theorem 1.1.23. *Let $A : \mathcal{H}_1 \to \mathcal{H}_2$ be a bounded linear operator. Then the operators AA^* and A^*A are nonnegative.*

A number $\lambda \in \mathbb{R}$ is called an *eigenvalue* of a bounded linear operator $A : \mathcal{H} \to \mathcal{H}$ if $Ax = \lambda x$ for some $x \neq 0$.

Theorem 1.1.24. *If λ is an eigenvalue of a bounded linear operator $A : \mathcal{H} \to \mathcal{H}$, then $|\lambda| \leq \|A\|$. In particular, if A is self-adjoint, then $|\lambda| \leq \sup_{\|x\| \leq 1} \langle Ax, x \rangle$.*

Definition 1.1.25. We say that a bounded linear operator $A : \mathcal{H}_1 \to \mathcal{H}_2$ is *compact* if for any bounded sequence $\{x^k\}_{k=0}^{\infty} \subseteq \mathcal{H}_1$, the sequence $\{Ax^k\}_{k=0}^{\infty} \subseteq \mathcal{H}_2$ contains a convergent subsequence.

If $\dim \mathcal{H}_1 < \infty$, then every bounded linear operator A is compact.

Theorem 1.1.26. *Let $A : \mathcal{H} \to \mathcal{H}$ be a nonnegative, nonzero compact linear operator. Then:*

(i) *There exists an eigenvalue $\lambda = \lambda_{\max}(A)$ of the operator A such that*

$$\lambda = \|A\| = \sup_{\|x\| \leq 1} \langle Ax, x \rangle, \tag{1.11}$$

consequently, at least one eigenvalue of the operator A is positive.
(ii) *If A is positive, then all of its eigenvalues are positive.*
(iii) *There exists $z \in \mathcal{H}$ with $\|z\| = 1$ such that*

$$\langle Az, z \rangle = \sup_{\|x\| \leq 1} \langle Ax, x \rangle.$$

Theorem 1.1.27. *Let $A : \mathcal{H}_1 \to \mathcal{H}_2$ be a linear and compact operator. Then*

$$\|A\| = \sqrt{\lambda_{\max}(A^* A)} = \sqrt{\lambda_{\max}(AA^*)} = \|A^*\| .$$

Definition 1.1.28. We say that an operator $T : \mathcal{H} \to \mathcal{H}$ is *monotone* if

$$\langle Tx - Ty, x - y \rangle \geq \eta \|x - y\|^2$$

for a constant $\eta \geq 0$ and for all $x, y \in X$. If $\eta > 0$, then we say that T is *strongly monotone* or *η-strongly monotone*.

Definition 1.1.29. We say that an operator $T : \mathcal{H} \to \mathcal{H}$ is *Lipschitz continuous* if

$$\|Tx - Ty\| \leq \kappa \|x - y\| .$$

for a positive constant κ. The constant κ is called a *Lipschitz constant*. We also say that T is *κ-Lipschitz continuous*.

Definition 1.1.30. We say that an operator $T : \mathcal{H}_1 \to \mathcal{H}_2$ is *differentiable* at a point $x \in \mathcal{H}_1$ if there exists a bounded linear operator $G \in L(\mathcal{H}_1, \mathcal{H}_2)$ such that

$$T(x + h) = T(x) + Gh + o(\|h\|).$$

The operator G is uniquely determined and is called a *derivative* or a *differential* of the operator T at the point x and is denoted by $DT(x)$ or $T'x$.

Theorem 1.1.31. *Let $T : \mathcal{H}_1 \to \mathcal{H}_2$ be continuously differentiable in a neighborhood of a point $x \in \mathcal{H}_1$ and $U : \mathcal{H}_2 \to \mathcal{H}_3$ be continuously differentiable in a neighborhood of $Tx \in \mathcal{H}_2$. Then there exists a differential $D(U \circ T)(x)$ and*

$$D(U \circ T)(x) = D(U(Tx))DT(x).$$

Definition 1.1.32. Let $T : X \to Y$, where $X \subseteq \mathcal{H}_1$ and $Y \subseteq \mathcal{H}_2$, and let $C \subseteq X$ and $D \subseteq Y$. The subset

$$T(C) := \{Tx : x \in C\} \subseteq Y$$

is called an *image* of C by the operator T. The subset

$$T^{-1}(D) := \{x \in X : Tx \in D\}$$

is called an *inverse image* of D by the operator T.

The *iteration* T^k of an operator $T : X \to X$, where $X \subseteq \mathcal{H}$, is defined by the recurrence: $T^1 = T$, $T^{k+1} = TT^k$, $k \geq 1$. For any $x \in X$ the sequence $\{T^k x\}_{k=0}^{\infty}$ is called an *orbit* of T. A point $z \in X$ is called a *fixed point* of an operator $T : X \to \mathcal{H}$, if $Tz = z$. The subset of all fixed points of the operator T is denoted by $\text{Fix } T$. A subset $C \subseteq X$ is called a *retract* of X if there exists a continuous operator $T : X \to C$ (called a *retraction*) with $\text{Fix } T = C$. We say that an operator $T : X \to X$ is *idempotent* or a *projection* if $TTx = Tx$ for all $x \in X$, i.e., $T(X) \subseteq \text{Fix } T$. Since the converse implication holds for all operators, we have: T is idempotent if and only if $T(X) = \text{Fix } T$. A continuous projection is a retraction. Let $V \subseteq \mathcal{H}$ be a closed subspace. The operator $P : \mathcal{H} \to V$ with the property $Px - x \in V^{\perp}$ for all $x \in \mathcal{H}$ is called the *orthogonal projection* onto V. The orthogonal projection P onto a closed subspace is defined uniquely and is a linear and bounded operator with the norm $\|P\| = 1$. Let $a \in \mathcal{H}$. The orthogonal projection onto $\text{Lin } a$ is denoted by P_a. We say that an operator $A : \mathcal{H}_1 \to \mathcal{H}_2$ is *affine* if for all $x, y \in \mathcal{H}_1$ and for any $\lambda \in \mathbb{R}$, $A((1 - \lambda)x + \lambda y) = (1 - \lambda)Ax + \lambda Ay$ holds. Let $A : \mathcal{H}_1 \to \mathcal{H}_2$ be a linear operator. The subset

$$\ker A := \{x \in \mathcal{H}_1 : Ax = 0\}$$

denotes a *kernel* or *null space* of A and the subset

$$\text{Im } A := \{y \in \mathcal{H}_2 : y = Ax \text{ for some } x \in \mathcal{H}_1\}$$

denotes an *image* or *range* of A. The subsets $\ker A \subseteq \mathcal{H}_1$ and $\text{Im } A \subseteq \mathcal{H}_2$ are subspaces of $\mathcal{H}_1$ and $\mathcal{H}_2$, respectively. It holds

$$(\ker A)^{\perp} = \text{Im } A^*,$$

consequently,

$$\mathcal{H}_1 = \ker A \oplus \text{Im } A^*.$$

1.1.2.5 Matrices

Definition 1.1.33. We say that a symmetric matrix A of type $n \times n$ is *positive definite*, if $x^{\top} Ax > 0$ for all nonzero vectors $x \in \mathbb{R}^n$. We say that A is *positive semi-definite* if $x^{\top} Ax \geq 0$ for all $x \in \mathbb{R}^n$.

Let $v = (v_1, \ldots, v_m) \in \mathbb{R}^m$. Then $\text{diag } v$ denotes a *diagonal matrix* of type $m \times m$ with $v_1, \ldots, v_m$ on its diagonal, i.e.,

$$\text{diag}\, v = \begin{bmatrix} v_1 & 0 & \dots & 0 \\ 0 & v_2 & \dots & 0 \\ \dots & \dots & \dots & \dots \\ 0 & 0 & \dots & v_m \end{bmatrix}.$$

Theorem 1.1.34. *For a symmetric matrix A the following conditions are equivalent:*

(i) *A is positive definite (positive semi-definite),*
(ii) *All eigenvalues of A are positive (nonnegative),*
(iii) *There exists an orthogonal matrix U and a diagonal matrix D with positive (nonnegative) elements on the diagonal such that $A = U^\top D U$,*
(iv) *There exists a positive definite (positive semi-definite) matrix B such that $A = B^2$.*

We denote by $A^{\frac{1}{2}}$ the matrix B with property (iv) above.

Definition 1.1.35. Let A be an $m \times n$ matrix. The matrix $A^\top A$ is called the *Gram matrix* of the columns of A.

The Gram matrix $A^\top A$ is positive semi-definite. Furthermore, it is positive definite if and only if A has full column rank, i.e., the columns of A are linearly independent.

Definition 1.1.36. The *Moore–Penrose pseudoinverse A^+* of an $m \times n$ matrix A is a uniquely determined $n \times m$ matrix satisfying the following conditions:

$$AA^+A = A, \; A^+AA^+ = A^+, \; (AA^+)^\top = AA^+, \; (A^+A)^\top = A^+A.$$

If A is a full column rank or a full row rank matrix, then

$$A^+ = (A^\top A)^{-1} A^\top$$

or

$$A^+ = A^\top (AA^\top)^{-1},$$

respectively.

1.1.2.6 Convex Subsets

Let $\mathcal{H}, \mathcal{H}_i, i = 1, 2, \dots, m$, be Hilbert spaces.

Definition 1.1.37. A subset $C \subseteq \mathcal{H}$ is said to be *convex*, if $(1 - \lambda)x + \lambda y \in C$ for all $x, y \in C$ and for all $\lambda \in [0, 1]$.

An intersection of arbitrary family of convex subsets is convex. In particular, a *polytope*, i.e., the intersection of a finite number of half-spaces, is convex. The ball

$B(x, \rho)$ is convex for any $x \in \mathcal{H}$ and for any $\rho \geq 0$. Let $A : \mathcal{H}_1 \to \mathcal{H}_2$ be an affine operator. If $C \subseteq \mathcal{H}_1$ is a convex subset, then the image $A(C) \subseteq \mathcal{H}_2$ is convex. If $D \subseteq \mathcal{H}_2$ is convex, then the inverse image $A^{-1}(D) \subseteq \mathcal{H}_1$ is convex. The product $C_1 \times \ldots \times C_m$ of convex subsets $C_i \subseteq \mathcal{H}_i, i = 1, 2, \ldots, m$, is a convex subset of the product space $\mathcal{H} := \mathcal{H}_1 \times \ldots \times \mathcal{H}_m$.

Theorem 1.1.38. *If $C \subseteq \mathcal{H}$ is convex, then its closure and its interior are convex subsets.*

Theorem 1.1.39. *A subset $C \subseteq \mathcal{H}$ is closed and convex if and only if C is an intersection of half-spaces.*

The following theorem follows from the Mazur theorem (see, e.g., [158, Chap. I, Sect. 1.2], [180, Sect. 2.3, Corollary 2], [267, Theorem 21.4], [300, Theorem 3.12] or [214, Theorem 8.48]).

Theorem 1.1.40. *If $C \subseteq \mathcal{H}$ is a closed convex subset, then C is weakly closed.*

Definition 1.1.41. The smallest convex subset containing $S \subseteq \mathcal{H}$ is called a *convex hull* of S and is denoted by conv S. The smallest affine subspace containing $S \subseteq \mathcal{H}$ is called an *affine hull* of S and is denoted by aff S. The subset

$$\text{ri } C := \{x \in C : B(x, \varepsilon) \cap \text{aff } C \subseteq C \text{ for some } \varepsilon > 0\}$$

is called a *relative interior* of a convex subset $C \subseteq \mathcal{H}$.

It holds

$$\text{ri } \Delta_m = \{u \in \mathbb{R}^m : u > 0, e^\top u = 1\},$$

where Δ_m denotes a standard simplex in $\mathbb{R}^n$.

Definition 1.1.42. A subset $K \subseteq \mathcal{H}$ is called a *convex cone* if $\alpha x + \beta y \in K$ for all $x, y \in K$ and for all $\alpha, \beta \geq 0$.

Definition 1.1.43. For a convex cone $C \subseteq \mathcal{H}$ the subset

$$C^* := \{x \in \mathcal{H} : \langle y, x \rangle \leq 0 \text{ for all } y \in C\}$$

is called a *polar cone* to $C \subseteq \mathcal{H}$.

Definition 1.1.44. We say that a convex cone $C \subseteq \mathcal{H}$ is *obtuse* if $-C^* \subseteq C$.

Theorem 1.1.45 (Moreau's decomposition). *Let $K \subseteq \mathcal{H}$ be a nonempty closed convex cone and $x, x^+, x^- \in \mathcal{H}$. Then the following conditions are equivalent:*

(i) $x = x^+ + x^-$, *where* $x^+ \in K$, $x^- \in K^*$ *and* $\langle x^+, x^- \rangle = 0$,
(ii) $x^+ = P_K x$ *and* $x^- = P_{K^*} x$.

Definition 1.1.46. The subset $N_C(x) := \{y \in \mathcal{H} : \langle y, z - x \rangle \leq 0 \text{ for all } z \in C\}$ is called a *normal cone* to a convex subset $C \subseteq \mathcal{H}$ at $x \in C$. The subset

$$T_C(x) := \left\{ s \in \mathcal{H} : \exists_{\{x^k\}_{k=0}^{\infty} \subseteq C, x^k \to x} \exists_{\{t_k\}_{k=0}^{\infty}, t_k \downarrow 0} \text{ with } s = \lim_k \frac{x^k - x}{t_k} \right\}$$

is called the *tangent cone* to a convex subset $C \subseteq \mathcal{H}$ at $x \in C$.

Theorem 1.1.47. *Let $C \subseteq \mathcal{H}$ be a closed convex subset and $x \in C$. The following equalities hold*

$$(N_C(x))^* = T_C(x) \text{ and } (T_C(x))^* = N_C(x).$$

1.1.2.7 Convex Functions

Let $X, Y \subseteq \mathcal{H}$ be convex subsets of a Hilbert space $\mathcal{H}$ and let $f : X \to \mathbb{R}$.

Definition 1.1.48. We say that f is *convex* if

$$f((1 - \lambda)x + \lambda y) \le (1 - \lambda)f(x) + \lambda f(y)$$

holds for all $x, y \in X$ and for all $\lambda \in [0, 1]$. If the inequality is strict for all $x, y \in X$, $x \ne y$ and for all $\lambda \in (0, 1)$, then we say that f is *strictly convex*. We say that f is α-*strongly convex*, where $\alpha > 0$ or, shortly, *strongly convex* if

$$f((1 - \lambda)x + \lambda y) \le (1 - \lambda)f(x) + \lambda f(y) - \frac{1}{2}\alpha\lambda(1 - \lambda)\|x - y\|^2$$

for all $x, y \in X$ and for all $\lambda \in [0, 1]$. A function $f : X \to \mathbb{R}$ is called *concave* if $-f$ is convex.

Properties of convex functions

1. If $f_i : X \to \mathbb{R}$, $i \in I$, are convex, then the function $f := \max_{i \in I} f_i$ is convex.
2. If $f : X \to \mathbb{R}$ is convex and $g : \mathbb{R} \to \mathbb{R}$ is convex and nondecreasing, then the composition $g \circ f$ is convex.
3. If $f_i : X \to \mathbb{R}$, $i = 1, 2, \ldots, m$, are convex and $F : \mathbb{R}^m \to \mathbb{R}$ is a convex function which is nondecreasing with respect to any coordinate, then the function $f := F(f_1, \ldots, f_m)$ is convex.
4. If $f : Y \to \mathbb{R}$ is a convex function and $A : X \to Y$ is an affine operator, then the function $f \circ A$ is convex.
5. If $f : X \times Y \to \mathbb{R}$ is convex, then the function $h : X \to \mathbb{R}$, $h(x) := \inf_{y \in Y} f(x, y)$ is convex.
6. Any sublevel set $S(f, \alpha)$, $\alpha \in \mathbb{R}$, of a convex function $f : X \to \mathbb{R}$ is convex.
7. A function $f : X \to \mathbb{R}$ is convex if and only if its epigraph epi f is a convex subset of $X \times \mathbb{R}$.
8. A convex function $f : \mathbb{R}^n \to \mathbb{R}$ is continuous.

9. A strongly convex function $f : \mathcal{H} \to \mathbb{R}$ is coercive.
10. A convex function $f : \mathcal{H} \to \mathbb{R}$ has directional derivatives $f'(\cdot, s)$ for all $s \in \mathcal{H}$, and for any $x, s \in \mathcal{H}$ it holds

$$f'(x, s) = \inf_{t>0} \frac{f(x + ts) - f(x)}{t}.$$

11. Let $x \in \mathcal{H}$. The directional derivative $f'(x, \cdot) : \mathcal{H} \to \mathbb{R}$ of a convex function $f : \mathcal{H} \to \mathbb{R}$ is convex and *positively homogeneous* of degree 1, i.e., for all $s \in \mathcal{H}$ and for all $\alpha \geq 0$ the equality $f'(x, \alpha s) = \alpha f'(x, s)$ holds.
12. If $f : \mathcal{H} \to \mathbb{R}$ is a convex function and $f'(x, s) \geq 0$ for all $s \in \mathcal{H}$, then f attains its minimum at $x \in \mathcal{H}$.
13. The norm $\|\cdot\| : \mathcal{H} \to \mathbb{R}$ is convex and differentiable on $\mathcal{H} \backslash \{0\}$ and $D\|x\| = \frac{x}{\|x\|}$ for any $x \neq 0$.
14. The function $f : \mathcal{H} \to \mathbb{R}$, $f(x) := \frac{1}{2}\|x\|^2$ is differentiable and 1-strongly convex, and $Df(x) = x$ for all $x \in \mathcal{H}$.
15. Let $C \subseteq \mathcal{H}$ be nonempty. A function $d(\cdot, C) : \mathcal{H} \to \mathbb{R}$ defined by $d(x, C) := \inf_{z \in C} \|z - x\|$ is called a *distance function* to the subset C. If C is convex, then $d(\cdot, C)$ is a convex function.
16. Let a function $f : \mathbb{R}^n \to \mathbb{R}$ have continuous partial derivatives of the second order. Then f is convex if and only if its Hessian $\nabla^2 f(x)$ is a positive semidefinite matrix for all $x \in \mathbb{R}^n$. The function f is α-strongly convex if and only if $s^{\mathsf{T}} \nabla^2 f(x) s \geq \alpha \|s\|^2$ for all $x, s \in \mathbb{R}^n$.

Theorem 1.1.49. *If a convex function $f : \mathcal{H} \to \mathbb{R}$ is bounded in a neighborhood of a point $x^0 \in \mathcal{H}$, then f is locally Lipschitz continuous on $\mathcal{H}$, i.e., for any $x \in \mathcal{H}$ there is $\rho > 0$ and $\kappa \geq 0$ such that $|f(u) - f(v)| \leq \kappa \|u - v\|$ for all $u, v \in B(x, \rho)$. In particular, a continuous convex function $f : \mathcal{H} \to \mathbb{R}$ is locally Lipschitz continuous on $\mathcal{H}$.*

If $\mathcal{H}$ is finite dimensional, then Theorem 1.1.49 can be strengthened. In this case the local Lipschitz continuity can be replaced by the global one on bounded subsets.

Theorem 1.1.50. *If a convex function $f : \mathcal{H} \to \mathbb{R}$ is κ-Lipschitz continuous in a neighborhood of a point $x \in \mathcal{H}$, then for any $s \in \mathcal{H}$ it holds*

$$f'(x, s) \leq \kappa \|s\|.$$

Theorem 1.1.51. *A lower semi-continuous convex function $f : \mathcal{H} \to \mathbb{R}$ is weakly lower semi-continuous.*

Theorem 1.1.52. *A continuous convex function $f : X \to \mathbb{R}$ defined on a closed bounded convex subset $X \subseteq \mathcal{H}$ attains its global minimum.*

Corollary 1.1.53. *If a continuous convex function $f : X \to \mathbb{R}$ defined on a closed convex subset $X \subseteq \mathcal{H}$ is coercive, then f attains its global minimum on X.*

Theorem 1.1.54. *A convex Gâteaux-differentiable function $f : \mathbb{R}^n \to \mathbb{R}$ is differentiable.*

Subdifferential of a Convex Function and Its Properties

Definition 1.1.55. Let $f : \mathcal{H} \to \mathbb{R}$ be convex. The subset

$$\partial f(x) := \{g \in \mathcal{H} : \langle g, y - x \rangle \leq f(y) - f(x) \quad \text{for all } y \in \mathcal{H}\}$$

is called a *subdifferential* of f at $x \in \mathcal{H}$. The function f is said to be *subdifferentiable* at x if $\partial f(x) \neq \emptyset$. An element of the subdifferential $\partial f(x)$ is called a *subgradient* of f at x and is denoted by $g_f(x)$. The affine function $\bar{f}_x : \mathcal{H} \to \mathbb{R}$ defined by

$$\bar{f}_x(y) := \langle g, y - x \rangle + f(x),$$

where $g \in \partial f(x)$ is called a *linearization* of f at x.

Theorem 1.1.56. *For any $x \in \mathcal{H}$ the subdifferential $\partial f(x)$ of a continuous convex function $f : \mathcal{H} \to \mathbb{R}$ is a nonempty, weakly closed and bounded convex set.*

Theorem 1.1.57. *A continuous convex function $f : \mathcal{H} \to \mathbb{R}$ is Gâteaux-differentiable at $x \in \mathcal{H}$ if and only if its subdifferential $\partial f(x)$ consists of one point. In this case $\partial f(x) = \{Df(x)\}$.*

Theorem 1.1.58. *Let $f : \mathcal{H} \to \mathbb{R}$ be continuous and convex and $x, s \in \mathcal{H}$. The following equality holds*

$$f'(x, s) = \sup\{\langle y, s \rangle : y \in \partial f(x)\}.$$

In particular, if f is differentiable, then $f'(x, s) = \langle Df(x), s \rangle$ If, moreover, $\mathcal{H}$ is a Euclidean space, then $f'(x, s) = s^\top \nabla f(x)$.

The subdifferential of the norm $\|\cdot\|$ has the form

$$\partial(\|x\|) = \begin{cases} \{\frac{x}{\|x\|}\} & \text{if } x \neq 0 \\ B(0, 1) & \text{if } x = 0. \end{cases}$$

   All the facts presented in this section can be found in the following references [33, 38, 40, 61, 133, 158, 180, 181, 209, 210, 214, 234, 262–264, 267, 298, 300, 317, 325].

1.2   Metric Projection

In this section we define the metric projection in a Hilbert space $\mathcal{H}$ and present its basic properties. This operator plays an important role in further parts of the book. Other properties of the metric projection will be presented in Chap. 2.

Definition 1.2.1. Let $C \subseteq \mathcal{H}$ be a nonempty subset and $x \in \mathcal{H}$. If there exists a point $y \in C$ such that

Fig. 1.1 Metric projection

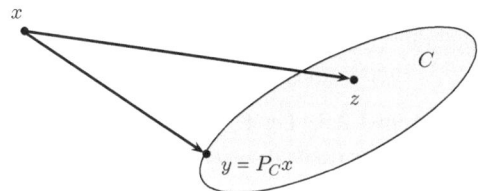

$$\|y - x\| \le \|z - x\|$$

for any $z \in C$, then y is called a *metric projection* of x onto C and is denoted by $P_C x$ (Fig. 1.1). The vector $s = P_C x - x$ is called *a projection vector* of x onto C. If $P_C x$ exists and is uniquely determined for all $x \in \mathcal{H}$, then the operator $P_C : \mathcal{H} \to C$ is called the *metric projection (onto C)*.

Remark 1.2.2. Let $C \subseteq \mathcal{H}$. If $x \in C$, then it follows from the definition of the metric projection that $P_C x = x$. If $x \notin C$ and there exists a metric projection $P_C x$, then $P_C x \in \operatorname{bd} C$. Indeed, if $P_C x \in \operatorname{int} C$, then $B(P_C x, \varepsilon) \subseteq C$ for a sufficiently small $\varepsilon > 0$. Then

$$z := x + (1 - \frac{\varepsilon}{\|P_C x - x\|})(P_C x - x) \in B(P_C x, \varepsilon) \subseteq C,$$

because $\|z - P_C x\| = \varepsilon$, and

$$\|z - x\| = (1 - \frac{\varepsilon}{\|P_C x - x\|}) \|P_C x - x\| < \|P_C x - x\|$$

which contradicts the definition of $P_C x$.

1.2.1 Existence and Uniqueness of the Metric Projection

Definition 1.2.1 does not yield the existence and the uniqueness of a metric projection $P_C x$ for a point $x \in \mathcal{H}$ and for a subset $C \subseteq \mathcal{H}$. If $C \subseteq \mathcal{H}$ is compact, then the existence of a metric projection follows from the Weierstrass theorem but it needs not to be defined uniquely. For example, both points $a, b \in \mathcal{H}$ are metric projections of the midpoint $x := \frac{a+b}{2}$ of the segment $[a, b]$ onto the subset $C = \{a, b\}$. Therefore, the convexity of C seems to be a natural assumption for the uniqueness of the metric projection $P_C x$ for any $x \in \mathcal{H}$. Furthermore, it follows from the definition of a metric projection and from the continuity of the norm that $P_C x$ (if it exists) belongs to the closure of C. Actually, $P_C x \in \operatorname{bd} C$ for $x \notin C$ (see Remark 1.2.2). Therefore, the closedness of C is a natural assumption for the existence of the metric projection $P_C x$ for any $x \in \mathcal{H}$. It turns out that in a Hilbert space these two assumptions (the closedness and the convexity of C) are sufficient for the existence and the uniqueness of the metric projection $P_C x$ for all $x \in C$. The following theorem can be found, e.g., in [61, Sect. 1.2.2, Theorem 1.7], [140, Theorem 3.4(2)], [173, Theorem 7.43], [209, Chap. III, Sect. 3.1] or [267, Theorem 8.25].

Theorem 1.2.3. *Let $C \subseteq \mathcal{H}$ be a nonempty closed convex subset. Then for any $x \in \mathcal{H}$ there exists a metric projection $P_C x$ and it is uniquely determined.*

Proof. First we prove the assertion for $x = 0$. Let $d := \inf\{\|y\| : y \in C\}$ and a sequence $\{y^k\}_{k=0}^{\infty} \subseteq C$ be such that $\|y^k\| \to d$. We split the proof onto three parts.

(a) We show that $\{y^k\}_{k=0}^{\infty}$ is a Cauchy sequence. Let $\varepsilon > 0$ and $k_0 \geq 0$ be such that $\|y^k\|^2 \leq d^2 + \varepsilon/4$ for $k \geq k_0$. Let $k, l \geq k_0$. Since C is convex, we have $\frac{1}{2} y^k + \frac{1}{2} y^l \in C$. Therefore, $\frac{1}{2} \|y^k + y^l\| \geq d$. The parallelogram law yields now

$$\left\| y^k - y^l \right\|^2 = 2 \left\| y^k \right\|^2 + 2 \left\| y^l \right\|^2 - \left\| y^k + y^l \right\|^2$$
$$\leq 2(d^2 + \varepsilon/4) + 2(d^2 + \varepsilon/4) - 4d^2 = \varepsilon,$$

i.e., $\{y^k\}_{k=0}^{\infty}$ is a Cauchy sequence.
(b) It follows from (a) that y^k converges to an element $y \in \mathcal{H}$, because $\mathcal{H}$ is a complete space. Furthermore, $y \in C$, because C is closed. Hence, by the continuity of the norm, it follows that $\|y\| = d$. Therefore, $y = P_C 0$.
(c) Now we show that the metric projection $P_C x$ is uniquely determined. Let $y' \in C$ with $\|y'\| = d$. It follows from the convexity of C that $\frac{1}{2} y + \frac{1}{2} y' \in C$. Furthermore,

$$d \leq \left\| \frac{1}{2} y + \frac{1}{2} y' \right\| \leq \frac{1}{2} \|y\| + \frac{1}{2} \|y'\| = d,$$

i.e., $\|y + y'\| = 2d$. Again, by the parallelogram law, we have

$$\left\| y - y' \right\|^2 = 2 \|y\|^2 + 2 \left\| y' \right\|^2 - \left\| y + y' \right\|^2 = 2d^2 + 2d^2 - 4d^2 = 0,$$

i.e., $y = y'$.

We have proved that $P_C 0$ exists and is uniquely defined. Of course, $C - x$ is closed convex, because C is closed convex. Therefore, $P_{C-x} 0$ also exists and is uniquely defined. It follows easily from the definition of the metric projection that

$$P_C x = x + P_{C-x} 0,$$

consequently, the assertion is true for every $x \in \mathcal{H}$. $\square$

One can ask if the converse theorem to Theorem 1.2.3 is also true. The answer is positive in the Euclidean space and is known as the Motzkin theorem (see [336, Theorem 7.5.5]). In general Hilbert spaces the problem is still open.

1.2.2 Characterization of the Metric Projection

The theorem below gives a criterion for $y \in C$ to be the metric projection of a point $x \in \mathcal{H}$ onto a convex set $C \subseteq \mathcal{H}$ and is often used in applications. The theorem can be found, e.g., in [140, Theorem 4.1], [173, Theorem 7.45], [209, Chap. III, Theorem 3.1.1] and in [185, Proposition 3.5] which contains several other equivalent conditions.

Theorem 1.2.4 (Characterization Theorem). *Let $x \in \mathcal{H}$, $C \subseteq \mathcal{H}$ be a convex subset and $y \in C$. The following conditions are equivalent:*

(i) $y = P_C x$,
(ii) $\langle x - y, z - y \rangle \le 0$ *for all $z \in C$.*

Proof. (i) $\Rightarrow$ (ii). Let $y = P_C(x)$, $z \in C$ and

$$z_\lambda = y + \lambda(z - y)$$

for $\lambda \in (0, 1)$. Obviously, $z_\lambda \in C$ because C is convex. By (i) and by the properties of the inner product, we have

$$\|x - y\|^2 \le \|x - z_\lambda\|^2 = \|x - y - \lambda(z - y)\|^2$$
$$= \|x - y\|^2 - 2\lambda \langle x - y, z - y \rangle + \lambda^2 \|z - y\|.$$

Since $\lambda > 0$, the above inequalities yield

$$\langle x - y, z - y \rangle \le \frac{\lambda}{2} \|z - y\|^2,$$

and for $\lambda \to 0$ we obtain (ii) in the limit.
(ii) $\Rightarrow$ (i). By the properties of the inner product and by (ii) we obtain for any $z \in C$

$$\|z - x\|^2 = \|z - y + y - x\|^2$$
$$= \|z - y\|^2 + \|y - x\|^2 + 2\langle z - y, y - x \rangle$$
$$\ge \|y - x\|^2,$$

which, by the definition of the metric projection, gives (i). $\square$

Condition (ii) of Theorem 1.2.4 says that $\sphericalangle(x - y, z - y) \ge \pi/2$ if $x - y$ and $z - y$ are nonzero vectors (see Fig. 1.2).
A simple proof of the following lemma is left to the reader.

Lemma 1.2.5. *Let $x, y, z \in \mathcal{H}$. The following conditions are equivalent:*

(a) $\langle x - y, z - y \rangle \le 0$,
(b) $\langle z - x, y - x \rangle \ge \|y - x\|^2$,
(c) $\|z - y\|^2 \le \|z - x\|^2 - \|y - x\|^2$,
(d) $\langle z - x, z - y \rangle \ge 0$.

Fig. 1.2 Characterization of
the metric projection

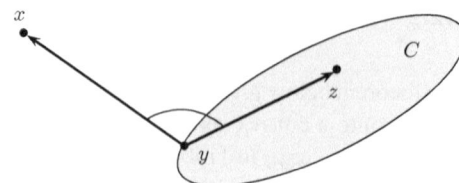

Fig. 1.3 Metric projection
onto a translated subset

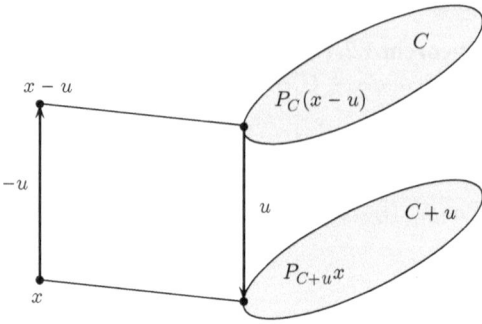

It follows from Lemma 1.2.5 that the inequality in condition (ii) of the Characterization Theorem 1.2.4 can be replaced by one of the conditions (a)–(d) of Lemma 1.2.5.

1.2.3   First Applications of the Characterization Theorem

In this section we give some useful facts which follow from the Characterization Theorem 1.2.4.

Lemma 1.2.6. *Let $x \in \mathcal{H}$, $C \subseteq \mathcal{H}$ be a convex subset and $u \in \mathcal{H}$. If a metric projection $P_C(x - u)$ exists, then a metric projection $P_{C+u}x$ also exists and*

$$P_{C+u}x = P_C(x - u) + u \tag{1.12}$$

holds (Fig. 1.3).

Proof. The proof of the lemma is a direct application of the Characterization Theorem 1.2.4 and is left to the reader.   □

Lemma 1.2.7. *Let $u \in \mathcal{H}$, $C \subseteq \mathcal{H}$ be convex and $A : \mathcal{H} \to \mathcal{H}$ be a unitary operator. If a metric projection $P_C(A^*u)$ exists, then a metric projection $P_{A(C)}u$ also exists and*

$$P_{A(C)}u = A(P_C(A^*u))$$

holds (Fig. 1.4).

Fig. 1.4 Metric projection
onto a unitary transformation
of a subset

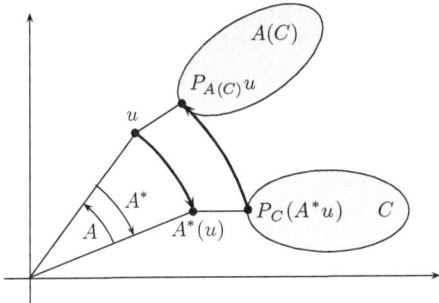

Proof. Let $y = P_C(A^*u)$ and $v = Ay$. We show that $v = P_{A(C)}u$. Since $y \in C$, we have $v \in A(C)$. Since A is a linear operator and C is convex, the subset $A(C)$ is convex. By the characterization of the metric projection (see Theorem 1.2.4), for all $w \in A(C)$ and for $z \in C$ with $Az = w$, we have

$$0 \geq \langle A^*u - y, z - y \rangle$$
$$= \langle A^*u - A^*Ay, z - y \rangle$$
$$= \langle u - v, Az - Ay \rangle$$
$$= \langle u - v, w - v \rangle.$$

Again by the characterization of the metric projection (see Theorem 1.2.4), we obtain $v = P_{A(C)}u$.   □

Let $\mathcal{H}_i$ be Hilbert spaces with the inner products $\langle \cdot, \cdot \rangle_{\mathcal{H}_i}$, $i \in I := \{1, 2, \ldots, m\}$, and $\mathcal{H} := \mathcal{H}_1 \times \ldots \times \mathcal{H}_m$ be the product Hilbert space.

Lemma 1.2.8. *Let $C_i \subseteq \mathcal{H}_i$ be convex, $i = 1, 2, \ldots, m$, $C := C_1 \times \ldots \times C_m \in \mathcal{H}$ and $x = (x_1, \ldots, x_m) \in \mathcal{H}$. If metric projections $P_{C_i} x_i$, $i = 1, 2, \ldots, m$, exist, then a metric projection $P_C x$ also exists and*

$$P_C x = (P_{C_1} x_1, \ldots, P_{C_m} x_m).$$

Proof. The proof of the lemma is a direct application of the Characterization Theorem 1.2.4 and is left to the reader.   □

Lemma 1.2.8 is illustrated in Fig. 1.5.

Lemma 1.2.9. *Let $C \subseteq \mathcal{H}$ be a nonempty closed convex set and $P_C : \mathcal{H} \to \mathcal{H}$ be a metric projection onto C. Then for any $x \in \mathcal{H}$ the following conditions are equivalent:*

(i) $y = P_C x$
(ii) $y \in C$ *and* $x - y \in N_C(y)$.

Fig. 1.5 Metric projection onto a product of subsets

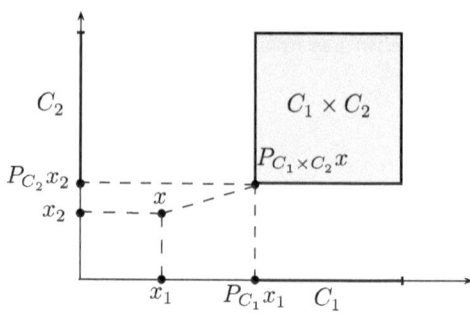

Fig. 1.6 Metric projection and normal cone

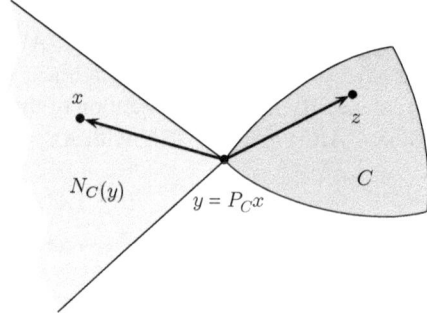

Proof. By the definition of the normal cone, we can write the second part of the right hand side of (ii) in the form $\langle x - y, z - y \rangle \leq 0$ for any $z \in C$ (Fig. 1.6). The lemma follows now from the characterization of the metric projection (see Theorem 1.2.4). $\qquad\qquad\square$

For $T : \mathcal{H} \to \mathcal{H}$ and $\lambda \in \mathbb{R}$ denote $T_\lambda := \mathrm{Id} + \lambda(T - \mathrm{Id})$.

Corollary 1.2.10. *Let* $T : \mathcal{H} \to \mathcal{H}$, $C \subseteq \mathcal{H}$ *be a nonempty closed convex subset. Then* $\mathrm{Fix}(P_C T_\lambda) = \mathrm{Fix}(P_C T)$ *for any* $\lambda > 0$.

Proof. Let $\lambda > 0$. It follows from Lemma 1.2.9 that

$$
\begin{aligned}
x \; &\in \; \mathrm{Fix}(P_C T_\lambda) \Longleftrightarrow P_C(x + \lambda(Tx - x)) = x \\
&\Longleftrightarrow \lambda(Tx - x) \in N_C(x) \\
&\Longleftrightarrow Tx - x \in N_C(x) \\
&\Longleftrightarrow P_C Tx = x \\
&\Longleftrightarrow x \in \mathrm{Fix}(P_C T)
\end{aligned}
$$

which completes the proof. $\qquad\qquad\square$

In Sect. 2.2.3 we give further properties of the metric projection.

1.3 Convex Optimization Problems

In this section we present several convex optimization problems: convex minimization, variational inequalities, convex feasibility problems and split feasibility
problems. These problems as well as the methods for solving them have applications
in various areas of mathematics, e.g., in solving linear equations, in probability
and statistics, in conformal mappings, in frame design problems and in many
other problems (for details, see, e.g., [139, 156, 182, 207, 329, 337]). Furthermore,
the mentioned above abstract problems can be treated as mathematical models
for many practical problems which arise in physical, medical, technical and
information sciences. The problems and methods presented in this book found
applications:

- In signal processing [192,275,316], in particular in resolution enhancement [109]
  and in signal synthesis [117, 123],
- In image processing [37,312,316], in particular in image recovery [118,309,315],
  in image restoration [354, 359], in image reconstruction [220, 221, 280, 289], in
  image reconstruction from projections [98, 100, 111, 224, 330], in demosaicking
  [248] and in color imaging [311],
- In discrete tomography [82],
- In medical imaging, in particular in computerized tomography [81, 189, 204, 205,
  270, 313], in proton computed tomography: [282] and in magnetic resonance
  imaging [301],
- In radiation therapy treatment planning [81, 83–85, 93, 110, 199, 206, 206, 212,
  241, 261, 324, 334, 335, 341],
- In physics [149, 240], in astrophysics [360], in acoustic [39], in crystallography
  [328], in mechanics [265], in materials science [225],
- In seismic tomography [60, 332], in meteorology [323],
- In optics [316], in particular in phase retrieval problem [26, 27, 161] and in
  holography [219, 310],
- In wavelet-based denoising [115], in watermarking [242], in adaptive filtering
  [314, 355],
- In control problems [192], in control design problems [193, 194],
- In antenna design [195], in antenna pattern synthesis problems [288],
- In data compression [245], in video coding [281],
- In graph matching [340],
- In learning process [319], in supervised learning process [222]
- And in finance [208].

In the next chapters we will show how to solve convex optimization problems by
iterative methods constructed by application of some algorithmic operators. We will
also show that the problem of finding a solution of a convex optimization problem
is equivalent to finding a fixed point of an algorithmic operator.

1.3.1 Convex Minimization Problems

Let $f : \mathcal{H} \to \mathbb{R}$ and $C \subseteq \mathcal{H}$. The *constrained minimization problem*

$$\begin{aligned} \text{minimize} \quad & f(x) \\ \text{with respect to } & x \in C \end{aligned} \tag{1.13}$$

is to find at least one *local minimizer* of the function f on C, i.e., a point $x^* \in C$ such that $f(x^*) \le f(x)$ for all $x \in C \cap B(x^*, \rho)$ and for some $\rho > 0$, if it exists. If $f(x^*) \le f(x)$ for all $x \in C$, then x^* is called a *global minimizer* of the function f on C. The point x^* is also called a (local or global) *optimal solution* of problem (1.13). The value $f^* = f(x^*)$ is called the minimum of the function f on the subset C.

Theorem 1.3.1. *Let $C \subseteq \mathcal{H}$ be a convex subset and $f : \mathcal{H} \to \mathbb{R}$ be a convex function. If x^* is a local minimizer of f on C, then it is also a global one. If, furthermore, f is strictly convex, then the minimizer is uniquely determined. If f is continuous and strongly convex and C is closed, then $\mathrm{Argmin}_{x \in C} f(x) \ne \emptyset$.*

Proof. Let $x^* \in C$ be a local minimizer of f on C. Suppose that there is a point $x' \in C$ with $f(x') < f(x^*)$. Let $x_\lambda = (1 - \lambda)x^* + \lambda x'$, where $\lambda \in (0, 1)$. It is clear that $x_\lambda \in C$, because C is convex. By the convexity of f, we have

$$f(x^*) \le f(x_\lambda) \le (1 - \lambda)f(x^*) + \lambda f(x') < f(x^*)$$

for a sufficiently small $\lambda > 0$. The contradiction shows that $f(x) \ge f(x^*)$ for all $x \in C$, i.e., x^* is a global minimizer of f on C. Suppose now that f is strictly convex and that $f(x') = f(x^*)$ holds for some $x' \ne x^*$. Then for $x = \frac{1}{2}x^* + \frac{1}{2}x'$, we obtain

$$f(x) < \frac{1}{2}f(x^*) + \frac{1}{2}f(x') = f(x^*) \le f(x),$$

a contradiction which shows that the minimizer is uniquely determined. Now suppose that f is continuous and strongly convex and that C is closed. Recall that a strongly convex function is coercive. Therefore, Corollary 1.1.53 yields $\mathrm{Argmin}_{x \in C} f(x) \ne \emptyset$. $\square$

The subset C in (1.13) is often given in the form

$$C := \{x \in \mathcal{H} : c_i(x) \le 0, i \in I\},$$

where $c_i : \mathcal{H} \to \mathbb{R}, i \in I := \{1, 2, \dots, m\}$. In this case (1.13) can be written as the following constrained minimization problem:

$$\text{minimize } f(x)$$
$$\text{subject to } c_i(x) \leq 0, i \in I, \tag{1.14}$$
$$x \in \mathcal{H}.$$

The functions c_i are called the *constraints* of problem (1.14). If the functions $f, c_i, i \in I$, are convex, then (1.14) is called a *convex minimization problem* (CMP). If $f, c_i, i \in I$, are differentiable, then (1.14) is called a *constrained differentiable minimization problem*. In order to solve problem (1.13) or (1.14) one can apply suitable methods (algorithms) which generate sequences approximating a solution x^* of the problem. These algorithms should be constructed in such a way that the sequences converge (in some sense) to a solution of the problem. Since an algorithm should terminate after a finite number of iterations, we cannot expect in general that the algorithm gives an exact solution. Let $\varepsilon \geq 0$. We say that $x_\varepsilon \in C$ is an *ε-optimal solution* of (1.13) or (1.14) if $f(x_\varepsilon) \leq f(x) + \varepsilon$ for all $x \in C$.

If f is a convex function and C is a closed convex subset (or $c_i, i \in I$, are convex functions), then problem (1.13) (or (1.14)) is called a *convex (constrained) minimization problem*.

For a constrained differentiable minimization problem (1.14) with $\mathcal{H} = \mathbb{R}^n$ (equipped with the standard inner product) a point $(x^*, y^*) \in \mathbb{R}^n \times \mathbb{R}^m$ satisfying the following system of equalities and inequalities

$$\nabla f(x) + y^T \nabla c(x) = 0$$
$$c(x) \leq 0$$
$$y \geq 0 \tag{1.15}$$
$$y^T c(x) = 0,$$

where $c(x) = (c_1(x), c_2(x), \dots, c_m(x))$, is called a *Karush–Kuhn–Tucker point* or, shortly, *KKT-point*. If f and $c_i, i \in I$, in (1.14) are convex (not necessarily differentiable), then a point $(x^*, y^*) \in \mathbb{R}^n \times \mathbb{R}^m$ satisfying the conditions

$$0 \in \partial f(x) + \sum_{i \in I} \lambda_i \partial c_i(x)$$
$$c(x) \leq 0$$
$$y \geq 0 \tag{1.16}$$
$$y^T c(x) = 0,$$

where $y = (\lambda_1, \dots \lambda_m) \in \mathbb{R}^m$, is also called a *KKT-point*. For convex differentiable minimization problem, (1.15) is, of course, a special case of (1.16).

Below we give necessary conditions and sufficient conditions for a point x to be an optimal solution of (1.13) or (1.14).

Theorem 1.3.2. *A continuous and convex function $f : \mathcal{H} \to \mathbb{R}$ attains its minimum at a point $x \in \mathcal{H}$ if and only if $0 \in \partial f(x)$.*

Proof. The theorem follows directly from the definition of the subdifferential. $\square$

Corollary 1.3.3. *A differentiable convex function* $f : \mathcal{H} \to \mathbb{R}$ *attains its minimum at a point* $x \in \mathcal{H}$ *if and only if* $Df(x) = 0$.

Theorem 1.3.4. *Let* $f : \mathcal{H} \to \mathbb{R}$ *be a continuous and convex function and* $C \subseteq \mathcal{H}$ *be a closed convex subset. The function* $f \mid_C$ *attains its minimum at a point* $x \in C$ *if and only if* $-\partial f(x) \subseteq N_C(x)$. *In particular, a convex differentiable function* f *attains its minimum at a point* $x \in C$ *if and only if* $-Df(x) \in N_C(x)$, *i.e.,*

$$f'(x, y - x) = \langle Df(x), y - x \rangle \geq 0$$

for all $y \in C$.

Proof. See [209, Chap. VII, Theorem 1.1.1] for finite dimensional case or [325, Sect. 7.1] for an infinite dimensional one. □

Corollary 1.3.5. *Let* $C \subseteq \mathcal{H}$ *be a closed convex subset and* $f : \mathcal{H} \to \mathbb{R}$ *be a differentiable convex function. Then for any* $\gamma > 0$ *it holds*

$$\operatorname*{Argmin}_{x \in C} f(x) = \operatorname{Fix} P_C (\operatorname{Id} - \gamma Df).$$

Proof. Let $\gamma > 0$. It follows from Theorem 1.3.4 and from Lemma 1.2.9 that

$$
\begin{aligned}
x \quad &\in \quad \operatorname*{Argmin}_{x \in C} f(x) \\
&\Longleftrightarrow -Df(x) \in N_C(x) \\
&\Longleftrightarrow -\gamma Df(x) \in N_C(x) \\
&\Longleftrightarrow x = P_C(x - \gamma Df(x)) \\
&\Longleftrightarrow x \in \operatorname{Fix} P_C(\operatorname{Id} - \gamma Df)
\end{aligned}
$$

which completes the proof. □

In the theorem below we assume that $\mathcal{H} = \mathbb{R}^n$ and that the constraints of the minimization problem (1.14) are regular in some sense, i.e., they satisfy a certain condition called a *constraints qualification*. If the constraints c_i are convex, $i = 1, 2, \ldots, m$, then one often supposes that they satisfy the *Slater constraints qualification*, i.e., there exists $\bar{x} \in \mathbb{R}^n$ such that $c_i(\bar{x}) < 0$ for all non-affine functions c_i. Other constraints qualifications are also considered in the literature. We omit the details and refer the reader to [169, Sect. 9.2], [209, Chap. VII, Sect. 2.2], [38, Sect. 3.3] or [179, Sect. 2.2].

Theorem 1.3.6 (Karush–Kuhn–Tucker). *Suppose that the constraints qualification for the convex minimization problem (1.14) is satisfied. If a point* $x^* \in \mathbb{R}^m$ *is an optimal solution of (1.14), then there exists* $y^* \in \mathbb{R}^m$ *such that* (x^*, y^*) *is a KKT-point.*

Proof. See [209, Chap. VII, Theorem 2.2.5]. □

Theorem 1.3.7. *Let* $f, c_i : \mathbb{R}^n \to \mathbb{R}, i \in I$ *be convex. If* $(x^*, y^*) \in \mathbb{R}^n \times \mathbb{R}^m$ *is a KKT-point of (1.14), then* x^* *is an optimal solution of this problem.*

Proof. See [169, Theorem 9.1.1] for the differentiable minimization problem or [209, Chap. VII, Theorem 2.1.4] for the convex minimization problem. □

1.3.2 Variational Inequality

Let $C \subseteq \mathcal{H}$ be closed convex and $\mathcal{F} : \mathcal{H} \to \mathcal{H}$. The *variational inequality problem* $(VIP(\mathcal{F}, C))$ is to find $x \in C$ such that

$$\langle \mathcal{F}x, y - x \rangle \geq 0 \qquad (1.17)$$

holds for all $y \in C$. Note that, by Theorem 1.3.4, the differentiable convex minimization problem (1.13) is a special case of the variational inequality. The proof of the following theorem is similar to the proof of Corollary 1.3.5 and is left to the reader.

Theorem 1.3.8. *Let* $\gamma > 0$. *A point* $x \in C$ *is a solution of variational inequality (1.17) if and only if* $x \in \text{Fix } P_C (\text{Id} - \gamma \mathcal{F})$.

For a general discussion on the existence of solutions of variational inequality problems we refer the reader to [165, 226, 358]. Below we only recall sufficient conditions for the existence and uniqueness of a solution of $VIP(\mathcal{F}, C)$.

Theorem 1.3.9. *If* $\mathcal{F} : \mathcal{H} \to \mathcal{H}$ *is a strongly monotone and Lipschitz continuous operator, then the variational inequality (1.17) has a unique solution* $x^* \in C$.

Proof. See, e.g., [358, Theorem 46.C]. □

1.3.3 Common Fixed Point Problem

Let a finite family of operators $U_i : \mathcal{H} \to \mathcal{H}$ with Fix $U_i \neq \emptyset, i \in I$, be given. If $F := \bigcap_{i \in I} \text{Fix } U_i \neq \emptyset$, then the *common fixed point problem* (CFPP) is to find a point $x \in \bigcap_{i \in I} \text{Fix } U_i$. In order to ensure that Fix U_i are closed and convex, some additional assumptions on the operators U_i are usually made. The details will be explained in Chap. 2.

1.3.4 Convex Feasibility Problem

Let a finite family of closed convex subsets $C_i, i \in I$, of a Hilbert space $\mathcal{H}$ be given. Denote $C := \bigcap_{i \in I} C_i$. If $C \neq \emptyset$, then the *convex feasibility problem* (CFP) is to find a point $x \in C$. We see that the CFP can be formulated as the CFPP with $U_i := P_{C_i}$, $i \in I$. On the other hand, the CFPP is a CFP with $C_i = \operatorname{Fix} U_i$. In applications concerning the CFP the subsets C_i are often called the *constraints sets* or, shortly, *constraints* and are given in the form

$$C_i := \{x \in \mathcal{H} : c_i(x) \leq 0\}, \tag{1.18}$$

where $c_i : \mathcal{H} \to \mathbb{R}$ are convex functions, called the *constraints functions* or, shortly, *constraints*, $i \in I$.

The CFP can also be considered without the assumption $C \neq \emptyset$. In this case we should define an appropriate convex *proximity function* $f : \mathcal{H} \to \mathbb{R}_+$ which measures a "distance" to the constraints. The proximity function should have the property

$$f(x) = 0 \text{ if and only if } x \in C. \tag{1.19}$$

The convex feasibility problem can be formulated as a minimization of the proximity function f. The proximity function $f : \mathcal{H} \to \mathbb{R}_+$ for the CFP can be defined by the following general form

$$f(x) := F(f_1(x), \ldots, f_m(x)), \tag{1.20}$$

where $f_i : \mathcal{H} \to \mathbb{R}_+$ are convex functions with the property

$$f_i(x) \begin{cases} = 0 \text{ for } x \in C_i \\ > 0 \text{ for } x \notin C_i, \end{cases} \tag{1.21}$$

$i \in I := \{1, 2, \ldots, m\}$, and the function $F : \mathbb{R}_+^m \to \mathbb{R}$ is convex and increasing (or at least nondecreasing) with respect to any variable and such that $F(v) = 0$ if and only if $v = 0$. In particular, if $F(v) = w^\top v$ for a vector $w = (\omega_1, \ldots, \omega_m) \in \mathbb{R}_{++}^m$, the proximity function f has the form of a weighted sum of functions f_i, $i \in I$,

$$f(x) := \sum_{i \in I} \omega_i f_i(x). \tag{1.22}$$

If $F(v) := \max_{i \in I} v_i$, where $v = (v_1, \ldots, v_m)$, the proximity function f has the form

$$f(x) = \max_{i \in I} f_i(x). \tag{1.23}$$

Two important examples of functions f_i with properties (1.21) are $f_i := d(\cdot, C_i)$ and $f_i := d^2(\cdot, C_i)$. We can also write $f_i(x) = \|P_{C_i} x - x\|$ in the first case or $f_i(x) = \frac{1}{2} \|P_{C_i} x - x\|^2$ in the other one. Note that both functions are convex and the other one is differentiable, as well. These facts will be proved in Sect. 2.2.3 (see Lemmas 2.2.27 and 2.2.28 and Corollary 2.2.29). The computation of these two functions requires, however, a simple structure of C_i. If we take

$f_i(x) = \frac{1}{2} \| P_{C_i} x - x \|^2$, $i \in I$, in (1.22), then we obtain the following important proximity function:

$$f(x) := \frac{1}{2} \sum_{i \in I} \omega_i \| P_{C_i} x - x \|^2 . \tag{1.24}$$

In Chap. 4 we show that determining a minimizer of the above function is equivalent to finding a fixed point of the operator $T = \sum_{i \in I} \omega_i P_{C_i}$ (see Theorem 4.4.6).

When the subset C_i is given by (1.18), the function $c_i^+ := \max\{0, c_i\}$ expresses a "distance" to the ith constraint, $i \in I$. In this case we can define a function f_i satisfying (1.21) in the form

$$f_i := h_i \circ c_i^+ \tag{1.25}$$

for a convex and increasing function $h_i : \mathbb{R}_+ \to \mathbb{R}_+$ such that $h_i(0) = 0$, $i \in I$. The function h_i has often the form $h_i(u) := u$ or $h_i(u) := \frac{1}{2} u^2$, $i \in I$. In the first case the proximity function f defined by (1.22) is a weighted sum of "distances" to the constraints, i.e.,

$$f(x) = \sum_{i \in I} \omega_i c_i^+(x),$$

and, in the other one, f has the form

$$f(x) = \frac{1}{2} \sum_{i \in I} \omega_i (c_i^+(x))^2,$$

where $w = (\omega_1, \ldots, \omega_m) \in \mathbb{R}_{++}^m$. One can additionally suppose that $\sum_{i \in I} \omega_i = 1$ which leads to the assumption that $w \in \mathrm{ri}\, \Delta_m$. This assumption does not change the minimization problem. Both proximity functions are convex and the other one is even differentiable.

As mentioned above, the CFP is, in general, to minimize a convex proximity function f. Denote $f^* := \inf_{x \in \mathcal{H}} f(x)$. It is clear that $f^* \geq 0$. We emphasize that the proximity function f needs not to attain its minimum. If $C \neq \emptyset$, we have $C = \mathrm{Argmin}_{x \in \mathcal{H}} f(x)$ for any proximity function with the above described properties. In this case, the problem of determining $x \in C$ and the problem of minimization of the proximity function f are equivalent. Furthermore, in this case, a significant information about f is available, namely $f^* := \min_{x \in \mathcal{H}} f(x) = 0$. One can say that in the general case, i.e., without the assumption $C \neq \emptyset$, the CFP is to find an element $x^* \in \mathcal{H}$, for which "distances" to the constraints C_i, $i \in I$, are minimal in some sense, if such x^* exists. For $x \in \mathcal{H}$, the value $f(x)$ expresses this "distance". If $C = \emptyset$ and the proximity function attains its minimum, then, of course, $f^* > 0$. Note, however, that the nonemptiness of C does not follow from the fact that $f^* := \inf_{x \in \mathcal{H}} f(x) = 0$.

In some applications, the CFP is to find an element $x \in C$ or to state that $C = \emptyset$. We can also say that the CFP is to find a zero of the proximity function or to state that the proximity function is positive for all arguments.

1.3.5 Linear Feasibility Problem

In the case where all C_i are half-spaces, i.e.,

$$C_i := H_-(a_i, \beta_i) = \{x \in \mathcal{H} : \langle a_i, x \rangle \le \beta_i\}$$

for $a_i \in \mathcal{H}$, $a_i \ne 0$ and $\beta_i \in \mathbb{R}$, $i \in I := \{1, 2, \ldots, m\}$, the CFP is called the *linear feasibility problem* (LFP). If $\mathcal{H} = \mathbb{R}^n$, the linear feasibility problem is to solve a system of linear inequalities

$$a_i^\top x \le \beta_i, i = 1, 2, \ldots, m, \tag{1.26}$$

where $a_i \in \mathbb{R}^n$, $a_i \ne 0$, $\beta_i \in \mathbb{R}$, $i \in I$, which can also be written in matrix form as

$$Ax \le b,$$

where A is a matrix with rows a_i, i.e., $A = [a_1, \ldots, a_m]^\top$ and $b = (\beta_1, \ldots, \beta_m)$. By duality theorems, the linear feasibility problem in $\mathbb{R}^n$ is equivalent to the linear programming problem (see, e.g., [4, Chap. 6]).

Similarly to the CFP, the linear feasibility problem can also be considered without the assumption $C \ne \emptyset$. In this case, the problem is to minimize a proximity function. One can apply general forms of a proximity function given by equalities (1.20)–(1.25). Since the constraints for the LFP have the form $a_i^\top x - \beta_i \le 0$ or, equivalently, $\|a_i\|^{-1} (a_i^\top x - \beta_i) \le 0$, as f_i one often takes functions f_i defined by

$$f_i(x) := h_i((a_i^\top x - \beta_i)_+) \tag{1.27}$$

or by

$$f_i(x) := h_i \left(\frac{(a_i^\top x - \beta_i)_+}{\|a_i\|} \right) \tag{1.28}$$

for some convex increasing functions $h_i : \mathbb{R}_+ \to \mathbb{R}_+$ with $h_i(0) = 0$, $i \in I$. In most cases the functions h_i are given by $h_i(u) := \omega_i u$ or $h_i(u) := \frac{1}{2}\omega_i u^2$ for $\omega_i > 0$, $i \in I$. If we apply these definitions of f_i and h_i to formulas (1.20)–(1.25) we obtain the following collection of proximity functions:

(a)

$$f(x) := \max_{i \in I}(a_i^\top x - \beta_i)_+, \tag{1.29}$$

in order to determine a point for which the value of the most violated constraint is minimal,

(b)

$$f(x) := \max_{i \in I} \frac{(a_i^\top x - \beta_i)_+}{\|a_i\|}, \tag{1.30}$$

in order to determine a point for which the distance to the furthest constraint is minimal,

(c)

$$f(x) := \sum_{i \in I} \omega_i (a_i^\top x - \beta_i)_+, \tag{1.31}$$

in order to determine a point for which the weighted sum of the values of the constraints is minimal,

(d)

$$f(x) := \sum_{i \in I} \omega_i \frac{(a_i^\top x - \beta_i)_+}{\|a_i\|}, \tag{1.32}$$

in order to determine a point for which the weighted sum of distances to the constraints is minimal,

(e)

$$f(x) := \frac{1}{2} \sum_{i \in I} \omega_i [(a_i^\top x - \beta_i)_+]^2, \tag{1.33}$$

in order to determine a point for which the weighted sum of squared values of the constraints is minimal, or

(f)

$$f(x) := \frac{1}{2} \sum_{i \in I} \omega_i \left[\frac{(a_i^\top x - \beta_i)_+}{\|a_i\|} \right]^2, \tag{1.34}$$

in order to determine a point for which the weighted sum of squared distances to the constraints is minimal.

All of the proximity functions presented herein are convex. The last two functions are even differentiable. If we denote the *residuum* of the ith constraint at a point x by $\rho_i(x)$, i.e., $\rho_i(x) := a_i^\top x - \beta_i$, $i \in I$, and by $r(x)$ the *residual vector*, i.e., $r(x) := (\rho_1(x), \ldots, \rho_m(x))$, then the proximity function (1.31) has the form $f(x) = w^\top r_+(x)$, where $w = (\omega_1, \ldots, \omega_m) \in \mathbb{R}^m_{++}$, and the proximity function (1.33) has the form $f(x) = \frac{1}{2} \|r_+(x)\|^2_W$ for $W := \operatorname{diag} w$. A system of linear equations $Ax = b$ is a special case of the LFP, because the system can be presented as

$$\begin{bmatrix} A \\ -A \end{bmatrix} x \le \begin{bmatrix} b \\ -b \end{bmatrix}.$$

In this case the proximity function f given by (1.34) obtains the form

$$f(x) := \frac{1}{2} \sum_{i \in I} \omega_i \left(\frac{a_i^\top x - \beta_i}{\|a_i\|} \right)^2. \tag{1.35}$$

Remark 1.3.10. If we multiply the inequality $a_i^\top x \le \beta_i$ by a constant $\alpha_i > 0$, $i \in I$, we obtain an equivalent linear feasibility problem, regardless of the proximity function under consideration, whenever $C \neq \emptyset$. In the case, where $C = \emptyset$, the new LFP defined by linear inequalities

$$\alpha_i a_i^\top x \leq \alpha_i \beta_i, i \in I, \tag{1.36}$$

is, in general, not equivalent to the original problem, even if we consider the same proximity function as for the original problem. However, there are some relationships between these two problems. The problems are equivalent if we use the proximity functions (1.30), (1.32) or (1.34). If we apply the proximity functions (1.31) or (1.33) with weights $\omega_i = \alpha_i^{-1} v_i$ or $\omega_i = \alpha_i^{-2} v_i$, respectively, $i \in I$, to problem (1.36), then we obtain a new LFP which is equivalent to the original one with the proximity function (1.31) or (1.33), respectively, and with weights $\omega_i = v_i$, $i \in I$. In particular, if we take $\alpha_i = \|a_i\|^{-1}, i \in I$, in (1.36), then the half-spaces in the new LFP are defined by normalized vectors. Therefore, if we apply the proximity function (1.33) with weights $\omega_i = \|a_i\|^2 v_i, i \in I$, to the new LFP, then the problem is equivalent to the original one with the proximity function (1.33) and with weights $\omega_i = v_i, i \in I$. Note that we can suppose, without loss of generality, that the vectors $a_i, i \in I$, are normalized. Otherwise, an appropriate change of the weights leads to an equivalent problem, because the weights $\omega_i, i \in I$, can be multiplied by a constant $\alpha > 0$ without changing the minimizers of the proximity function. Therefore, we can suppose that $w \in \operatorname{ri} \Delta_m$ instead of $w \in \mathbb{R}^m_{++}$.

Example 1.3.11. (*Linear least squares problem*) Given a linear system $Ax = b$, where A is a matrix of type $m \times n$, $x \in \mathbb{R}^n$ and $b \in \mathbb{R}^m$, the problem

$$\begin{aligned} &\text{minimize } \tfrac{1}{2} \|Ax - b\|^2 \\ &\text{subject to } x \in \mathbb{R}^n \end{aligned} \tag{1.37}$$

is called a *linear least squares problem* (LLSP). This problem is equivalent to the following compatible system of linear equations

$$A^\top A x = A^\top b \tag{1.38}$$

Actually, any solution of (1.38) has the form $x = x^N + \hat{x}$, where $x^N = A^+ b$, $\hat{x} \in \ker A$ and A^+ denotes the Moore–Penrose pseudoinverse of A. System (1.38) is called a *normal equation*. The equivalence of (1.37) and (1.38) follows from the necessary and sufficient condition for convex minimization (see Corollary 1.3.3). In particular, any solution of a compatible system $Ax = b$ is a solution of the LLSP. In general, one can solve system (1.38) by so called SVD-decomposition of the matrix $A^\top A$ (for details see, e.g., in [181, 234, 317]). We show that a solution of (1.37) (or, equivalently, (1.38)) with a minimal norm has the form $x := x^N = A^+ b$. This solution is called the *normal solution* of problem (1.37). Consider the following differentiable convex constrained minimization problem

$$\text{minimize } \tfrac{1}{2}\|x\|^2$$
$$\text{subject to } A^\top(Ax - b) = 0 \qquad (1.39)$$
$$x \in \mathbb{R}^n.$$

It follows from the definition of the metric projection that a solution x^* of (1.39) is equal to $P_L 0$, where $L := \{x \in \mathbb{R}^n : A^\top A x = A^\top b\}$. The uniqueness of the metric projection yields the uniqueness of x^*. The Lagrange function $L : \mathbb{R}^n \times \mathbb{R}^n \to \mathbb{R}$ has the form

$$L(x, y) = \frac{1}{2}\|x\|^2 + y^\top A^\top(Ax - b)$$

and the KKT-system has the form

$$\nabla_x L(x, y) = x + A^\top A y = 0$$
$$\nabla_y L(x, y) = A^\top(Ax - b) = 0. \qquad (1.40)$$

By the KKT-theorem, this system has a solution $(x, y) \in \mathbb{R}^n \times \mathbb{R}^n$. It follows from the definition of the Moore–Penrose pseudoinverse that

$$A^\top A A^+ b = A^\top (A A^+)^\top b = (A A^+ A)^\top b = A^\top b.$$

Therefore, $x^N := A^+ b$ satisfies the second equation of the KKT-system (1.40). By sufficient optimality conditions for the differentiable convex minimization problem (see Theorem 1.3.7), x^N is a solution of (1.39). Therefore, $x^* = x^N$. Subsuming, $x^N := A^+ b$ is the unique solution of (1.37).

Note that the LLSP is a special case of the LFP with the proximity function defined by (1.35).

1.3.6   General Convex Feasibility Problem

In many applications of the convex feasibility problems one seeks a point, for which some constraints are satisfied and a proximity function defined for the rest of constraints is minimal. The constraints which should be satisfied are called *hard constraints* and the rest of constraints are called *soft constraints*. We call the problem a *general convex feasibility problem* (GCFP). In other words, the GCFP is to find a point satisfying the hard constraints for which "distances" to soft constraints are minimal in some sense. Such model was described in details in [123].

Let $\{C_i\}_{i \in I}$ be a family of closed convex subsets. Let C_i for $i \in I_h$ be the hard constraints and C_i for $i \in I_s$ be the soft constraints, where $I = I_h \cup I_s$ and $I_h \cap I_s = \emptyset$. Furthermore, let $C_h := \bigcap_{i \in I_h} C_i$ and $C_s := \bigcap_{i \in I_s} C_i$. If $C_h \cap C_s \neq \emptyset$, then GCFP is to find $x^* \in C_h \cap C_s$. If we do not suppose $C_h \cap C_s \neq \emptyset$, then, similarly as for the CFP, one defines a convex proximity function $f : C_h \to \mathbb{R}_+$, defined on the subset C_h of hard constraints, which measures "distances" to the soft constraints. Similarly as for the CFP, we require the proximity function f to have the property

Fig. 1.7 Split feasibility
problem

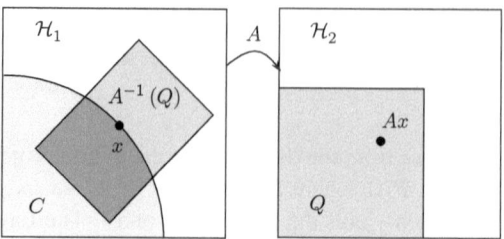

$f(x) = 0$ if and only if $x \in C_h \cap C_s$. As a proximity function one can use functions
described in Sect. 1.3.4, e.g., the function $f : \mathcal{H} \rightarrow \mathbb{R}_+$ defined by (1.24) with I
replaced by I_s.

1.3.7   Split Feasibility Problem

In some applications the general convex feasibility problem has the form: Given
closed convex subsets $C \subseteq \mathcal{H}_1$, $Q \subseteq \mathcal{H}_2$ of Hilbert spaces $\mathcal{H}_1, \mathcal{H}_2$ and a bounded
linear operator $A : \mathcal{H}_1 \rightarrow \mathcal{H}_2$, find a point $x \in C$ such that $Ax \in Q$ (if such a point
exists) (see Fig. 1.7). If $\mathcal{H}_1, \mathcal{H}_2$ are Euclidean spaces, then the above problem has
the form: Given closed convex subsets $C \subseteq \mathbb{R}^n$, $Q \subseteq \mathbb{R}^m$ and a matrix A of type
$m \times n$, find a point $x \in C$ such that $Ax \in Q$ (if such a point exists).

The subset C is a hard constraint and the subset $D := A^{-1}(Q) = \{x \in \mathcal{H}_1 :
Ax \in Q\}$ is a soft constraint. It is clear that D is closed convex, because for a
bounded linear operator A, the inverse image of a closed convex subset Q is closed
and convex. This problem was introduced by Censor and Elfving [88] and was called
a *split feasibility problem* (SFP). If we do not suppose that $C \cap D \neq \emptyset$, then,
similarly as for the GCFP, we introduce a convex proximity function $f : C \rightarrow \mathbb{R}_+$
and we present the SFP as minimization of f. Byrne [55, Sect. 2] has proposed a
proximity function defined by

$$f(x) := \frac{1}{2} \left\| P_Q(Ax) - Ax \right\|^2 . \tag{1.41}$$

The function f is convex as a composition of a linear operator A and a convex
function $\frac{1}{2}d^2(\cdot, Q)$. Note that $d(Ax, Q) = \left\| P_Q(Ax) - Ax \right\|$. The function defined
by (1.41) measures the square distance of the image of a point $x \in C$ by the operator
A to the subset Q. The function f has the required property:

$$x \in C \cap D \Leftrightarrow f(x) = 0.$$

Practical applications of the SFP require a simple structure of C and Q, which
allows an easy computation of $P_C(x)$ and $P_Q(y)$ for any $x \in \mathcal{H}_1$ and $y \in \mathcal{H}_2$.

1.3.8   Linear Split Feasibility Problem

Now consider the SFP in the finite dimensional case. If the subset $Q \subseteq \mathbb{R}^m$ has the form

$$Q := \{y \in \mathbb{R}^m : y \leq b\}, \tag{1.42}$$

where $b \in \mathbb{R}^m$, then the problem is called a *linear split feasibility problem* (LSFP). In other words, the problem can be written as follows: find $x \in C$ such that $Ax \leq b$ (if such a point exists). In general, we do not suppose, however, that $C \cap \{x \in \mathbb{R}^n : Ax \leq b\} \neq \emptyset$. Now we present a few examples of a proximity function for the LSFP which are often used in applications. In Sect. 4 we will show that for the subset Q defined by (1.42) we have

$$P_Q(Ax) - Ax = b - Ax.$$

(see Example 4.1.4). Therefore, the proximity function $f : C \to \mathbb{R}$ defined by (1.41), which is applied to the LSFP can be written in the form

$$f(x) := \frac{1}{2} \|r_+(x)\|^2 = \frac{1}{2} \sum_{i=1}^{m} [(a_i^\top x - \beta_i)_+]^2.$$

Note that f is differentiable. Alternatively, one can also apply a proximity function given by

$$f(x) := w^\top r_+(x) = \sum_{i=1}^{m} \omega_i (a_i^\top x - \beta_i)_+$$

or by

$$f(x) := \|r_+(x)\|_W^2 = \frac{1}{2} \sum_{i=1}^{m} \omega_i [(a_i^\top x - \beta_i)_+]^2, \tag{1.43}$$

where $W := \operatorname{diag} w$ for $w = (\omega_1, \ldots, \omega_m) \in \mathbb{R}_{++}^m$. The latter proximity function is used in many applications.

1.3.9   Multiple-Sets Split Feasibility Problem

The following problem, called a *multiple-sets split feasibility problem*, has been studied in the literature (see, e.g., [93, 103, 257, 344]): Given closed convex subsets $C_i \subseteq \mathcal{H}_1$, $i = 1, 2, \ldots, p$, $Q_j \subseteq \mathcal{H}_2$, $j = 1, 2, \ldots, r$, of Hilbert spaces $\mathcal{H}_1, \mathcal{H}_2$, respectively, and a bounded linear operator $A : \mathcal{H}_1 \to \mathcal{H}_2$, find a point $x \in C := \bigcap_{i=1}^{p} C_i$ such that $Ax \in Q := \bigcap_{j=1}^{r} Q_j$.

This problem can be considered as a split feasibility problem with the proximity function given by (1.41). The application of this proximity function requires, however, a simple structure of C and Q, which allows an easy computation of $P_C x$

and $P_Q y$ for any $x \in \mathbb{R}^n$ and $y \in \mathbb{R}^m$. If the structure of C_i, $i = 1, 2, \ldots, p$, and Q_j, $j = 1, 2, \ldots, r$, is simple, one can also use the following proximity function $f : \mathbb{R}^n \to \mathbb{R}$ proposed in [93]:

$$f(x) := \frac{1}{2} \sum_{i=1}^{p} \| P_{C_i} x - x \|^2 + \frac{1}{2} \sum_{j=1}^{r} \beta_j \| P_{Q_j} Ax - Ax \|^2.$$

Note that in this case the problem is a special case of the general convex feasibility problem, where all subsets C_i, $i = 1, 2, \ldots, p$, and Q_j, $j = 1, 2, \ldots, r$, are treated as soft constraints. If C_h has a simple structure allowing an easy computation of $P_{C_h} x$ for any $x \in \mathbb{R}^n$, one can also use the proximity function $f : C_h \to \mathbb{R}$ given by

$$f(x) := \frac{1}{2} \sum_{j=1}^{r} \beta_j \| P_{Q_j} Ax - Ax \|^2.$$

In this case the problem is a special case of the general convex feasibility problem, where C_i, $i = 1, 2, \ldots, p$, are hard constraints and Q_j, $j = 1, 2, \ldots, r$, are soft constraints.

1.4 Exercises

Exercise 1.4.1. Prove that in a Hilbert space the Cauchy–Schwarz inequality and the triangle inequality are equivalent.

Exercise 1.4.2. Let $w = (\omega_1, \omega_2, \ldots, \omega_m) \in \Delta_m$. Prove that

$$\left\| \sum_{i=1}^{m} \omega_i x_i \right\|^2 = \sum_{i=1}^{m} \omega_i \| x_i \|^2 - \frac{1}{2} \sum_{i,j=1}^{m} \omega_i \omega_j \| x_i - x_j \|^2. \tag{1.44}$$

Exercise 1.4.3. Prove the parallelogram law (see equality (1.8)).

Exercise 1.4.4. Prove Theorem 1.1.1.

Exercise 1.4.5. Let G be a positive definite matrix of type $m \times m$. Prove that the function $\langle \cdot, \cdot \rangle_G$ defined by

$$\langle w, v \rangle_G := w^\top G v$$

is an inner product in $\mathbb{R}^m$.

Exercise 1.4.6. Let $\mathcal{H} = l_2$ and $e_k = (e_{k1}, e_{k2}, \ldots)$ with

$$e_{kj} = \begin{cases} 1 \text{ if } j = k \\ 0 \text{ if } j \neq k, \end{cases}$$

$j, k \geq 0$. Prove that $\{e_k\}_{k=1}^{\infty}$ converges weakly in l_2 but does not converge strongly.

Exercise 1.4.7. Let $f : \mathcal{H} \to \mathbb{R}$ be a convex function. Prove that:

(a) The sublevel set $S(f, \alpha)$ is convex for any $\alpha \in \mathbb{R}$.
(b) The subset $\text{Argmin}_{x \in \mathcal{H}} f(x)$ is convex.

Exercise 1.4.8. Prove that the epigraph of a function $f : \mathcal{H} \to \mathbb{R}$ is a convex subset if and only if f is convex.

Exercise 1.4.9. Let $A : \mathcal{H}_1 \to \mathcal{H}_2$ be an affine operator, $C \subseteq \mathcal{H}_1$ and $D \subseteq \mathcal{H}_2$ be convex subsets. Show that $A(C)$ and $A^{-1}(D) \subseteq \mathcal{H}_1$ are convex.

Exercise 1.4.10. Evaluate derivatives of the following functions $f : \mathcal{H} \to \mathbb{R}$:

(a) $f(x) = \langle a, x \rangle$,
(b) $f(x) = \frac{1}{2} \langle Ax, x \rangle$, where $A : \mathcal{H} \to \mathcal{H}$ is a bounded linear operator.

Exercise 1.4.11. Evaluate derivatives of the following functions $f : \mathcal{H}_1 \to \mathbb{R}$:

(a) $f(x) = \frac{1}{2} \| Ax - b \|^2$, where $A : \mathcal{H}_1 \to \mathcal{H}_2$ is a bounded linear operator and $b \in \mathcal{H}_2$,
(b) $f(x) = \frac{1}{2} \| P_Q(Ax) - Ax \|^2$, where $A : \mathcal{H}_1 \to \mathcal{H}_2$ is a bounded linear operator and $Q \subseteq \mathcal{H}_2$ is a closed convex subset.

Exercise 1.4.12. Let A be an $m \times n$ matrix. Prove that the matrix $AA^\top$ is positive semi-definite and that $AA^\top$ is positive definite if and only if A has full row rank, i.e., the rows of A are linearly independent.

Exercise 1.4.13. Let A be a positive definite matrix of type $n \times n$ and $\lambda_{\min}(A)$ denote the smallest eigenvalue of A. Prove that:

(a) $\frac{1}{2} x^\top A x \geq \lambda_{\min}(A) \| x \|^2$ for any $x \in \mathbb{R}^n$,
(b) The function $f : \mathbb{R}^n \to \mathbb{R}$ defined by $f(x) := \frac{1}{2} x^\top A x$ is convex.

Exercise 1.4.14. Let $C \subseteq \mathcal{H}$ be a polytope, i.e., $C = \{x \in \mathcal{H} : \langle a_i, x \rangle \leq \beta_i, i \in I\}$ and let $x \in C$. Show that

$$N_C x = \text{cone}\{a_i : i \in I(x)\},$$

where $I(x) = \{i \in I : \langle a_i, x \rangle = \beta_i\}$.

Exercise 1.4.15. Let $\Delta_m = \{u \in \mathbb{R}^m : u \geq 0, e^\top u = 1\}$ denote a standard simplex in $\mathbb{R}^n$. Show that
$$\text{ri } \Delta_m = \{u \in \mathbb{R}^m : u > 0, e^\top u = 1\}.$$

Exercise 1.4.16. Let $C \subseteq \mathcal{H}$ be a nonempty convex subset. Show that the distance function $d(\cdot, C) : \mathcal{H} \to \mathbb{R}$ defined by $d(x, C) := \inf_{y \in C} \| x - y \|$ is convex.

Exercise 1.4.17. Prove Lemma 1.2.5.

Exercise 1.4.18. Let $f : \mathcal{H} \to \mathbb{R}$ be a differentiable convex function. Show that the function $h : \mathcal{H} \to \mathbb{R}_+$ defined by $h(x) = ((f(x))_+)^2$ is differentiable and convex.

Exercise 1.4.19. Let $f_i : X \to \mathbb{R}$, $i = 1, 2, \ldots, m$, be convex functions and $F : \mathbb{R}^m \to \mathbb{R}$ be a convex function which is nondecreasing with respect to any coordinate. Show that the function $f := F(f_1, \ldots, f_m)$ is convex.

Exercise 1.4.20. Let A be a positive definite matrix of type $n \times n$. Show that the function $f : \mathbb{R}^n \to \mathbb{R}$, $f(x) = \frac{1}{2} x^\top A x$ is strongly convex.

Exercise 1.4.21. Consider the following problem: Find a point $x \in \mathbb{R}^n_+$ for which the sum of squares of distances to the balls $C_i := B(a_i, \rho_i)$, where $a_i \in \mathbb{R}^n$, $\rho_i \geq 0$, $i = 1, 2, \ldots, m$, is minimal. Present the problem as a convex feasibility problem and as a generalized convex feasibility problem, where $\mathbb{R}^n_+$ is a subset of hard constraints. Write corresponding proximity functions.

Exercise 1.4.22. Prove Lemma 1.2.6.

Exercise 1.4.23. Prove Lemma 1.2.8.

Exercise 1.4.24. Prove Theorem 1.3.8.

Chapter 2
Algorithmic Operators

In Chap. 5 we will present several methods for solving convex optimization problems. We will focus our study on *iterative methods* (we also call them *iterative procedures* or *algorithms*) which are given in the form of the following recurrence

$$x^{k+1} = T_k x^k \tag{2.1}$$

defined on a closed convex subset $X \subseteq \mathcal{H}$, where $T_k : X \to X$ is a sequence of operators. We suppose that the starting point x^0 is an element of a starting subset $X_0 \subseteq X$. Usually, one supposes that $X_0 = X$. A sequence $\{x^k\}_{k=0}^{\infty}$ generated by the iterative method (2.1) is called an *approximating sequence*. If $T_k = T$ for all $k \geq 0$, then this sequence is called an *orbit* of T. Any iterative method for solving a convex optimization problem is constructed in such a way that the approximating sequences $\{x^k\}_{k=0}^{\infty}$ generated by this method converge (at least weakly) to a solution of the optimization problem. As we will see, the solution is a fixed point of an operator $S : X \to \mathcal{H}$, which is usually a nonexpansive one. The form of this operator depends on the considered optimization problem. A sequence of operators T_k which defines the iterative method is usually constructed in such a way that Fix $S \subseteq \bigcap_{k=0}^{\infty}$ Fix T_k.

In this chapter we deal with general properties of operators which define algorithms for solving convex optimization problems. In one iteration of the algorithm an appropriate operator $T : X \to X$ defines an *actualization*, also called an *update* x^+ of the current approximation x of a solution of the convex optimization problem. Usually, this actualization has the form $x^+ = Tx$. We call T an *algorithmic operator*. One can also consider algorithms, where the actualization has the form $x^+ \in Tx$ for a mapping (multifunction) $T : X \rightrightarrows X$. In this case, T is called an *algorithmic mapping*.

Operators defining iterations of an algorithm usually depend on some parameters which are constant or vary during the iteration process. The properties of approximating sequences depend on the properties of algorithmic operators defining the iterative method as well as on the choice of parameters defining these operators.

A. Cegielski, *Iterative Methods for Fixed Point Problems in Hilbert Spaces*, Lecture Notes in Mathematics 2057, DOI 10.1007/978-3-642-30901-4_2, © Springer-Verlag Berlin Heidelberg 2012

2.1 Basic Definitions and Properties

Let $\mathcal{H}$ be a Hilbert space. In what follows, we consider operators which are defined on a nonempty closed convex subset $X \subseteq \mathcal{H}$.

Remark 2.1.1. Let $U_i : X \to X$, $i \in I := \{1, 2, \ldots, m\}$. If (i) $U = \sum_{i \in I} \omega_i U_i$, where $w = (\omega_1, \omega_2, \ldots, \omega_m) \in \Delta_m$, or (ii) $U = U_m U_{m-1} \ldots U_1$, then the following obvious inclusion holds

$$\bigcap_{i \in I} \operatorname{Fix} U_i \subseteq \operatorname{Fix} U$$

The converse inclusion needs not to be true even if all U_i, $i \in I$, have a common fixed point (see Example 2.1.27).

Definition 2.1.2. Let $T : X \to \mathcal{H}$ and $\lambda \in [0, 2]$. The operator $T_\lambda : X \to \mathcal{H}$ defined by

$$T_\lambda := (1 - \lambda)\operatorname{Id} + \lambda T$$

is called a *λ-relaxation* or, shortly, *relaxation* of the operator T. Obviously, $T_\lambda = \operatorname{Id} + \lambda(T - \operatorname{Id})$. We call λ a *relaxation parameter*. If $\lambda \in (0, 1)$, then T_λ is called an *under-relaxation* of T. If $\lambda \in (1, 2)$, then T_λ is called an *over-relaxation* of T and if $\lambda = 2$, then T_λ is called the *reflection* of T. If $\lambda \in (0, 2)$, then T_λ is called a *strict relaxation* of T.

A relaxation T_λ of an operator T can be defined for any $\lambda \in \mathbb{R}$. However, if we do not extend explicitly the range of λ, we assume that $\lambda \in [0, 2]$.

Remark 2.1.3. Note that the equality $(T_\lambda)_\mu = T_{\lambda\mu}$ holds for all $\lambda, \mu \in \mathbb{R}$, consequently $(T_\lambda)_{\lambda^{-1}} = T$ for $\lambda \neq 0$.

Remark 2.1.4. It is clear that $\operatorname{Fix} T = \operatorname{Fix} T_\lambda$ whenever $\lambda \neq 0$.

Let $U_i : X \to X$, $i \in I := \{1, 2, \ldots, m\}$, $U := U_m U_{m-1} \ldots U_1$ and $Q_i := U_i U_{i-1} \ldots U_1 U_m \ldots U_{i+1}$, $i = 1, 2, \ldots, m$. Denote $Q_0 := Q_m = U$ and $U_0 := U_m$. There exists a relationship among the subsets of fixed points of operators Q_i, which is expressed by the following theorem.

Theorem 2.1.5. *For $i = 1, 2, \ldots, m$ there holds*

$$\operatorname{Fix} Q_i = U_i(\operatorname{Fix} Q_{i-1}). \tag{2.2}$$

Proof. Let $i \in I$. First we prove the inclusion

$$\operatorname{Fix} Q_i \supseteq U_i(\operatorname{Fix} Q_{i-1}). \tag{2.3}$$

Let $z^{i-1} \in \operatorname{Fix} Q_{i-1}$ and $z^i = U_i z^{i-1}$. Then we have

$$z^i = U_i z^{i-1} = U_i Q_{i-1} z^{i-1} = U_i U_{i-1} \ldots U_1 U_m \ldots U_{i+1} U_i z^{i-1}$$
$$= U_i U_{i-1} \ldots U_1 U_m \ldots U_{i+1} z^i = Q_i z^i,$$

which proves that (2.3) holds for any $i \in I$. Consequently,

$$\text{Fix } Q_i \supseteq U_i(\text{Fix } Q_{i-1}) \supseteq U_i U_{i-1}(\text{Fix } Q_{i-2})$$

$$\supseteq \ldots \supseteq U_i U_{i-1} \ldots U_1(\text{Fix } Q_0) = U_i U_{i-1} \ldots U_1(\text{Fix } Q_m)$$

$$\supseteq U_i U_{i-1} \ldots U_1 U_m(\text{Fix } Q_{m-1})$$

$$\supseteq \ldots \supseteq U_i U_{i-1} \ldots U_1 U_m U_{m-1} \ldots U_{i+1}(\text{Fix } Q_i)$$

$$= Q_i(\text{Fix } Q_i) = (\text{Fix } Q_i),$$

$i \in I$, and all inclusions are, actually, equations. In particular, $\text{Fix } Q_i = U_i(\text{Fix } Q_{i-1})$, i.e., (2.2) is satisfied for all $i \in I$.   □

2.1.1 Nonexpansive Operators

Definition 2.1.6. We say that an operator $T : X \to \mathcal{H}$ is:

(i) *Nonexpansive* (NE), if

$$\|Tx - Ty\| \leq \|x - y\|$$

for all $x, y \in X$,

(ii) *Strictly nonexpansive* if

$$\|Tx - Ty\| < \|x - y\| \text{ or } x - y = Tx - Ty$$

for all $x, y \in X$,

(iii) An α-*contraction*, where $\alpha \in (0, 1)$ or, shortly, a *contraction* if

$$\|Tx - Ty\| \leq \alpha \|x - y\|$$

for all $x, y \in X$.

The theorem below, called the *Banach fixed point theorem* or the *Banach theorem on contractions*, is widely applied in various areas of mathematics. The theorem holds for any complete metric space, and hence, in particular, for every closed subset of a Hilbert space.

Theorem 2.1.7 (Banach, 1922). *Let $\mathcal{X}$ be a complete metric space and $T : \mathcal{X} \to \mathcal{X}$ be a contraction. Then T has exactly one fixed point $x^* \in \mathcal{X}$. Furthermore, for any $x \in \mathcal{X}$, the orbit $\{T^k x\}_{k=0}^{\infty}$ converges to x^* with a rate of geometric progression.*

Proof. See, e.g., original paper of Banach [15], [185, Theorem 1.1], [267, Theorem 24.2], [184, Theorem 2.1], [183, Theorem 2.1] or [36, Theorem 2.1].   □

The Banach fixed point theorem is a widely applied tool for an iterative approximation of fixed points. Unfortunately, its application is restricted to contractions. We will need, however, appropriate tools for an iterative approximation of fixed points of nonexpansive operators T with Fix $T \neq \emptyset$.

Below, we present several classical fixed points theorems.

Theorem 2.1.8 (Brouwer, 1912). *Let* $X \subseteq \mathbb{R}^n$ *be nonempty compact and convex and* $T : X \to X$ *be continuous. Then* T *has a fixed point.*

Proof. See, e.g., original paper of Brouwer [43] or [191, Chap. II, §5, Theorem 7.2] or [183, Theorem 7.6].   □

The Brouwer fixed point theorem was generalized by Juliusz Schauder.

Theorem 2.1.9 (Schauder, 1930). *Let* X *be a nonempty compact and convex subset of a Banach space and* $T : X \to X$ *be continuous. Then* T *has a fixed point.*

Proof. See, e.g., original paper of Schauder [302] or [191, Chap. II, §6, Theorem 3.2] or [183, Theorem 8.1].   □

For nonexpansive operators in a Hilbert space $\mathcal{H}$ the compactness of $X \subseteq \mathcal{H}$ in the Schauder theorem can be replaced by the boundedness of X. The following theorem was proved independently by Browder [45, Theorem 1], Göhde [188] and by Kirk [227]. The proof can also be found, e.g., in [185, Theorem 5.1], [191, Chap. I, §4, Theorem 1.3], [183, Theorem 4.1] or [36, Theorem 3.1].

Theorem 2.1.10 (Browder–Göhde–Kirk, 1965). *Let* X *be a nonempty closed, convex and bounded subset of a uniformly convex Banach space (e.g., of a Hilbert space* $\mathcal{H}$*) and* $U : X \to X$ *be nonexpansive. Then* U *has a fixed point.*

Contrary to the Banach fixed point theorem, the theorems of Brouwer, Schauder and of Browder–Göhde–Kirk are only of existential nature. In Chap. 3 we present theorems which can be applied to iterative methods for determining fixed points of nonexpansive operators.

Below, we present some properties of the subset of fixed points of a nonexpansive operator. The following result can be found in [185, Proposition 5.3].

Proposition 2.1.11. *The subset of fixed points of a nonexpansive operator* $T :$ $X \to \mathcal{H}$ *is closed and convex.*

Proof. (cf. [185, Proposition 5.3]) Let $x^k \in$ Fix T and $x^k \to x$. We have $x \in X$ because X is closed. By the continuity of T,

$$x = \lim x^k = \lim_k T x^k = T x,$$

i.e., Fix T is a closed subset. Now we show the convexity of Fix T. Let $x, y \in$ Fix T, $x \neq y$ and $z = (1 - \lambda)x + \lambda y$ for $\lambda \in (0, 1)$. By the nonexpansivity of T and by the positive homogeneity of the norm we have

$$\|x - Tz\| = \|Tx - Tz\| \leq \|x - z\| = \lambda \|x - y\| \tag{2.4}$$

and
$$\|Tz - y\| = \|Tz - Ty\| \le \|z - y\| = (1 - \lambda)\|x - y\|. \tag{2.5}$$

Now, the triangle inequality yields
$$\begin{aligned}
\|x - y\| &\le \|x - Tz\| + \|Tz - y\| \\
&\le \lambda\|x - y\| + (1 - \lambda)\|x - y\| \\
&= \|x - y\|.
\end{aligned}$$

Consequently,
$$\|x - y\| = \|x - Tz\| + \|Tz - y\|.$$

By the strict convexity of the norm, the vectors $x - Tz$ and $Tz - y$ are positive linearly dependent. Therefore, $\alpha(x - Tz) + \beta(y - Tz) = 0$ for some $\alpha, \beta \ge 0$. Since $x \ne y$, it follows that $\alpha + \beta > 0$, and hence, $Tz = \frac{\alpha}{\alpha+\beta}x + \frac{\beta}{\alpha+\beta}y$. Now, the nonexpansivity of T and inequalities (2.4) and (2.5) yield

$$\frac{\beta}{\alpha + \beta}\|x - y\| = \|x - Tz\| = \|Tx - Tz\| \le \|x - z\| = \lambda\|x - y\| \tag{2.6}$$

and

$$\frac{\alpha}{\alpha + \beta}\|x - y\| = \|Tz - y\| = \|Tz - Ty\| \le \|z - y\| = (1 - \lambda)\|x - y\|. \tag{2.7}$$

If at least one inequality in (2.6) and (2.7) is strict, then by summing up (2.6) and (2.7) we would obtain a contradiction. Therefore, $\frac{\beta}{\alpha+\beta} = \lambda$ and $\frac{\alpha}{\alpha+\beta} = (1-\lambda)$, consequently $Tz = (1 - \lambda)x + \lambda y = z$.   □

The closedness and convexity of the subset of fixed points of a nonexpansive operator follows also from a property presented in Sect. 2.2 (see Corollary 2.2.48).

Lemma 2.1.12. *Let $S_i : X \to X$, $i \in I := \{1, 2, \ldots, m\}$, be nonexpansive. Then:*

(i) *A convex combination $S := \sum_{i \in I} \omega_i S_i$, where $w = (\omega_1, \ldots, \omega_m) \in \Delta_m$, is nonexpansive. If, furthermore, at least one operator S_i is a contraction and the corresponding weight $\omega_i > 0$, then S is a contraction;*

(ii) *A composition $S := S_m S_{m-1} \ldots S_1$ is nonexpansive. If, furthermore, at least one operator S_i is a contraction, then S is a contraction.*

Proof. Let $x, y \in X$ and S_i be nonexpansive, i.e., $\|S_i x - S_i y\| \le \alpha_i\|x - y\|$, where $\alpha_i \in (0, 1]$, $i \in I$.

(i) Let $w \in \Delta_m$, $S := \sum_{i \in I} \omega_i S_i$ and $\alpha = \sum_{j \in I} \omega_j \alpha_j$. It is clear that $\alpha \in (0, 1]$. By the convexity of the norm and the nonexpansivity of S_i, $i \in I$, we have

$$\|Sx - Sy\| = \left\| \sum_{i \in I} \omega_i (S_i x - S_i y) \right\|$$

$$\leq \sum_{i \in I} \omega_i \|S_i x - S_i y\|$$

$$\leq \sum_{i \in I} \omega_i \alpha_i \|x - y\|$$

$$= \sum_{i \in I} \frac{\omega_i \alpha_i}{\sum_{j \in I} \omega_j \alpha_j} \alpha \|x - y\|$$

$$= \alpha \|x - y\|,$$

i.e., S is a nonexpansive operator. Now suppose that S_{i_0} is a contraction, i.e., $\alpha_{i_0} < 1$ and that $\omega_{i_0} > 0$, for some $i_0 \in I$. Then $\alpha \in (0,1)$, i.e., S is a contraction.

(ii) We have

$$\|Sx - Sy\| = \|S_m S_{m-1} \ldots S_1 x - S_m S_{m-1} \ldots S_1 y\| \leq \alpha \|x - y\|,$$

where $\alpha = \alpha_m \alpha_{m-1} \ldots \alpha_1 \in (0, 1]$. If S_{i_0} is a contraction for some $i_0 \in I$, i.e., $\alpha_{i_0} \in (0,1)$, then, of course, $\alpha \in (0,1)$ and S is a contraction.   □

Theorem 2.1.13. *Let $U_i : X \to X$ be nonexpansive for all $i \in I := \{1, 2, \ldots, m\}$, and $U := U_m U_{m-1} \ldots U_1$. If $U_j(X)$ is bounded for at least one $j \in I$, then* Fix $U \neq \emptyset$.

Proof. Let $U_j(X)$ be bounded for some $j \in I$. Since U_i are nonexpansive, $i \in I$, the boundedness of $U_j(X)$ yields the boundedness of $U(X)$. Therefore, $Y := \operatorname{cl} \operatorname{conv} U(X)$ is closed, convex and bounded. Since $U(X) \subseteq X$ and X is closed and convex, we have $Y \subseteq X$. The operator $U \mid_Y$ maps a closed, convex and bounded subset Y into itself. By the Browder–Göhde–Kirk Fixed Point Theorem, the operator $U \mid_Y$ has a fixed point $z \in Y$. Of course, $Uz = U \mid_Y (z) = z$.   □

Theorem 2.1.14. *Let $U_i : X \to \mathcal{H}$, $i \in I := \{1, 2, \ldots, m\}$, be nonexpansive operators with a common fixed point and $U := \sum_{i \in I} \omega_i U_i$ with $w \in \operatorname{ri} \Delta_m$. Then*

$$\operatorname{Fix} U = \bigcap_{i \in I} \operatorname{Fix} U_i.$$

Proof. The inclusion $\bigcap_{i \in I} \operatorname{Fix} U_i \subseteq \operatorname{Fix} U$ is always true (see Remark 2.1.1). Now we show that the converse inclusion also holds. Let $z \in \operatorname{Fix} U$ and $u \in \bigcap_{i \in I} \operatorname{Fix} U_i$. If $z = u$, then, of course, $z \in \bigcap_{i \in I} \operatorname{Fix} U_i$. Otherwise, for $z \neq u$, by the convexity of the norm and by the nonexpansivity of U_i, $i \in I$, we have

$$\|z - u\| = \|U z - u\|$$

$$= \left\| \sum_{i \in I} \omega_i U_i z - u \right\| = \left\| \sum_{i \in I} \omega_i (U_i z - u) \right\|$$

$$\leq \sum_{i \in I} \omega_i \|U_i z - u\| = \sum_{i \in I} \omega_i \|U_i z - U_i u\|$$

$$\leq \sum_{i \in I} \omega_i \|z - u\| = \|z - u\|.$$

Consequently,

$$\left\| \sum_{i \in I} \omega_i (U_i z - u) \right\| = \sum_{i \in I} \omega_i \|U_i z - u\| = \sum_{i \in I} \omega_i \|z - u\|. \tag{2.8}$$

Since $\omega_i > 0$ for all $i \in I$, the first equality in (2.8) yields a positive linear dependence of all pairs of vectors $U_i z - u$ and $U_j z - u$, $i, j \in I$, $i \neq j$, i.e.,

$$\|U_i z - u\| (U_j z - u) = \|U_j z - u\| (U_i z - u). \tag{2.9}$$

The second equality in (2.8), together with the inequality $\|U_i z - u\| \leq \|z - u\|$, $i \in I$, and the assumption $\omega_i > 0$, $i \in I$, yield

$$\|U_i z - u\| = \|z - u\| \tag{2.10}$$

for all $i \in I$. Since $z \neq u$, we have $U_i z \neq u$, $i \in I$. Now, it follows from (2.9) and (2.10) that $U_i z = v$ for all $i \in I$ and for some $v \in \mathcal{H}$. Consequently,

$$z = U z = \sum_{j \in I} \omega_j U_j z = \sum_{j \in I} \omega_j v = v = U_i z,$$

for all $i \in I$, i.e., $z \in \bigcap_{i \in I} \text{Fix } U_i$. $\square$

2.1.2   Quasi-nonexpansive Operators

Definition 2.1.15.   We say that an operator $T : X \to \mathcal{H}$ is:

(i) *Fejér monotone* (FM) with respect to a nonempty subset $C \subseteq X$ if

$$\|T x - z\| \leq \|x - z\|$$

for all $x \in X$ and $z \in C$,

Fig. 2.1 Equivalence (2.11)

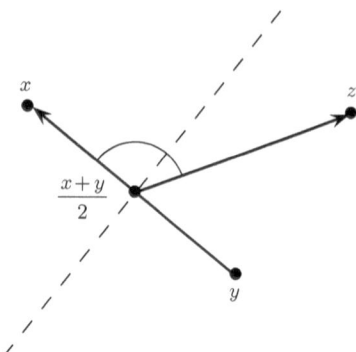

(ii) *Strictly Fejér monotone* with respect to a nonempty subset $C \subseteq X$ if

$$\|Tx - z\| < \|x - z\|$$

for all $x \notin C$ and $z \in C$.

Remark 2.1.16. Because of the following obvious equivalence

$$\|z - y\| \le \|z - x\| \iff \left\langle z - \frac{y + x}{2}, y - x \right\rangle \ge 0 \tag{2.11}$$

for arbitrary $x, y, z \in \mathcal{H}$ (see Fig. 2.1), an operator $T : X \to \mathcal{H}$ is Fejér monotone with respect to C if and only if

$$\left\langle z - \frac{Tx - x}{2}, Tx - x \right\rangle \ge 0. \tag{2.12}$$

Furthermore, T is strictly Fejér monotone if and only if inequality (2.12) is strict for all $x \notin C$. We have not supposed that C is closed convex in Definition 2.1.15. Inequality (2.12) yields, however, that if T is (strictly) Fejér monotone with respect to C, then T is (strictly) Fejér monotone with respect to $\operatorname{conv} C$. Furthermore, the continuity of the norm yields that if T is Fejér monotone with respect to C, then T is Fejér monotone with respect to $\operatorname{cl} C$. Therefore, we can suppose, without loss of generality, that C is closed convex in Definition 2.1.15 (i) and that C is convex in Definition 2.1.15 (ii).

There exists the largest subset, with respect to which an operator T is Fejér monotone. This subset is closed and convex, as follows from the following lemma.

Lemma 2.1.17. *Let $T : X \to \mathcal{H}$. If the subset*

$$\operatorname{Fej} T := \bigcap_{x \in X} \left\{ z \in X : \left\langle z - \frac{Tx + x}{2}, Tx - x \right\rangle \ge 0 \right\} \tag{2.13}$$

is nonempty, then $\operatorname{Fej} T$ is the largest subset, with respect to which T is Fejér monotone.

Proof. The assertion follows directly from the equivalence (2.11).   □

Remark 2.1.18. Because of frequent use we state some obvious properties of Fejér monotone operators:

(i) If T is (strictly) Fejér monotone with respect to a nonempty subset $C \subseteq \mathcal{H}$, then for an arbitrary $\lambda \in (0, 1)$ its relaxation T_λ is also (strictly) Fejér monotone with respect to C.

(ii) If T is (strictly) Fejér monotone with respect to a nonempty subset $C \subseteq \mathcal{H}$, then T is (strictly) Fejér monotone with respect to any nonempty subset $D \subseteq C$.

(iii) Every composition and every convex combination of operators which are Fejér monotone with respect to a nonempty subset $C \subseteq \mathcal{H}$ is Fejér monotone with respect C.

Definition 2.1.19. We say that an operator $T : X \to \mathcal{H}$ having a fixed point is:

(i) *Quasi-nonexpansive* (QNE) if T is Fejér monotone with respect to Fix T, i.e.,

$$\|Tx - z\| \leq \|x - z\|$$

for all $x \in X$ and $z \in$ Fix T,

(ii) *Strictly quasi-nonexpansive* (sQNE) if T is strictly Fejér monotone with respect to Fix T, i.e.,

$$\|Tx - z\| < \|x - z\|$$

for all $x \notin$ Fix T and $z \in$ Fix T,

(iii) *C-strictly quasi-nonexpansive* (C-sQNE), where $C \neq \emptyset$ and $C \subseteq$ Fix T, if T is quasi-nonexpansive and

$$\|Tx - z\| < \|x - z\|$$

for all $x \notin$ Fix T and $z \in C$.

For an operator T having a fixed point the following relation is clear:

$$T \text{ is sQNE} \Longrightarrow T \text{ is } C\text{-sQNE}$$

where $C \subseteq$ Fix T. Furthermore, by definition,

$$T \text{ is Fix } T\text{-sQNE} \Longrightarrow T \text{ is sQNE.}$$

The metric projection onto a closed convex subset is a typical example of a strictly quasi-nonexpansive operator.

A nonexpansive and strictly Fejér monotone operator is also called *attracting* (see [22, Definition 2.1]). Yamada and Ogura use the name an *attracting quasi-nonexpansive* operator for a strictly quasi-nonexpansive one (see [346, page 623]). Vasin and Ageev call these operators *strongly Q-quasi-nonexpansive* (see [333, Definition 2.2]). Reich and Zaslavski define a more general operator than the strictly quasi-nonexpansive one and call it an *F-attracting mapping*, where $F = \operatorname{Fix} T$ (see [297, Sect. 1]). A continuous strictly quasi-nonexpansive operator is also called a *paracontraction* (see, [164, Definition 1]). The class of quasi-nonexpansive operators is denoted in [126, page 161] by $\mathcal{F}^0$. Properties of quasi-nonexpansive operators in metric spaces have been intensively studied since 1969 (see, e.g., [145, 148, 283], [50, Sect. 1], [113]), but the name quasi-nonexpansive was introduced by Dotson [147].

Lemma 2.1.20. *A nonexpansive operator $U : X \to \mathcal{H}$ with a fixed point is quasi-nonexpansive.*

Proof. Let U be nonexpansive and $z \in \operatorname{Fix} U$. Then

$$\|Ux - z\| = \|Ux - Uz\| \le \|x - z\|,$$

i.e., U is quasi-nonexpansive. □

It is clear that the class of nonexpansive operators having a fixed point is an essential subclass of quasi-nonexpansive operators, because a quasi-nonexpansive operator needs not to be continuous. Moreover, a quasi-nonexpansive operator needs not to be nonexpansive even if it is continuous (see Exercise 2.5.2). In this section we present properties of the family of quasi-nonexpansive operators. In further parts of the book we show that these operators play an important role in iterative methods for fixed point problems.

The following lemma gives a relation between the subset $\operatorname{Fej} T$ and the subset $\operatorname{Fix} T$ for an operator $T : X \to \mathcal{H}$ (cf. [24, Proposition 2.6 (ii)]).

Lemma 2.1.21. *For any operator $T : X \to \mathcal{H}$ the inclusion $\operatorname{Fej} T \subseteq \operatorname{Fix} T$ holds. If $\operatorname{Fix} T \ne \emptyset$ and T is quasi-nonexpansive, then the converse inclusion also holds. Consequently, the subset of fixed points of a quasi-nonexpansive operator is closed and convex.*

Proof. If $\operatorname{Fej} T = \emptyset$, then the first part of the assertion is obvious. Now let $\operatorname{Fej} T \ne \emptyset$ and $w \in \operatorname{Fej} T$. Then, for $z = x = w$ in (2.13), we obtain

$$0 \le \langle w - \frac{Tw + w}{2}, Tw - w \rangle$$

$$= -\frac{1}{2} \|Tw - w\|^2 \le 0,$$

i.e., $Tw = w$. Therefore, $\operatorname{Fej} T \subseteq \operatorname{Fix} T$. Now suppose that $\operatorname{Fix} T \ne \emptyset$ and that T is quasi-nonexpansive, i.e., T is Fejér monotone with respect to $\operatorname{Fix} T$. Then, Lemma 2.1.17 yields the inclusion $\operatorname{Fix} T \subseteq \operatorname{Fej} T$, which together with the first part of the lemma gives $\operatorname{Fix} T = \operatorname{Fej} T$. The convexity and the closedness of $\operatorname{Fix} T$ follows now from Lemma 2.1.17 and from the fact that the intersection of closed half-spaces is closed and convex. □

Fig. 2.2 Nonconvex Fix T
for a Fejér monotone
operator T

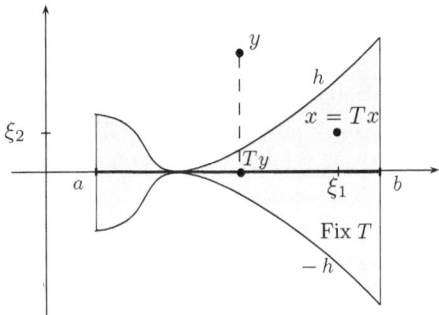

Remark 2.1.22. It follows from Remark 2.1.18 (ii), Lemmas 2.1.17 and 2.1.21 that a quasi-nonexpansive operator $T : X \to X$ is Fejér monotone with respect to any nonempty subset of Fix T. Therefore, we will restrict our further consideration of Fejér monotone operators to quasi-nonexpansive ones. Note, however, that without the quasi nonexpansivity of T the equality Fix T = Fej T needs not to be true. In this case, Fix T needs not to be convex, even if T is Fejér monotone.

Example 2.1.23. Let $\mathcal{H} = \mathbb{R}^2$, $X := [a,b] \times \mathbb{R}$ for $-\infty \leq a \leq b \leq +\infty$ and $h : X \to \mathbb{R}_+$ be a function with $\inf_{x\in[a,b]} h(x) = 0$. Define the operator $T : X \to \mathbb{R}^2$ by

$$Tx := \begin{cases} x & \text{if } |\xi_2| \leq h(\xi_1) \\ (\xi_1, 0) & \text{if } |\xi_2| > h(\xi_1), \end{cases}$$

where $x = (\xi_1, \xi_2)$ (see Fig. 2.2).
The reader may check that Fej $T = [a,b] \times \{0\}$ and that Fix $T = \{x \in X : |\xi_2| \leq h(\xi_1)\}$. If h is positive in at least one point, then Fej $T \neq$ Fix T. If, moreover, h is not concave, then Fix T is not convex.

Let $U_i : X \to X$, $i \in I := \{1, 2, \ldots, m\}$, and:

(i) $U := \sum_{i\in I} \omega_i U_i$, where $w = (\omega_1, \omega_2, \ldots, \omega_m) \in \Delta_m$ or
(ii) $U := U_m U_{m-1} \ldots U_1$.

As we observed before, the following inclusion holds

$$\bigcap_{i\in I} \text{Fix } U_i \subseteq \text{Fix } U \tag{2.14}$$

(see Remark 2.1.1) and the converse inclusion holds in case (i) when all U_i, $i \in I$, are nonexpansive operators with a common fixed point and $w \in \text{ri } \Delta_m$ (see Theorem 2.1.14). It turns out that, in both cases (i) and (ii), the inclusion converse to (2.14)) is true for strictly quasi-nonexpansive operators (see [22, Proposition 2.12], where the property was formulated for attracting operators). In case (i) this property is also true for a more general form of the operator $U = \sum_{i\in I} \omega_i U$, where the weights ω_i, $i \in I$, may depend on x.

Definition 2.1.24. A function $w : X \to \Delta_m$, with $w(x) = (\omega_1(x), \ldots, \omega_m(x))$ is called a *weight function*.

Definition 2.1.25. Let $U_i : X \to \mathcal{H}, i \in I$. We say that the weight function $w :$ $X \to \Delta_m$ is *appropriate with respect to the family* $\{U_i\}_{i \in I}$ or, shortly, *appropriate*, if for any $x \notin \bigcap_{i \in I} \text{Fix } U_i$ there exists an index $j \in I$ such that

$$\omega_j(x) \, \| \, U_j x - x \, \| \neq 0. \tag{2.15}$$

Denote

$$I(x) := \{i \in I : x \notin \text{Fix } U_i\} \tag{2.16}$$

for a family of operators $U_i : X \to \mathcal{H}, i \in I$. The subset $I(x)$ is called a subset of *violated constraints*. Note that w is appropriate if and only if

$$w_j(x) > 0 \text{ for some } j \in I(x) \tag{2.17}$$

and for any $x \notin \bigcap_{i \in I} \text{Fix } U_i$ (or, equivalently, for any $x \in \mathcal{H}$ such that $I(x) \neq \emptyset$).
   A weight function $w : X \to \text{ri } \Delta_m$ is appropriate with respect to any family of operators $\{U_i\}_{i \in I}$ if:

(i) $w \in \text{ri } \Delta_m$ is a vector of constant weights (this case was considered in [22, Proposition 2.12]), or if
(ii) $w_i(x) > 0$ for all $x \notin \text{Fix } U_i$ and for all $i \in I$.

   It is clear that property (2.15) is weaker than conditions (i) and (ii) above.
   The following theorem extends important results of [22, Proposition 2.12], where $C = \bigcap_{i \in I} \text{Fix } U_i$ and only constant weights are considered (see also [25, Proposition 2.5] for a related result). These extended results will be applied in further parts of the book.

Theorem 2.1.26. *Let the operators* $U_i : X \to X, i \in I$, *with* $\bigcap_{i \in I} \text{Fix } U_i \neq \emptyset$, *be* C-*strictly quasi-nonexpansive, where* $C \subseteq \bigcap_{i \in I} \text{Fix } U_i, C \neq \emptyset$. *If* U *has one of the following forms:*

(i) $U := \sum_{i \in I} \omega_i U_i$ *and the weight function* $w : X \to \Delta_m$ *is appropriate,*
(ii) $U := U_m U_{m-1} \ldots U_1,$

   then

$$\text{Fix } U = \bigcap_{i \in I} \text{Fix } U_i \tag{2.18}$$

and U *is* C-*strictly quasi-nonexpansive.*

Proof. The inclusion $\bigcap_{i \in I} \text{Fix } U_i \subseteq \text{Fix } U$ is obvious. Now we show that $\text{Fix } U \subseteq$ $\bigcap_{i \in I} \text{Fix } U_i$. This inclusion is clear if $\bigcap_{i \in I} \text{Fix } U_i = X$. Now suppose that $x \notin$ $\bigcap_{i \in I} \text{Fix } U_i$. Let $z \in \bigcap_{i \in I} \text{Fix } U_i$. If $z \in C$, then the C-strict quasi nonexpansivity of $U_i, i \in I$, yields

$$\|U_i x - z\| < \|x - z\| \text{ for any } i \in I(x). \tag{2.19}$$

(i) Let $Ux = \sum_{i\in I} \omega_i(x)U_i x$, where the weight function $w : X \to \Delta_m$ is appropriate. Then the convexity of the norm, (2.19) and (2.15) yield

$$\|Ux - z\| = \left\| \sum_{i\in I} \omega_i(x)(U_i x - z) \right\|$$

$$\leq \sum_{i\in I} \omega_i(x) \|U_i x - z\| \leq \sum_{i\in I} \omega_i(x) \|x - z\| = \|x - z\|,$$

where the second inequality is strict if $z \in C$.

(ii) Let $j := \min\{i \in I : x \notin \text{Fix } U_i\}$. Then we have $U_j U_{j-1} \ldots U_1 x = U_j x$ and (2.19) yields

$$\|Ux - z\| = \|U_m \ldots U_1 x - z\|$$
$$= \|U_m \ldots U_j x - z\|$$
$$\leq \|U_{m-1} \ldots U_j x - z\|$$
$$\leq \ldots \leq \|U_j x - z\| \leq \|x - z\|,$$

where the latter inequality is strict if $z \in C$.

Now it is clear that $x \notin \text{Fix } U$ because, otherwise, for $z \in C$, in both cases (i) and (ii) we would obtain

$$\|x - z\| = \|Ux - z\| < \|x - z\|,$$

a contradiction. We have proved that $\text{Fix } U \subseteq \bigcap_{i\in I} \text{Fix } U_i$. Hence, (2.18) holds and, in both cases (i) and (ii), U is C-strictly quasi-nonexpansive.   □

Note that equality (2.18) needs not to be true for nonexpansive operators, even if they have a common fixed point.

Example 2.1.27.   (cf. [22, Remark 2.11]) Let $X \subseteq \mathcal{H}$ be a subspace with dim $X > 0$. Let $U_i : X \to X$, $U_i := -\text{Id}$, $i = 1, 2$. We have $U_2 U_1 = \text{Id}$, consequently, $\text{Fix}(U_2 U_1) = X$, but $\text{Fix } U_1 \cap \text{Fix } U_2 = \{0\}$.

The assumption on the C-strict quasi nonexpansivity in Theorem 2.1.26 (i) can be weakened. In this case it suffices to suppose that all U_i are quasi-nonexpansive, $i \in I$, and at least one of them is C-strictly quasi-nonexpansive. The assumption that the weight function w is appropriate should be replaced in this case by a stronger one, namely: $w_j(x) > 0$ for all x such that $I(x) \neq \emptyset$ and for all $j \in I(x)$. We leave the proof of this fact to the reader.

A stronger version of the first part of Theorem 2.1.26 (ii) for two operators is stated below (cf. [346, Proposition 1(d) (i)]).

Theorem 2.1.28. *Let* $S : X \rightarrow X$ *be quasi-nonexpansive,* $T : X \rightarrow X$ *be strictly quasi-nonexpansive and* Fix $S \cap$ Fix $T \neq \emptyset$. *Then* Fix $ST =$ Fix $TS =$ Fix $S \cap$ Fix T. *Furthermore,* ST *is quasi-nonexpansive and* TS *is strictly quasi-nonexpansive.*

Proof. The inclusions Fix $S \cap$ Fix $T \subseteq$ Fix ST and Fix $S \cap$ Fix $T \subseteq$ Fix TS are clear.

(i) We prove that Fix $ST \subseteq$ Fix $S \cap$ Fix T. The inclusion is obvious if Fix $ST = \emptyset$. Suppose that Fix $ST \neq \emptyset$ and let $x \in$ Fix ST be such that $x \notin$ Fix $S \cap$ Fix T. We consider two cases:

   (a) $x \in$ Fix T. Then $x = STx = Sx$, i.e., $x \in$ Fix S. Therefore, $x \in$ Fix $S \cap$ Fix T.
   (b) $x \notin$ Fix T. Let $z \in$ Fix $S \cap$ Fix T. By the quasi nonexpansivity of S and by the strict quasi nonexpansivity of T, we have

$$\|x - z\| = \|STx - z\| \leq \|Tx - z\| < \|x - z\|.$$

   In both cases we obtain a contradiction, which proves that Fix $ST \subseteq$ Fix $S \cap$ Fix T.

(ii) We prove that Fix $TS \subseteq$ Fix $T \cap$ Fix S. The inclusion is obvious if Fix $TS = \emptyset$. Suppose that Fix $TS \neq \emptyset$ and let $x \in$ Fix TS be such that $x \notin$ Fix $T \cap$ Fix S. Consider two cases:

   (a) $Sx \in$ Fix T. Then $x = TSx = Sx$, consequently, $x \in$ Fix S. Now we have $x = Sx \in$ Fix T, i.e., $x \in$ Fix $T \cap$ Fix S.
   (b) $Sx \notin$ Fix T. Let $z \in$ Fix $T \cap$ Fix S. By the strict quasi nonexpansivity of T and by the quasi nonexpansivity of S, we have

$$\|x - z\| = \|TSx - z\| < \|Sx - z\| \leq \|x - z\|.$$

   In both cases we obtain a contradiction, which proves that Fix $TS \subseteq$ Fix $T \cap$ Fix S.

Let now $z \in$ Fix $TS =$ Fix $T \cap$ Fix S and $x \in X$. We have

$$\|STx - z\| \leq \|Tx - z\| \leq \|x - z\|,$$

i.e., ST is quasi-nonexpansive. Furthermore,

$$\|TSx - z\| \leq \|Sx - z\| \leq \|x - z\|,$$

where the second inequality is strict if $x \notin$ Fix S and the first one is strict if $x \in$ Fix S and $x \notin$ Fix T. Hence, TS is strictly quasi-nonexpansive.   □

Fig. 2.3 $P_{\mathrm{cl\,conv}\,C}$ is a
separator of C

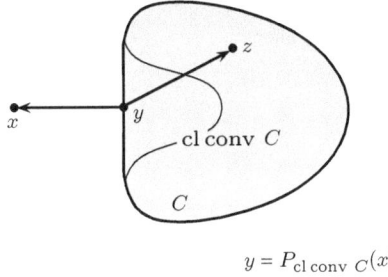

$$y = P_{\mathrm{cl\,conv}\,C}(x)$$

Corollary 2.1.29. *Let* $U := U_m U_{m-1} \ldots U_1$, *where* $U_1, U_2, \ldots, U_{m-1} : X \to X$ *are quasi-nonexpansive,* $U_m : X \to X$ *is strictly quasi-nonexpansive and* $\bigcap_{i \in I} \mathrm{Fix}\, U_i \neq \emptyset$. *Then* $\mathrm{Fix}\, U = \bigcap_{i \in I} \mathrm{Fix}\, U_i$ *and* U *is strictly quasi-nonexpansive.*

Proof. The corollary follows from Theorem 2.1.28. We leave to the reader an easy proof by induction with respect to m. □

The assumption of Theorem 2.1.28 that T is strictly quasi-nonexpansive is essential. Note that the composition of quasi-nonexpansive operators needs not to be quasi-nonexpansive (see Example 2.1.52). Furthermore, the assumptions of Theorem 2.1.28 do not yield the strict quasi nonexpansivity of the operator ST (see Example 2.1.54).

2.1.3 Cutters and Strongly Quasi-nonexpansive Operators

Definition 2.1.30. Let $x \in \mathcal{H}$. We say that $y \in \mathcal{H}$ *separates a subset* $C \subseteq \mathcal{H}$ from x if

$$\langle x - y, z - y \rangle \leq 0$$

for all $z \in C$. We say that an operator $T : X \to \mathcal{H}$ is a *separator of a subset* $C \subseteq X$ or T *separates a subset* C, if $y := Tx$ separates C from x for all $x \in \mathcal{H}$. We say that T is an α-*relaxed separator* of C, where $\alpha \in [0, 2]$, if T is an α-relaxation of a separator of C. Let T have a fixed point. We say that T is a *cutter* if T is a separator of $\mathrm{Fix}\, T$, i.e.,

$$\langle x - Tx, z - Tx \rangle \leq 0 \qquad (2.20)$$

for all $x \in X$ and all $z \in \mathrm{Fix}\, T$. We say that T is an α-*relaxed cutter*, where $\alpha \in [0, 2]$, if T is an α-relaxed separator of $\mathrm{Fix}\, T$.

For any nonempty $C \subseteq \mathcal{H}$ the projection $P_{\mathrm{cl\,conv}\,C}$ is a separator of C (see Fig. 2.3). In general, a separator of C is not uniquely determined.

The name *cutter* expresses the fact that, for any $x \notin \mathrm{Fix}\, T$, the hyperplane $H(x - Tx, \langle Tx, x - Tx \rangle)$ cuts the space into two half-spaces, one of which contains the point x while the other one contains the subset $\mathrm{Fix}\, T$ (see Fig. 2.4). In the literature one can find different names for cutters. Bauschke and Combettes call the class of

Fig. 2.4 A cutter and an
α-relaxed cutter

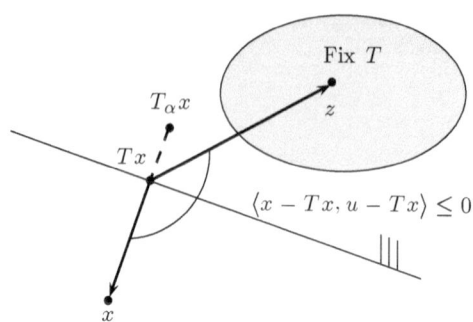

cutters a $\mathcal{T}$-*class* (see [24, Definition 2.2] and [121, Definition 2.1]). Yamada and Ogura (see [346, Sect. B]) and Mărușter (see [254]) call the operators *firmly quasi-nonexpansive*. Zaknoon, Segal and Censor denoted cutters as *directed operators* (see [104–106, 307, 356]). In [69] these operators were called *separating operators*. The name *cutter* was proposed by Cegielski and Censor in [70].

Note that a separator and, in particular, a cutter need not to be continuous operators.

Remark 2.1.31. Let $T : X \to \mathcal{H}$ have a fixed point. Then, by Lemma 1.2.5, the operator T is a cutter if and only if

$$\langle Tx - x, z - x \rangle \geq \|Tx - x\|^2 \tag{2.21}$$

holds for all $x \in X$ and for all $z \in \text{Fix}\, T$ (cf. [121, Proposition 2.3 (ii)]), and T is an α-relaxed cutter, where $\alpha \in [0, 2]$, if and only if

$$\alpha \langle Tx - x, z - x \rangle \geq \|Tx - x\|^2 \tag{2.22}$$

holds for all $x \in X$ and for all $z \in \text{Fix}\, T$. Furthermore, if T is a cutter, then $T \mid_{\text{Fix}\, T} = \text{Id}$. Therefore, T is a cutter (respectively, an α-relaxed cutter) if and only if inequality (2.21) (respectively, (2.22)) is satisfied for all $x \notin \text{Fix}\, T$ and for all $z \in \text{Fix}\, T$. Relaxed cutters were also studied in [253, 255, 346] and in [249], where they were called *averaged quasi-nonexpansive* mappings.

Remark 2.1.32. Let $T : X \to \mathcal{H}$ be a separator of a subset $C \subseteq X$. Then the following obvious properties of T hold:

(i) T is a separator of the closed convex hull of C.
(ii) T is a separator of any subset $D \subseteq C$.
(iii) For an arbitrary $\lambda \in [0, 1]$, the relaxation T_λ of T is a separator of C.
(iv) $C \subseteq \text{Fix}\, T$.

Corollary 2.1.33. *Let* $U : X \to \mathcal{H}$, $T := \frac{1}{2}(U + \text{Id})$ *and* $C \subseteq X$. *Then:*

(i) U *is Fejér monotone with respect to* C *if and only if* T *is a separator of* C.

(ii) If Fix $U \neq \emptyset$, *then U is quasi-nonexpansive if and only if T is a cutter.*

Proof. The corollary follows easily from equivalence (2.11) (see [24, Proposition 2.3 (v)⇔(vi)] for a different proof). □

Corollary 2.1.34. *Let $T : X \rightarrow \mathcal{H}$ and $C \subseteq X$. If T is Fejér monotone with respect to C, then T is Fejér monotone with respect to the closed convex hull of C.*

Proof. The corollary follows directly from Corollary 2.1.33 (i) and from Remark 2.1.32 (i). □

By Remark 2.1.32 (iii), the right hand side of the equivalence in Corollary 2.1.33 (i) can be written in the form: U_λ *is a separator of C for all $\lambda \in [0, \frac{1}{2}]$.* Similarly, the right hand side of the equivalence in Corollary 2.1.33 (ii) can be written in the form: U_λ *is a cutter for all $\lambda \in [0, \frac{1}{2}]$.* Corollary 2.1.33 (ii) can also be written equivalently as follows:

U is quasi-nonexpansive if and only if there is a cutter $S : X \rightarrow H$ and $\mu \in [0, 2]$ such that $U = S_\mu$.

A subset $C \subseteq X$ for which the operator $T : X \rightarrow \mathcal{H}$ is a separator needs not to be convex. However there exists the largest subset for which T is a separator, which is closed and convex. This fact follows from the following lemma.

Lemma 2.1.35. *Let $T : X \rightarrow \mathcal{H}$. If the subset*

$$\text{Sep}\, T := \bigcap_{x \in X} \{z \in X : \langle z - Tx, x - Tx \rangle \leq 0\}$$

is nonempty, then Sep T *is the largest subset for which T is a separator. Furthermore,* Sep T *is a closed convex subset.*

Proof. The first part of the lemma follows directly from Definition 2.1.30. The second part follows from the fact that an intersection of closed convex subsets is closed and convex. □

If T is nonexpansive, then Fix T is a closed convex subset (see Proposition 2.1.11). It turns out that cutters have the same property. The second part of the following lemma was proved in [24, Proposition 2.6 (i)–(ii)].

Lemma 2.1.36. *Let $T : X \rightarrow \mathcal{H}$. The following inclusion holds*

$$\text{Sep}\, T \subseteq \text{Fix}\, T. \tag{2.23}$$

If T is a cutter, then a converse inclusion is also true. Hence, the subset of fixed points of a cutter is closed and convex.

Proof. Let $y \in$ Sep T, i.e., $\langle x - Tx, y - Tx \rangle \leq 0$ for all $x \in X$. If we take $x = y$, we get $\|y - Ty\| \leq 0$, and hence, $Ty = y$, i.e., $y \in$ Fix T. Now suppose that T is a cutter and that $y \in$ Fix T. Then for any $x \in X$ we have $\langle y - Tx, x - Tx \rangle \leq 0\}$, i.e.,

$y \in \text{Sep } T$. Therefore, we have $\text{Sep } T = \text{Fix } T$. The subset $\text{Fix } T$ is closed convex as an intersection of closed convex subsets. $\qquad \square$

It follows from Remark 2.1.32 (ii) and from Lemmas 2.1.35 and 2.1.36 that a cutter $T : X \rightarrow X$ is a separator of any nonempty subset of $\text{Fix } T$. Therefore, we will restrict our further considerations of separators to cutters. Note, however, that the converse inclusion of (2.23) is not true in general (see Example 2.2.7). Hence, there is a separator with a fixed point which is not a cutter.

It is an immediate consequence of the characterization of the metric projection (see Theorem 1.2.4) that an operator $T : \mathcal{H} \rightarrow \mathcal{H}$ is a metric projection onto a closed convex subset if and only if $T^2 = T$ and T is a cutter (a more general fact will be presented in Theorem 2.2.5). In this case, we have $T = P_{\text{Fix } T}$. Even if a cutter T is not idempotent, T is closely related to the metric projection. The following corollary was proved in [121, Proposition 2.3 (iii)].

Corollary 2.1.37. *Let $T : X \rightarrow \mathcal{H}$ be a cutter. Then, for any $x \in X$, it holds*

$$\|Tx - x\| \le \|P_{\text{Fix } T} x - x\|. \tag{2.24}$$

Proof. If $x \in \text{Fix } T$, then inequality (2.24) is obvious. Now let $x \notin \text{Fix } T$. Then it follows from inequality (2.21) for $z := P_{\text{Fix } T} x$ together with the Cauchy–Schwarz inequality that

$$\|Tx - x\| \le \frac{\langle Tx - x, P_{\text{Fix } T} x - x \rangle}{\|Tx - x\|} \le \|P_{\text{Fix } T} x - x\|$$

which completes the proof. $\qquad \square$

Definition 2.1.38. Let $\alpha \ge 0$ and assume that $T : X \rightarrow \mathcal{H}$ has a fixed point. We say that T is α-*strongly quasi-nonexpansive* (α-SQNE), if

$$\|Tx - z\|^2 \le \|x - z\|^2 - \alpha \|Tx - x\|^2 \tag{2.25}$$

for all $x \in X$ and $z \in \text{Fix } T$. If T satisfies (2.25) with $\alpha > 0$, then T is called *strongly quasi-nonexpansive* *(SQNE)*.

A property which is more general than the strong nonexpansivity was introduced by Halperin [198, Sect. 2] and was called φ-property, where $\varphi : [0, \infty) \rightarrow [0, \infty)$ is a nondecreasing function. If $\varphi(t) = t^2$ for all $t \in [0, \infty)$, then φ-property is equivalent to the strong quasi-nonexpansivity. The notion strong quasi nonexpansivity was introduced by Bruck [50, Sect. 1] for operators defined on a metric space. Strongly quasi-nonexpansive operators are widely studied in the literature. Bauschke and Borwein use the name *strongly attracting operators* for operators which are NE and SQNE (see [22, Definition 2.1]). Reich and Zaslavski define a more general operator and call it a *uniformly F-attracting mapping*, where $F = \text{Fix } T$ (see [297, Sect. 1]). Vasin and Ageev call the α-SQNE operators, where $\alpha \in (0, 1)$, *Q-pseudocontractive operators* (see [333, Definition 2.3]). Yamada and Ogura use the notation α-*attracting quasi-nonexpansive* for the α-SQNE operators [346]. Crombez denotes the class of α-SQNE operators by $\mathcal{F}^\alpha$ (see [126, pages 160–161]) and gives several equivalent conditions for $T \in \mathcal{F}^\alpha$ (see [126, Theorem 2.1]).

It follows easily from the equivalence (a)$\Leftrightarrow$(c) of Lemma 1.2.5 that an operator T which has a fixed point is a cutter if and only if it is 1-strongly quasi-nonexpansive. The following theorem extends this property to relaxations of T (cf. [121, Proposition 2.3 (ii)]).

Theorem 2.1.39. *Assume that* $T : X \to \mathcal{H}$ *has a fixed point and let* $\lambda \in (0, 2]$. *Then T is a cutter if and only if its relaxation T_λ is $\frac{2-\lambda}{\lambda}$-strongly quasi-nonexpansive, i.e.,*

$$\|T_\lambda x - z\|^2 \le \|x - z\|^2 - \frac{2 - \lambda}{\lambda} \|T_\lambda x - x\|^2 \qquad (2.26)$$

for all $x \in X$ and for all $z \in$ Fix T.

Proof. Since

$$T_\lambda x - x = \lambda(Tx - x),$$

the properties of the inner product yield

$$\begin{aligned}
\|T_\lambda x - z\|^2 &- \|x - z\|^2 + \frac{2 - \lambda}{\lambda} \|T_\lambda x - x\|^2 \\
&= \|x - z + \lambda(Tx - x)\|^2 - \|x - z\|^2 + \lambda(2 - \lambda) \|Tx - x\|^2 \\
&= 2\lambda(\|Tx - x\|^2 - \langle z - x, Tx - x\rangle) \\
&= 2\lambda\langle z - Tx, x - Tx\rangle
\end{aligned}$$

for all $x \in X$ and for all $z \in C$. The assertion follows directly from the equalities above. $\qquad\square$

The following corollary is an equivalent formulation of Theorem 2.1.39.

Corollary 2.1.40. *Assume that* $U : X \to \mathcal{H}$ *has a fixed point and let* $\alpha \in (0, 2]$. *Then U is an α-relaxed cutter if and only if U is $\frac{2-\alpha}{\alpha}$-strongly quasi-nonexpansive.*

In general, a relaxation T_λ of a cutter T with $\lambda \ge 2$ needs not to be strongly quasi-nonexpansive. Nevertheless, the following proposition holds.

Proposition 2.1.41. *Let* $T : X \to \mathcal{H}$ *be a cutter with* int Fix $T \ne \emptyset$ *and* $\lambda > 0$. *Then for any $z \in$ int Fix T and $x \notin$ Fix T it holds*

$$\|T_\lambda x - z\|^2 \le \|x - z\|^2 - \lambda(2 + \frac{2\delta}{\|Tx - x\|} - \lambda) \|Tx - x\|^2, \qquad (2.27)$$

where $\delta > 0$ is such that $B(z, \delta) \subseteq$ Fix T. If X is bounded, then T_λ is int Fix T-strictly quasi-nonexpansive for any $\lambda \in (0, 2]$.

Proof. Let $z \in \text{int Fix } T$ and $x \notin \text{Fix } T$. Then $w := z - \delta \frac{Tx - x}{\|Tx - x\|} \in \text{Fix } T \subseteq X$ and inequality (2.21) yields

$$
\begin{aligned}
\|T_\lambda x - z\|^2 &= \|x + \lambda(Tx - x) - z\|^2 \\
&= \|x - z\|^2 + \lambda^2 \|Tx - x\|^2 - 2\lambda\langle z - x, Tx - x \rangle \\
&= \|x - z\|^2 + \lambda^2 \|Tx - x\|^2 \\
&\quad -2\lambda\langle z - w, Tx - x \rangle - 2\lambda\langle w - x, Tx - x \rangle \\
&\le \|x - z\|^2 + \lambda^2 \|Tx - x\|^2 \\
&\quad -2\lambda\delta \|Tx - x\| - 2\lambda \|Tx - x\|^2 \\
&= \|x - z\|^2 - \lambda(2 + \frac{2\delta}{\|Tx - x\|} - \lambda) \|Tx - x\|^2 .
\end{aligned}
$$

Let X be bounded and $d > 0$ be such that $\|Tu - u\| \le d$ for any $u \in X$. The existence of such d follows from Corollary 2.1.37. Denote $\varepsilon := \frac{2\delta}{d}$. Then (2.27) yields

$$
\|T_\lambda x - z\|^2 \le \|x - z\|^2 - \lambda(2 + \varepsilon - \lambda) \|Tx - x\|^2 .
$$

Consequently, T_λ is int Fix T-strictly quasi-nonexpansive for any $\lambda \in (0, 2]$. $\square$

The corollary below follows immediately from Proposition 2.1.41 and from Theorem 2.1.26.

Corollary 2.1.42. *Let* $U_i : X \to \mathcal{H}$, $i \in I$ *be quasi-nonexpansive with* $C := \bigcap_{i \in I} \text{Fix } U_i \ne \emptyset$ *and let* $U := U_m U_{m-1} \ldots U_1$. *If* int $C \ne \emptyset$, *then* Fix $U = \bigcap_{i \in I} \text{Fix } U_i$ *and* U *is int* C-*strictly quasi-nonexpansive.*

An equivalent formulation of the following result appeared in [127, Theorem 3.2 (iii)].

Corollary 2.1.43. *Assume that* $U : X \to \mathcal{H}$ *has a fixed point and let* $\beta \ge 0$. *Then* U *is* β-*strongly quasi-nonexpansive if and only if* U *is a* $\frac{2}{\beta+1}$-*relaxed cutter.*

Proof. It suffices to take $\alpha = \frac{2}{\beta+1}$ in Corollary 2.1.40. $\square$

Remark 2.1.44. Assume that $T : X \to \mathcal{H}$ has a fixed point and is α-strongly quasi-nonexpansive, where $\alpha \ge 0$.

(i) If $\alpha = 0$, then T is quasi-nonexpansive.
(ii) T is γ-strongly quasi-nonexpansive for all $\gamma \in [0, \alpha]$.
(iii) If $\alpha > 0$, then T is strictly quasi-nonexpansive. Therefore, all properties of strictly quasi-nonexpansive operators are also valid for strongly quasi-nonexpansive operators and for cutters.

Fig. 2.5 Solution of (2.28) as a function of $\lambda, \mu \in (0, 2)$

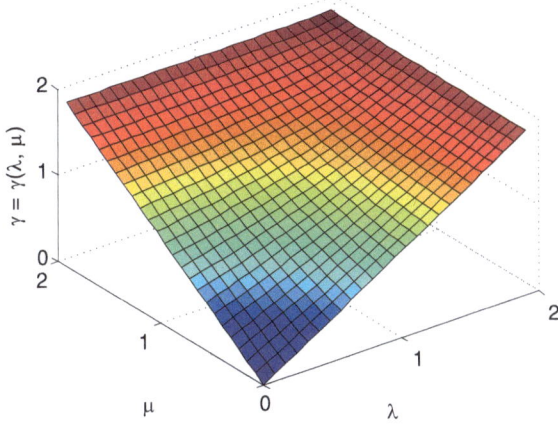

Cutters and strongly quasi-nonexpansive operators play an important role in methods presented in further parts of the book. Therefore, we focus our attention on the properties of these operators which enable us to construct new cutters or strongly quasi-nonexpansive operators. Below, we show that a family of relaxed cutters is closed under composition and under convex combination of operators having a common fixed point. The first property of relaxed cutters follows from the lemma below whose proof is left to the reader.

Lemma 2.1.45. *Let* $\lambda, \mu \in (0, 2)$. *The unique solution* γ *of the equation*

$$\left(\frac{1 - \frac{2}{\gamma}}{2} \right)^2 = (\frac{1}{\lambda} - \frac{1}{\gamma})(\frac{1}{\mu} - \frac{1}{\gamma}) \tag{2.28}$$

is

$$\gamma = \frac{2}{(\frac{\lambda}{2-\lambda} + \frac{\mu}{2-\mu})^{-1} + 1} = \frac{4(\lambda + \mu - \lambda\mu)}{4 - \lambda\mu}. \tag{2.29}$$

Moreover,

$$0 < \min\{\lambda, \mu\} < \frac{4\min\{\lambda, \mu\}}{\min\{\lambda, \mu\} + 2} \leq \gamma \leq \frac{4\max\{\lambda, \mu\}}{\max\{\lambda, \mu\} + 2} < 2.$$

A solution of (2.28) is illustrated in Fig. 2.5.

Theorem 2.1.46. *Let* $T : X \to X$ *be a* λ-*relaxed cutter,* $U : X \to X$ *be a* μ-*relaxed cutter, where* $\lambda, \mu \in (0, 2]$, *and let* $\text{Fix } T \cap \text{Fix } U \neq \emptyset$. *If* $\lambda, \mu \in (0, 2)$, *then* UT *is a* γ-*relaxed cutter, where* γ *is given by* (2.29). *If* $\lambda = 2$ *and* $\mu < 2$ *or* $\mu = 2$ *and* $\lambda < 2$, *then* UT *is a quasi-nonexpansive operator or, equivalently,* UT *is a 2-relaxed cutter.*

Proof. Suppose that T is a λ-relaxed cutter and that U is a μ-relaxed cutter. Take $a := Tx - x$ and $b := UTx - Tx$. Then it follows from inequality (2.22) that $\langle z - x, a \rangle \geq \frac{1}{\lambda} \|a\|^2$ and $\langle z - Tx, b \rangle \geq \frac{1}{\mu} \|b\|^2$ for any $z \in \text{Fix } U \cap \text{Fix } T$.

Let $\lambda, \mu \in (0, 2)$ and γ be defined by (2.29). Then Lemma 2.1.45 yields

$$\langle z - x, UTx - x \rangle - \frac{1}{\gamma} \|UTx - x\|^2$$

$$= \langle z - x, a + b \rangle - \frac{1}{\gamma} \|a + b\|^2$$

$$= \langle z - x, a \rangle + \langle z - x, b \rangle - \frac{1}{\gamma} \|a + b\|^2$$

$$= \langle z - x, a \rangle + \langle z - Tx, b \rangle + \langle a, b \rangle - \frac{1}{\gamma} \|a + b\|^2$$

$$\geq \frac{1}{\lambda} \|a\|^2 + \frac{1}{\mu} \|b\|^2 + \langle a, b \rangle - \frac{1}{\gamma} \|a + b\|^2$$

$$= (\frac{1}{\lambda} - \frac{1}{\gamma}) \|a\|^2 + (\frac{1}{\mu} - \frac{1}{\gamma}) \|b\|^2 + (1 - \frac{2}{\gamma}) \langle a, b \rangle$$

$$= \left\| \sqrt{\frac{1}{\lambda} - \frac{1}{\gamma}} a - \sqrt{\frac{1}{\mu} - \frac{1}{\gamma}} b \right\|^2 \geq 0.$$

Applying inequality (2.22) we obtain that UT is a γ-relaxed cutter. If $\lambda = 2$ and $\mu < 2$ or $\mu = 2$ and $\lambda < 2$, then UT is quasi-nonexpansive by Theorem 2.1.28.   □

The following result is due to Yamada and Ogura (see [346, Proposition 1(d)]).

Corollary 2.1.47. *Let* $T, U : X \to X$ *have a common fixed point and* $\rho, \sigma > 0$. *If* T *is* ρ-SQNE *and* U *is* σ-SQNE, *then* UT *is* δ-SQNE, *where*

$$\delta = \frac{1}{\frac{1}{\rho} + \frac{1}{\sigma}}. \tag{2.30}$$

Proof. Suppose that T is ρ-SQNE and U is σ-SQNE. It follows from Corollary 2.1.43 that T is a λ-relaxed cutter and that U is a μ-relaxed cutter, where $\lambda = \frac{2}{1+\rho}$ and $\mu = \frac{2}{1+\sigma}$. By Theorem 2.1.46 the operator UT is a γ-relaxed cutter, where

$$\gamma = \frac{2}{(\frac{\lambda}{2-\lambda} + \frac{\mu}{2-\mu})^{-1} + 1} = \frac{2}{(\frac{1}{\rho} + \frac{1}{\sigma})^{-1} + 1}.$$

Corollary 2.1.40 yields now that UT is δ-SQNE, where δ is given by (2.30).   □

Theorem 2.1.48. *Let* $T_i : X \to X$ *be an* α_i-*relaxed cutter, where* $\alpha_i \in (0,2)$, $i \in I := \{1,2,\ldots,m\}$, *or, equivalently,* T_i *be* β_i-*strongly quasi-nonexpansive, where* $\beta_i = \frac{2-\alpha_i}{\alpha_i} \in (0,+\infty)$, $i \in I$. *Let* $\bigcap_{i \in I} \operatorname{Fix} T_i \neq \emptyset$ *and* $U_m := T_m T_{m-1} \ldots T_1$. *Then:*

(i) The operator U_m *is a* γ_m-*relaxed cutter, with*

$$\gamma_m = \frac{2}{(\frac{\alpha_1}{2-\alpha_1} + \frac{\alpha_2}{2-\alpha_2} + \ldots + \frac{\alpha_m}{2-\alpha_m})^{-1} + 1}. \tag{2.31}$$

(ii) The operator U_m *is* δ_m-*strongly quasi-nonexpansive, with*

$$\delta_m = \frac{1}{\frac{1}{\beta_1} + \frac{1}{\beta_2} + \ldots + \frac{1}{\beta_m}}. \tag{2.32}$$

Moreover,

$$0 < \min_{i \in I} \alpha_i < \frac{2m \min_{i \in I} \alpha_i}{(m-1) \min_{i \in I} \alpha_i + 2} \leq \gamma_m \leq \frac{2m \max_{i \in I} \alpha_i}{(m-1) \max_{i \in I} \alpha_i + 2} < 2 \tag{2.33}$$

and

$$0 < \frac{\min_{i \in I} \beta_i}{m} \leq \delta_m \leq \frac{\max_{i \in I} \beta_i}{m}. \tag{2.34}$$

Proof. The assertion is obvious for $m = 1$. Note that $\gamma_m = \frac{2}{\delta_m + 1}$ and that Corollary 2.1.43 yields the equivalence of conditions (i) and (ii). We prove by induction with respect to m that these conditions hold for any $m \geq 2$.

1^0 If $m = 2$, then conditions (i) and (ii) follow directly from Theorem 2.1.46 and from Corollary 2.1.47.

2^0 Suppose that (ii) is true for some $m = k$. Consequently, U_k is δ_k-SQNE. It follows now from Corollary 2.1.47 that the operator $U_{k+1} = T_{k+1} U_k$ is δ-SQNE, where

$$\delta = \frac{1}{\frac{1}{\delta_k} + \frac{1}{\beta_{k+1}}} = \frac{1}{\frac{1}{\beta_1} + \frac{1}{\beta_2} + \ldots + \frac{1}{\beta_k} + \frac{1}{\beta_{k+1}}} = \delta_{k+1}.$$

Now, for $m = k + 1$, equality (2.31) follows from the above mentioned equivalence of (i) and (ii).

Hence, we have proved that conditions (i) and (ii) hold for all $m \geq 1$. Both inequalities in (2.34) follow immediately from equality (2.32). Now we have

$$\gamma_m = \frac{2}{\delta_m + 1} \leq \frac{2}{\frac{\min_{i \in I} \beta_i}{m} + 1} = \frac{2}{\frac{\min_{i \in I} \frac{2-\alpha_i}{\alpha_i}}{m} + 1} = \frac{2m \max_{i \in I} \alpha_i}{(m-1) \max_{i \in I} \alpha_i + 2} < 2.$$

In a similar way one can prove that

$$\gamma_m \geq \frac{2m \min_{i \in I} \alpha_i}{(m-1) \min_{i \in I} \alpha_i + 2} > \min_{i \in I} \alpha_i > 0$$

which completes the proof.   □

Bauschke and Borwein proved that a composition of β_i-SQNE operators with a common fixed point is β-SQNE for $\beta := \frac{\min_{i \in I} \beta_i}{2^m - 1}$ (see [22, Theorem 2.10 (ii)]). It is clear that this result is weaker than Theorem 2.1.48 (ii), because $\beta \leq \frac{\min_{i \in I} \beta_i}{m} \leq \delta_m$. Note that the first inequality is strict for $m > 2$ and that the other one is strict if $\beta_i \neq \beta_j$ for at least one pair $i, j \in I$.

Corollary 2.1.49. *Let $U_i : X \to \mathcal{H}$ be cutters with a common fixed point, $i \in I := \{1, 2, \ldots, m\}$, and $w : X \to \Delta_m$ be an appropriate weight function. Then the operator $U := \sum_{i \in I} w_i U_i$ is a cutter.*

Proof. Let $U := \sum_{i \in I} w_i U_i$. It is clear that a cutter is strictly quasi-nonexpansive (see Remark 2.1.44 (iii)). Therefore, it follows from Theorem 2.1.26 (i) that $\operatorname{Fix} U = \bigcap_{i \in I} \operatorname{Fix} U_i$. By Remark 2.1.31 and by the convexity of the function $\|\cdot\|^2$, we have

$$\langle Ux - x, z - x \rangle = \sum_{i \in I} w_i(x) \langle U_i x - x, z - x \rangle$$

$$\geq \sum_{i \in I} w_i(x) \|U_i x - x\|^2$$

$$\geq \left\| \sum_{i \in I} w_i(x) U_i x - x \right\|^2$$

$$= \|Ux - x\|^2$$

for all $x \in X$ and all $z \in \operatorname{Fix} U$. Again, by Remark 2.1.31, U is a cutter.   □

Theorem 2.1.50. *Let $T_i : X \to \mathcal{H}$ be an α_i-relaxed cutter, where $\alpha_i \in (0, 2)$, $i \in I := \{1, 2, \ldots, m\}$, or, equivalently, T_i be β_i-strongly quasi-nonexpansive, where $\beta_i = \frac{2 - \alpha_i}{\alpha_i} \in (0, +\infty)$, $i \in I$. Let $\bigcap_{i \in I} \operatorname{Fix} T_i \neq \emptyset$ and $w \in \Delta_m$. Then the operator $T := \sum_{i \in I} w_i T_i$ is an α-relaxed cutter with*

$$\alpha := \sum_{i \in I} w_i \alpha_i. \tag{2.35}$$

Consequently, T is β-SQNE, with

$$\beta := \left(\sum_{i \in I} \frac{w_i}{\beta_i + 1} \right)^{-1} - 1. \tag{2.36}$$

Moreover,

$$0 < \min_{i \in I} \alpha_i \le \alpha \le \max_{i \in I} \alpha_i < 2 \qquad (2.37)$$

and

$$0 < \min_{i \in I} \beta_i \le \beta \le \max_{i \in I} \beta_i. \qquad (2.38)$$

Proof. Without loss of generality we suppose that $w \in \text{ri } \Delta_m$. Let $U_i := (T_i)_{\alpha_i^{-1}}$, i.e.,

$$U_i = \text{Id} + \frac{1}{\alpha_i}(T_i - \text{Id}).$$

It is clear that U_i are cutters, $i \in I$. Let α be defined by (2.35) and $v_i := \frac{\omega_i \alpha_i}{\alpha}$, $i \in I$. Note that $v = (v_1, v_2, \ldots, v_m) \in \text{ri } \Delta_m$, consequently, v is appropriate. Define $U := \sum_{i \in I} v_i U_i$. By Corollary 2.1.49, the operator U is a cutter. We have

$$U = \sum_{i \in I} v_i U_i = \text{Id} + \sum_{i \in I} \frac{v_i}{\alpha_i}(T_i - \text{Id}) = \text{Id} + \frac{1}{\alpha} \sum_{i \in I} \omega_i(T_i - \text{Id}) = \text{Id} + \frac{1}{\alpha}(T - \text{Id}),$$

i.e., $T = \text{Id} + \alpha(U - \text{Id})$ and T is an α-relaxed cutter. The second part of the theorem follows now immediately from Corollaries 2.1.40 and 2.1.43. Inequalities in (2.37) are obvious and inequalities in (2.38) follow easily from (2.36). $\square$

Bauschke and Borwein proved that a convex combination of β_i-SQNE operators, $i \in I$, with a common fixed point is β-SQNE, where $\beta := \min_{i \in I} \beta_i$ (see [22, Proposition 2.12]). By inequality (2.38) this result is weaker than Theorem 2.1.50. Note that this inequality is strict if $\beta_i \ne \beta_j$ for at least one pair $i, j \in I$ for which ω_i and ω_j are nonzero. The second part of Theorem 2.1.50 for $m = 2$ was proved by Yamada and Ogura (see [346, Proposition 1(c)]).

The following important result extends Theorem 2.1.39.

Theorem 2.1.51. *Let $S : \mathcal{H} \to X$ be nonexpansive, $T : X \to \mathcal{H}$ be a cutter and $\lambda \in (0, 2)$. If $\text{Fix } S \cap \text{Fix } T \ne \emptyset$, then, for any $x \in \text{Fix } S$ and $z \in \text{Fix } S \cap \text{Fix } T$, the following estimations hold*

$$\|S T_\lambda x - z\|^2 \le \|x - z\|^2 - \lambda(2 - \lambda)\|Tx - x\|^2 \qquad (2.39)$$

and

$$\|S T_\lambda x - z\|^2 \le \|x - z\|^2 - \frac{2 - \lambda}{\lambda}\|S T_\lambda x - x\|^2. \qquad (2.40)$$

Consequently, the operator $S T_\lambda |_{\text{Fix } S}$ is $\frac{2-\lambda}{\lambda}$-strongly quasi-nonexpansive.

Proof. Let $x \in \text{Fix } S$ and $z \in \text{Fix } S \cap \text{Fix } T$. Then the assumptions that S is nonexpansive and T is a cutter yield

Fig. 2.6 Composition of
cutters needs not to be a cutter

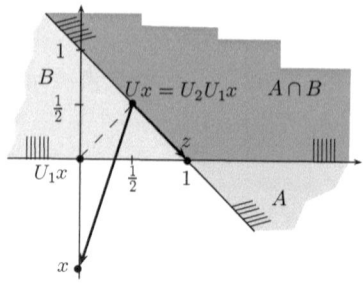

$$\|ST_\lambda x - z\|^2 = \|ST_\lambda x - Sz\|^2 \leq \|T_\lambda x - z\|^2$$
$$= \|x - z\|^2 + \lambda^2 \|Tx - x\|^2 - 2\lambda \langle z - x, Tx - x\rangle$$
$$\leq \|x - z\|^2 - \lambda(2 - \lambda) \|Tx - x\|^2$$
$$= \|x - z\|^2 - \frac{2 - \lambda}{\lambda} \|T_\lambda x - x\|^2$$
$$\leq \|x - z\|^2 - \frac{2 - \lambda}{\lambda} \|ST_\lambda x - Sx\|^2$$
$$= \|x - z\|^2 - \frac{2 - \lambda}{\lambda} \|ST_\lambda x - x\|^2$$

which completes the proof. $\square$

Below we give several examples which show that a composition of quasi-nonexpansive operators does not need to be quasi-nonexpansive, that a composition of a strictly quasi-nonexpansive operator and a quasi-nonexpansive one does not need to be strictly quasi-nonexpansive and that a composition of cutters does not need to be a cutter, even if they have a common fixed point.

Example 2.1.52. Let $X := [-1, 1] \subseteq \mathbb{R}$, $S, T : X \to X$, $S := -\operatorname{Id}$ and

$$Tx := \begin{cases} -x & \text{if } x = 1 \\ \frac{1}{2}x & \text{otherwise.} \end{cases}$$

One can easily check that S, T are quasi-nonexpansive, $\operatorname{Fix} S = \operatorname{Fix} T = \{0\}$ and $\operatorname{Fix} ST = \{0, 1\}$. The operator ST is not quasi-nonexpansive, because a subset of fixed points of a quasi-nonexpansive operator is convex (see Lemma 2.1.21).

Example 2.1.53. Let $X = \mathcal{H} := \mathbb{R}^2$, $A := \{x \in \mathbb{R}^2 : \langle e, x\rangle \geq 1\}$, $B := \{x \in \mathbb{R}^2 : \xi_2 \geq 0\}$, $U_1 := P_B$, $U_2 := P_A$ and $U := U_2 U_1$. Then U_1 and U_2 are cutters and it follows from Theorem 2.1.26 that $\operatorname{Fix} U = \operatorname{Fix} U_1 \cap \operatorname{Fix} U_2 = A \cap B \neq \emptyset$. For $x = (0, -1)$ and $z = (1, 0) \in A \cap B$, we have $Ux = (\frac{1}{2}, \frac{1}{2})$ and $\langle x - Ux, z - Ux\rangle = \frac{1}{2}$ (see Fig. 2.6). Therefore, U is not a cutter.

Example 2.1.54. Let $A, B \subseteq \mathcal{H}$ be nonempty closed convex subsets and $A \subsetneqq B$. Define $S := 2P_A - \mathrm{Id}$ and $T := P_B$. We have $\mathrm{Fix}\, S \cap \mathrm{Fix}\, T = A$. It follows easily from the characterization of the metric projection that P_A and P_B are cutters. By Theorem 2.1.39, T is strictly quasi-nonexpansive and S is quasi-nonexpansive. By Theorem 2.1.28, the operator ST is quasi-nonexpansive. Unfortunately, ST is not strictly quasi-nonexpansive, because for any $x \in B \backslash A$ and for $z := P_A x$ it holds

$$\|STx - z\| = \|Sx - z\| = \|x - z\|.$$

2.2 Firmly Nonexpansive Operators

Definition 2.2.1. We say that an operator $T : X \to \mathcal{H}$ is *firmly nonexpansive* (*FNE*), if

$$\langle Tx - Ty, x - y \rangle \geq \|Tx - Ty\|^2 \qquad (2.41)$$

for all $x, y \in X$. Let $\lambda \in [0, 2]$. We say that $T : X \to \mathcal{H}$ is *λ-relaxed firmly nonexpansive* (*λ-RFNE*) or, shortly, *relaxed firmly nonexpansive* (*RFNE*) if T is a λ-relaxation of a firmly nonexpansive operator U, i.e., $T = U_\lambda = (1 - \lambda)\,\mathrm{Id} + \lambda U$. If, furthermore, $\lambda \in (0, 2)$, then we say that T is *strictly relaxed firmly nonexpansive*.

The definition of a firmly nonexpansive operator in a Hilbert space is due to Browder (see [46]), who called it a *firmly contractive operator*. Bruck introduced the name firmly nonexpansive for operators in a Banach space (see [49, Definition 6]). In Hilbert spaces both definitions coincide, as we will show in Theorem 2.2.10. Condition (vi) of this theorem is, actually, the definition of a firmly nonexpansive operator proposed by Bruck.

The following lemma is obvious.

Lemma 2.2.2. *Let $T : X \to \mathcal{H}$ and $x, y \in X$. The following inequalities are equivalent:*

(i) $\langle Tx - Ty, x - y \rangle \geq \|Tx - Ty\|^2$,
(ii) $\langle Tx - Ty, (x - Tx) - (y - Ty) \rangle \geq 0$,
(iii) $\langle Ty - Tx, x - Tx \rangle + \langle Tx - Ty, y - Ty \rangle \leq 0$,
(iv) $\langle Ty - x, Tx - x \rangle + \langle Tx - y, Ty - y \rangle \geq \|Tx - x\|^2 + \|Ty - y\|^2$.

It follows from Lemma 2.2.2 that inequality (2.41) defining a firmly nonexpansive operator can be replaced by any inequality in (i)–(iv).

Corollary 2.2.3. *Let $\lambda > 0$. An operator $S : X \to \mathcal{H}$ is λ-RFNE if and only if*

$$\langle y - x, Sx - x \rangle + \langle x - y, Sy - y \rangle \geq \frac{1}{\lambda} \|(Sx - x) - (Sy - y)\|^2. \qquad (2.42)$$

Fig. 2.7 NE and monotone
operator which is not FNE

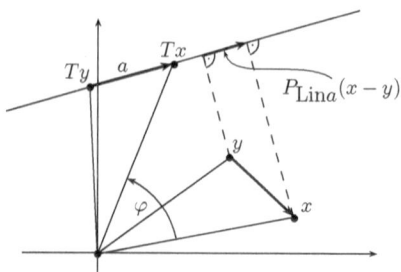

Proof. Let $S := T_\lambda = \text{Id} + \lambda(T - \text{Id})$ for a firmly nonexpansive operator T :
$X \to \mathcal{H}$. Let $x, y \in X$. It follows from the equivalence (i)$\Leftrightarrow$(iv) in Lemma 2.2.2
and from the equality $T = S_{\lambda^{-1}}$ (see Remark 2.1.3) that S is λ-RFNE if and only if

$$\langle Ty - x, Sx - x \rangle + \langle Tx - y, Sy - y \rangle \geq \frac{1}{\lambda}(\|Sx - x\|^2 + \|Sy - y\|^2).$$

Since $Ty - x = y - x + \frac{1}{\lambda}(Sy - y)$ and $Tx - y = x - y + \frac{1}{\lambda}(Sx - x)$, the last
inequality is equivalent to

$$\langle y-x, Sx-x \rangle + \langle x-y, Sy-y \rangle + \frac{2}{\lambda}\langle Sx-x, Sy-y \rangle \geq \frac{1}{\lambda}(\|Sx - x\|^2 + \|Sy - y\|^2).$$

The latter inequality is equivalent to (2.42).   □

2.2.1 Basic Properties of Firmly Nonexpansive Operators

Theorem 2.2.4. *A firmly nonexpansive operator* $T : X \to \mathcal{H}$ *is monotone and*
nonexpansive.

Proof. Let T be firmly nonexpansive. By the Cauchy–Schwarz inequality, we have

$$\|Tx - Ty\| \cdot \|x - y\| \geq \langle Tx - Ty, x - y \rangle \geq \|Tx - Ty\|^2 \geq 0,$$

for all $x, y \in X$, which yields the monotonicity and the nonexpansivity of T.   □

The converse of Theorem 2.2.4 is not true, e.g., the operator $T : \mathbb{R}^2 \to \mathbb{R}^2$,

$$Tx := (\xi_1 \cos \varphi - \xi_2 \sin \varphi, \xi_1 \sin \varphi + \xi_2 \cos \varphi)$$

is nonexpansive and monotone for $\varphi \in (0, \pi/2)$, but T is not firmly nonexpansive
(see Fig. 2.7).
Now we prove a property of firmly nonexpansive operators, which also appears
in Theorem 1.2.4. In particular, the characterization of the metric projection is,

actually, a corollary of the following theorem which is due to Goebel and Reich (see [185, pp. 43–44]).

Theorem 2.2.5. *Let* $T : X \to \mathcal{H}$ *be an operator with a fixed point.*

  (i) If T *is firmly nonexpansive, then* T *is a cutter, i.e.,*

$$\langle z - Tx, x - Tx \rangle \leq 0 \tag{2.43}$$

    for all $x \in X$ *and* $z \in \operatorname{Fix} T$.

 (ii) *If* T *is a projection, i.e.,* $T(X) = \operatorname{Fix} T$, *then the implication converse to* (i) *is also true. In this case,* $T = P_{\operatorname{Fix} T}$.

Proof. (i) Let T be firmly nonexpansive, $x \in X$ and $z \in \operatorname{Fix} T$. By the equivalence (i)$\Leftrightarrow$(iii) in Lemma 2.2.2, we have

$$\langle Ty - Tx, x - Tx \rangle + \langle Tx - Ty, y - Ty \rangle \leq 0,$$

and for $y = z \in \operatorname{Fix} T$ we obtain (2.43).

(ii) Suppose that T is a projection and that inequality (2.43) holds for all $x \in X$ and $z \in \operatorname{Fix} T$. Let $u, v \in X$. Taking $x = u$ and $z = Tv$ in (2.43) we get

$$\langle Tv - Tu, u - Tu \rangle \leq 0, \tag{2.44}$$

and, taking $x = v$ and $z = Tu$ in (2.43), we get

$$\langle Tu - Tv, v - Tv \rangle \leq 0. \tag{2.45}$$

Note that, in both cases, $z \in \operatorname{Fix} T$ because $T(X) = \operatorname{Fix} T$. Therefore, the characterization of the metric projection yields that $T = P_{\operatorname{Fix} T}$. Summing up inequalities (2.44) and (2.45) we get

$$\langle Tu - Tv, (Tu - Tv) - (u - v) \rangle \leq 0,$$

i.e., T is firmly nonexpansive (see equivalence (i)$\Leftrightarrow$(ii) in Lemma 2.2.2). $\quad\square$

Suppose that $\operatorname{Fix} T \neq \emptyset$. It follows from the equivalence of (i) and (iii) in Lemma 2.2.2 that inequality (2.41) for $y = z \in \operatorname{Fix} T$ gives (2.43). Therefore, for T being a cutter, inequality (2.41) is required for all $x \in X$ and all $y \in \operatorname{Fix} T$, while for T being firmly nonexpansive this inequality should hold for all $x, y \in X$.

Remark 2.2.6. Neither a projection nor a separator of a nonempty subset $C \subseteq \mathcal{H}$ need to be nonexpansive (note that a separator can even be discontinuous). Furthermore, a nonexpansive separator and even a nonexpansive cutter need not to be firmly nonexpansive (see Examples 2.2.7 and 2.2.8 below).

Fig. 2.8 NE separator which
is not FNE

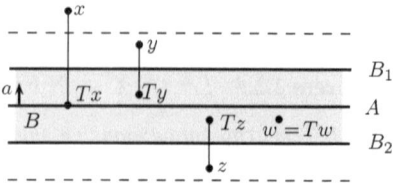

Example 2.2.7. (cf. [204] and [78, Sect. 4.10]) Let $a \in \mathcal{H}$, $\|a\| = 1$ and $\alpha > 0$.
Furthermore, let $A := \{x \in \mathcal{H} : \langle a, x \rangle = 0\}$, $B_1 := \{x \in \mathcal{H} : \langle a, x \rangle = \alpha\}$,
$B_2 := \{x \in \mathcal{H} : \langle a, x \rangle = -\alpha\}$ and $B := \{x \in \mathcal{H} : |\langle a, x \rangle| \leq \alpha\}$. The subset
B is a band with a width of 2α and is bounded by two hyperplanes B_1 and B_2.
The hyperplane A cuts the band B into two bands bounded by A and B_1 and by A
and B_2. Define the operator $T : \mathcal{H} \to \mathcal{H}$ as follows

$$
Tx = \begin{cases}
P_A x & \text{if } |\langle a, x \rangle| \geq 2\alpha \\
2 P_{B_1} x - x & \text{if } \alpha < \langle a, x \rangle < 2\alpha \\
2 P_{B_2} x - x & \text{if } -2\alpha < \langle a, x \rangle < -\alpha \\
x & \text{if } |\langle a, x \rangle| \leq \alpha
\end{cases}
\tag{2.46}
$$

Note that T projects onto A all points with the distance to A equal at least 2α, T
reflects (with respect to the closest hyperplane B_1 or B_2) the points which do not
belong to the band B with the distance to A less than 2α and T does not move
the elements of the band B (see Fig. 2.8). The reader can easily check that T is
nonexpansive and that T is a separator of A but T is not firmly nonexpansive. Note
that Fix $T = B$ and that T is not a cutter, i.e., it does not separate Fix T, but T
separates A (see Fig. 2.8).

Example 2.2.8. Let $A := \mathbb{R} \times \{0\}$ and $B := \{0\} \times \mathbb{R}$ be two subspaces of $\mathbb{R}^2$ and
$T : \mathbb{R}^2 \to \mathbb{R}^2$ be defined by

$$
Tx := [1 - \lambda(x)] P_A x + \lambda(x) P_B x,
$$

where $\lambda(x) = \frac{\xi_1^2}{\xi_1^2 + \xi_2^2}$ for $x = (\xi_1, \xi_2) \in \mathbb{R}^2$ (see Fig. 2.9). We have $P_A x = (\xi_1, 0)$,
$P_B x = (0, \xi_2)$. Consequently,

$$
Tx = \left(\frac{\xi_1 \xi_2^2}{\xi_1^2 + \xi_2^2}, \frac{\xi_1^2 \xi_2}{\xi_1^2 + \xi_2^2} \right)
$$

for $x \neq (0, 0)$. Note that $z := (0, 0)$ is the unique fixed point of T.
The operator T is a cutter, because

$$
\langle z - Tx, x - Tx \rangle = -\frac{\xi_1^2 \xi_2^2}{\xi_1^2 + \xi_2^2} \leq 0
$$

Fig. 2.9 NE cutter which is
not FNE

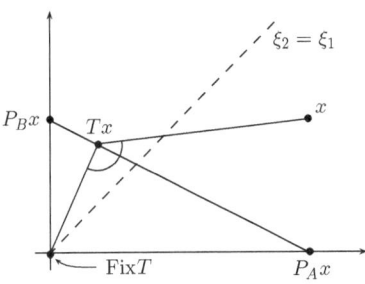

for all $x \neq z$. (Since the weight function $w : \mathbb{R}^2 \to \Delta_2$, $w(x) := (1 - \lambda(x), \lambda(x))$
is appropriate, this fact follows also from Corollary 2.1.49). Let $x, y \neq (0,0)$.
A straightforward calculation shows that

$$\frac{\|Tx - Ty\|^2}{\|x - y\|^2} = \frac{\xi_2^2 \eta_2^2 (\xi_1 - \eta_1)^2 + \xi_1^2 \eta_1^2 (\xi_2 - \eta_2)^2}{(\xi_1^2 + \xi_2^2)(\eta_1^2 + \eta_2^2)[(\xi_1 - \eta_1)^2 + (\xi_2 - \eta_2)^2]}$$

holds for all $x = (\xi_1, \xi_2) \in \mathbb{R}^2$ and for all $y = (\eta_1, \eta_2) \in \mathbb{R}^2$, $x \neq y$. If $\xi_1 \eta_1 = \xi_2 \eta_2 = 0$, then, of course $Tx = Ty = (0,0)$. Suppose that $0 < \xi_1^2 \eta_1^2 \leq \xi_2^2 \eta_2^2$. Then
we have

$$\frac{\|Tx - Ty\|^2}{\|x - y\|^2} = \frac{(\xi_1 - \eta_1)^2 + \frac{\xi_1^2 \eta_1^2}{\xi_2^2 \eta_2^2}(\xi_2 - \eta_2)^2}{(1 + \frac{\xi_1^2}{\xi_2^2})(1 + \frac{\eta_1^2}{\eta_2^2})[(\xi_1 - \eta_1)^2 + (\xi_2 - \eta_2)^2]} \leq 1.$$

If $0 < \xi_2^2 \eta_2^2 \leq \xi_1^2 \eta_1^2$, then we have

$$\frac{\|Tx - Ty\|^2}{\|x - y\|^2} = \frac{\frac{\xi_2^2 \eta_2^2}{\xi_1^2 \eta_1^2}(\xi_1 - \eta_1)^2 + (\xi_2 - \eta_2)^2}{(1 + \frac{\xi_2^2}{\xi_1^2})(1 + \frac{\eta_2^2}{\eta_1^2})[(\xi_1 - \eta_1)^2 + (\xi_2 - \eta_2)^2]} \leq 1.$$

Therefore, T is nonexpansive. If we take $x = (\frac{1}{2}, 1)$ and $y = (1, \frac{1}{2})$, then $Tx = (\frac{2}{5}, \frac{1}{5})$, $Ty = (\frac{1}{5}, \frac{2}{5})$ and

$$\langle Tx - Ty, x - y \rangle = -\frac{1}{5} < \frac{2}{25} = \|Tx - Ty\|^2.$$

Therefore, T is not firmly nonexpansive.

The following property of firmly nonexpansive operators (cf. [22, Lemma 2.4
(iv)]) is often used in applications.

Corollary 2.2.9. *Let* $T : X \rightarrow \mathcal{H}$ *be an operator with a fixed point and* $\lambda \in$
$(0, 2]$. *If* T *is firmly nonexpansive, then its relaxation* T_λ *is* $\frac{2-\lambda}{\lambda}$-*strongly quasi-nonexpansive, i.e.,*

$$\|T_\lambda x - z\|^2 \leq \|x - z\|^2 - \frac{2 - \lambda}{\lambda} \|T_\lambda x - x\|^2 \qquad (2.47)$$

for all $x \in X$ *and* $z \in \text{Fix } T$.

Proof. It follows from the first part of Theorem 2.2.5 that a firmly nonexpansive operator having a fixed point is a cutter. Therefore, T_λ is $\frac{2-\lambda}{\lambda}$-strongly quasi-nonexpansive (see Theorem 2.1.39).   □

2.2.2 Relationships Between Firmly Nonexpansive and Nonexpansive Operators

One can find in the literature several equivalent definitions of firmly nonexpansive operators. The properties of these operators were studied by Zarantonello [357, Sect. 1], Bruck [49, Sects. 2 and 3], Rockafellar [299], Bruck and Reich [51, Sect. 1], Goebel and Reich [185, Chap. 1, Sect. 11], Reich and Shafrir [296], Goebel and Kirk [184, Chap. 12], Bauschke and Borwein [22, Sects. 2 and 3], Byrne [56, Sect. 2], and by Crombez [127, Sect. 2].

The class of firmly nonexpansive operators is included in the class of nonexpansive ones (see Theorem 2.2.4). Further important relationships between these two classes are also useful for the investigation of firmly nonexpansive operators. These relationships are given in the following theorem.

Theorem 2.2.10. *Let* $T : X \rightarrow \mathcal{H}$. *Then the following conditions are equivalent:*

  (i) T *is firmly nonexpansive.*
  (ii) T_λ *is nonexpansive for any* $\lambda \in [0, 2]$.
  (iii) T *has the form* $T = \frac{1}{2}(S + \text{Id})$, *where* $S : X \rightarrow \mathcal{H}$ *is a nonexpansive operator.*
  (iv) $\text{Id} - T$ *is firmly nonexpansive.*
  (v) *For all* $x, y \in X$ *it holds*

$$\|Tx - Ty\|^2 \leq \|x - y\|^2 - \|(x - Tx) - (y - Ty)\|^2. \qquad (2.48)$$

  (vi) *For all* $x, y \in X$ *and for any* $\alpha \geq 0$ *it holds*

$$\|Tx - Ty\| \leq \|\alpha(x - y) + (1 - \alpha)(Tx - Ty)\|.$$

Proof. The equivalence (i)⇔(iv) in Theorem 2.2.10 is obvious, because both conditions can be written in the form $\langle Tx - Ty, (x - Tx) - (y - Ty) \rangle \geq 0$ for all $x, y \in X$. Nevertheless, we prove the following relations among (i)-(vi):

$$\text{(i)} \Rightarrow \text{(ii)} \Rightarrow \text{(iii)} \Rightarrow \text{(iv)} \Rightarrow \text{(v)} \Rightarrow \text{(i)} \Leftrightarrow \text{(vi)}.$$

(i)$\Rightarrow$(ii) Let T be firmly nonexpansive and $x, y \in X$. By the definition of a firmly nonexpansive operator, the Cauchy–Schwarz inequality and the nonexpansivity of T (see Theorem 2.2.4), we have

$$
\begin{aligned}
\| T_\lambda x - T_\lambda y \|^2 &= \| \lambda T x + (1 - \lambda)x - \lambda T y - (1 - \lambda)y \|^2 \\
&= \| \lambda(T x - T y) + (1 - \lambda)(x - y) \|^2 \\
&= \lambda^2 (\| T x - T y \|^2 - \langle T x - T y, x - y \rangle) \\
&\quad + (2\lambda - \lambda^2)\langle T x - T y, x - y \rangle + (1 - \lambda)^2 \| x - y \|^2 \\
&\leq (2\lambda - \lambda^2)\langle T x - T y, x - y \rangle + (1 - \lambda)^2 \| x - y \|^2 \\
&\leq (2\lambda - \lambda^2) \| T x - T y \| \, \| x - y \| + (1 - \lambda)^2 \| x - y \|^2 \\
&\leq (2\lambda - \lambda^2) \| x - y \|^2 + (1 - \lambda)^2 \| x - y \|^2 \\
&= \| x - y \|^2,
\end{aligned}
$$

i.e., T_λ is nonexpansive.

(ii)$\Rightarrow$(iii) This implication is obvious. It suffices to take $S = T_\lambda$ for $\lambda = 2$.

(iii)$\Rightarrow$(iv) Let S be nonexpansive, $T := \frac{1}{2}(S + \mathrm{Id})$ and $G := \mathrm{Id} - T$. Then we have $G = \frac{1}{2}(\mathrm{Id} - S)$ and

$$
\begin{aligned}
\| Gx - Gy \|^2 &= \langle Gx - Gy, x - y \rangle + \langle Gx - Gy, (Gx - Gy) - (x - y) \rangle \\
&= \langle Gx - Gy, x - y \rangle \\
&\quad + \frac{1}{4}\langle (Sx - Sy) - (x - y), (Sx - Sy) + (x - y) \rangle \\
&= \langle Gx - Gy, x - y \rangle + \frac{1}{4}(\| Sx - Sy \|^2 - \| x - y \|^2) \\
&\leq \langle Gx - Gy, x - y \rangle,
\end{aligned}
$$

for all $x, y \in X$.

(iv)$\Rightarrow$(v) Let $G := \mathrm{Id} - T$ be firmly nonexpansive. Then, for all $x, y \in X$ we have

$$
\begin{aligned}
&\| T x - T y \|^2 + \| (\mathrm{Id} - T)x - (\mathrm{Id} - T)y \|^2 \\
&\leq \| T x - T y \|^2 + \langle (\mathrm{Id} - T)x - (\mathrm{Id} - T)y, x - y \rangle \\
&= \| T x - T y \|^2 - \langle T x - T y, x - y \rangle + \| x - y \|^2 \\
&= -\langle T x - T y, (x - T x) - (y - T y) \rangle + \| x - y \|^2 \\
&\leq \| x - y \|^2,
\end{aligned}
$$

i.e., (2.48) holds.

(v)⇒(i) Let $x, y \in X$. If (2.48) holds, then, by the properties of the inner product, we have

$$\begin{aligned}
\|Tx - Ty\|^2 &\leq \|x - y\|^2 - \|(x - y) - (Tx - Ty)\|^2 \\
&= -\|Tx - Ty\|^2 + 2\langle Tx - Ty, x - y \rangle,
\end{aligned}$$

i.e., T is firmly nonexpansive.

(i)⇔(vi) Let $x, y \in X$. The function $h : \mathbb{R}_+ \to \mathbb{R}_+$ defined by

$$h(\alpha) = \frac{1}{2} \|\alpha(x - y) + (1 - \alpha)(Tx - Ty)\|^2$$

is convex as a composition of the convex function $f(\cdot) = \frac{1}{2}\|\cdot\|^2$ and an affine function $A : \mathbb{R} \to \mathcal{H}$, $A(\alpha) = \alpha(x - y) + (1 - \alpha)(Tx - Ty)$. Note that $h(0) = \frac{1}{2}\|Tx - Ty\|^2$, h is differentiable and

$$h'(0) = \langle Tx - Ty, (x - y) - (Tx - Ty) \rangle.$$

Since h is convex, we have

$$h(0) \leq h(\alpha) \iff h'(0) \geq 0$$

for all $\alpha \geq 0$, i.e.,

$$\begin{aligned}
\|Tx - Ty\|^2 &\leq \|\alpha(x - y) + (1 - \alpha)(Tx - Ty)\|^2 \\
&\iff \langle Tx - Ty, x - y \rangle \geq \|Tx - Ty\|^2
\end{aligned}$$

which completes the proof.   □

The same kind of correspondences between firmly nonexpansive operators and nonexpansive ones (the equivalence (i)⇔(iii) in Theorem 2.2.10) and between cutters and quasi-nonexpansive operators (Corollary 2.1.33 (ii)) explain the name *firmly quasi-nonexpansive operators* for cutters (see [346, page 624]).

Condition (ii) in Theorem 2.2.10 can be formulated equivalently as follows: (ii') $T_2 := 2T - \mathrm{Id}$ is nonexpansive.

The nonexpansivity of T_2 and Lemma 2.1.12 (i) yield the nonexpansivity of T_λ for all $\lambda \in [0, 2]$, because $T_\lambda = (1 - \frac{\lambda}{2})\,\mathrm{Id} + \frac{\lambda}{2}T_2$. Moreover, the assumption that T_2 is nonexpansive is sufficient in the implication (ii)⇒(iii) as follows from the proof.

Now we present a series of corollaries of Theorem 2.2.10.

Corollary 2.2.11. *Let $T : X \to \mathcal{H}$. The operator T is firmly nonexpansive if and only if its relaxation T_λ is firmly nonexpansive for all $\lambda \in [0, 1]$.*

Proof. Let T be firmly nonexpansive and $\lambda \in [0, 1]$. By the implication (i)⇒(iii) in Theorem 2.2.10 we obtain

$$T_\lambda = (1-\lambda)\,\mathrm{Id} + \frac{\lambda}{2}(\mathrm{Id}+S) = \frac{1}{2}[\mathrm{Id}+(1-\lambda)\,\mathrm{Id}+\lambda S]$$

for a nonexpansive operator S. Note that $(1-\lambda)\,\mathrm{Id}+\lambda S$ is nonexpansive as a convex combination of nonexpansive operators (see Lemma 2.1.12 (ii)). Therefore, T_λ is firmly nonexpansive by the implication (iii)$\Rightarrow$(i) in Theorem 2.2.10. The sufficiency of the condition is obvious. $\qquad\square$

Corollary 2.2.12. *Let $U : X \to \mathcal{H}$ and $\lambda \in [0,2]$. Then U is λ-RFNE if and only if U is μ-RFNE for all $\mu \in [\lambda, 2]$.*

Proof. Let $U := T_\lambda = \mathrm{Id}+\lambda(T-\mathrm{Id})$, where $T : X \to \mathcal{H}$ is a firmly nonexpansive operator, and $\mu \in [\lambda, 2]$. It is easy to see that

$$U = \mathrm{Id} + \mu(T_{\lambda/\mu} - \mathrm{Id}).$$

The corollary follows now from the fact that $T_{\lambda/\mu}$ is firmly nonexpansive (see Corollary 2.2.11). $\qquad\square$

Corollary 2.2.13. *Let $X \subseteq \mathcal{H}$ be a closed convex subset and $S : X \to \mathcal{H}$. The following conditions are equivalent:*

(i) S is nonexpansive,
(ii) $S = 2T - \mathrm{Id}$, where $T : X \to \mathcal{H}$ is a firmly nonexpansive operator.

Proof. (ii)$\Rightarrow$(i) Let $S := 2T - \mathrm{Id}$ for a firmly nonexpansive operator T. It follows from the implication (i)$\Rightarrow$(ii) in Theorem 2.2.10 that S is nonexpansive.

(i)$\Rightarrow$(ii) Let S be nonexpansive and $T := \frac{1}{2}(S + \mathrm{Id})$. By the implication (iii)$\Rightarrow$(i) in Theorem 2.2.10 the operator F is firmly nonexpansive. Furthermore, $S = 2T - \mathrm{Id}$. $\qquad\square$

Definition 2.2.14. (cf. [127, Definition 2.1]) We say that an operator $U : X \to \mathcal{H}$ is *ν-firmly nonexpansive* (*ν-FNE*), where $\nu > 0$, if

$$\|Ux - Uy\|^2 \le \|x - y\|^2 - \nu \|(x - Ux) - (y - Uy)\|^2.$$

Vasin and Ageev call a ν-firmly nonexpansive operator for $\nu \in (0,1)$, a *pseudo-contractive operator* (see [333, Definition 2.5]). In [127, Theorem 2.3] several equivalent conditions for U to be ν-FNE are presented.

By the equivalence (i)$\Leftrightarrow$(v) of Theorem 2.2.10, an operator is firmly nonexpansive if and only if it is 1-firmly nonexpansive. Note, however, that there is a difference between a λ-RFNE operator and a λ-FNE operator. Below, we present the relationship between these two notions.

Corollary 2.2.15. *Let $\lambda \in (0,2)$. An operator $U : X \to \mathcal{H}$ is λ-relaxed firmly nonexpansive if and only if U is $\frac{2-\lambda}{\lambda}$-firmly nonexpansive, i.e.,*

$$\|Ux - Uy\|^2 \leq \|x - y\|^2 - \frac{2 - \lambda}{\lambda} \|(x - Ux) - (y - Uy)\|^2$$

for all $x, y \in X$. *If, furthermore,* Fix $U \neq \emptyset$, *then*

$$\|Ux - z\|^2 \leq \|x - z\|^2 - \frac{2 - \lambda}{\lambda} \|Ux - x\|^2$$

for all $x \in X$ *and* $z \in$ Fix U, *i.e.,* U *is* $\frac{2-\lambda}{\lambda}$*-strongly quasi-nonexpansive.*

Proof. Let $U := T_\lambda$ for a firmly nonexpansive operator T and $x, y \in X$. Applying the properties of the inner product we get for $G := \mathrm{Id} - T$

$$\begin{aligned}
\|Ux - Uy\|^2 &= \|(1 - \lambda)x + \lambda Tx - (1 - \lambda)y - \lambda Ty\|^2 \\
&= \|x - y - \lambda(Gx - Gy)\|^2 \\
&= \|x - y\|^2 - 2\lambda\langle x - y, Gx - Gy\rangle + \lambda^2 \|Gx - Gy\|^2.
\end{aligned}$$

Since $x - Ux = x - T_\lambda x = \lambda Gx$, the equalities above yield

$$\begin{aligned}
&\|Ux - Uy\|^2 - \|x - y\|^2 + \frac{2 - \lambda}{\lambda} \|(x - Ux) - (y - Uy)\|^2 \\
&= \|Ux - Uy\|^2 - \|x - y\|^2 + \lambda(2 - \lambda) \|Gx - Gy\|^2 \\
&= -2\lambda(\langle x - y, Gx - Gy\rangle - \|Gx - Gy\|^2).
\end{aligned}$$

The first part of the corollary follows now from the equivalence (i)⇔(iv) in Theorem 2.2.10, and now the other part follows directly from the definition of an α-strongly quasi-nonexpansive operator.   □

Definition 2.2.16. Let $\alpha \in (0, 1)$. We say that an operator $T : X \to \mathcal{H}$ is *α-averaged* or, shortly, *averaged (AV)* if

$$T = (1 - \alpha)\,\mathrm{Id} + \alpha S$$

holds for a nonexpansive operator $S : X \to \mathcal{H}$.

Averaged operators were studied, e.g., by Mann [252], Krasnosel'skiĭ [238], Baillon et al. [14, Sect. 2]. In [56, Sect. 2], Byrne gives relationships between averaged operators and *inverse strongly monotone operators*, i.e., operators $G : X \to \mathcal{H}$ such that

$$\langle Gx - Gy, x - y\rangle \geq \nu \|Gx - Gy\|^2$$

for all $x, y \in X$ and for some constant $\nu > 0$.

Definition 2.2.16 states that an operator is averaged if and only if it is an underrelaxation of a nonexpansive operator.

Corollary 2.2.17. *Let* $\lambda \in (0,2)$ *and* $\alpha = \lambda/2$. *An operator* $U : X \to \mathcal{H}$ *is* λ-*relaxed firmly nonexpansive if and only if* U *is* α-*averaged.*

Proof. ($\Rightarrow$) Let $T : X \to \mathcal{H}$ be firmly nonexpansive and $U := T_\lambda = (1 - \lambda)$ Id $+\lambda T$. By the implication (i)$\Rightarrow$(iii) in Theorem 2.2.10 we have $T = \frac{1}{2}(S + \text{Id})$ for a nonexpansive operator $S : X \to \mathcal{H}$. Hence, $U = (1 - \alpha)$ Id $+\alpha S$, i.e., U is α-averaged.

($\Leftarrow$) Let U be α-averaged, i.e., $U = (1-\alpha)$ Id $+\alpha S$ for a nonexpansive operator S and for $\alpha = \lambda/2 \in (0,1)$. By Corollary 2.2.13 we have

$$U = (1 - \alpha)\,\text{Id} + \alpha(2T - \text{Id})$$
$$= (1 - 2\alpha)\,\text{Id} + 2\alpha T$$

for a firmly nonexpansive operator T. Hence, U is the λ-relaxation of $T : X \to \mathcal{H}$ with $\lambda = 2\alpha \in (0,2)$. $\qquad\square$

Corollary 2.2.18. *Let* $G : X \to \mathcal{H}$. *Then* G *is firmly nonexpansive if and only if* Id $-\mu G$ *is averaged for any* $\mu \in (0,2)$.

Proof. Necessity. Let G be firmly nonexpansive. We have

$$\text{Id} - \mu G = (1 - \mu/2)\,\text{Id} + (\mu/2)[2(\text{Id} - G) - \text{Id}].$$

By the implications (i)$\Rightarrow$(iv) and (i)$\Rightarrow$(ii) in Theorem 2.2.10 the operator $2(\text{Id} - G) - \text{Id}$ is nonexpansive. Consequently, the operator Id $-\mu G$ is averaged.

Sufficiency. Let Id $-\mu G$ be averaged for any $\mu \in (0,2)$. Then Id $-\mu G$ is nonexpansive for any $\mu \in (0,2)$ and Id $-2G$ is nonexpansive as a limit of nonexpansive operators. Now, it follows from the implication (ii)$\Rightarrow$(i) in Theorem 2.2.10 that G is firmly nonexpansive. $\qquad\square$

Corollary 2.2.19. *Let* $U : X \to \mathcal{H}$ *and* $\lambda \in (0,2]$. *The operator* U *is* λ-*relaxed firmly nonexpansive if and only if its relaxation* U_μ *is firmly nonexpansive for* $\mu \in [0, \frac{1}{\lambda}]$.

Proof. Take $U := T_\lambda$ for a firmly nonexpansive operator $T : X \to \mathcal{H}$. Then the claim follows from the equality $U_{\lambda^{-1}} = T$ (see Remark 2.1.3) and Corollary 2.2.11. The converse implication is obvious. $\qquad\square$

The following corollary shows that the family of firmly nonexpansive operators is closed under convex combination.

Corollary 2.2.20. *Let* $T_i : X \to \mathcal{H}$, $i \in I := \{1, 2, \ldots, m\}$, *be firmly nonexpansive and* $w = (\omega_1, \omega_2, \ldots, \omega_m) \in \Delta_m$. *Then the operator* $T := \sum_{i \in I} \omega_i T_i$ *is firmly nonexpansive.*

Proof. Let $T := \sum_{i \in I} \omega_i T_i$. By the implication (i)$\Rightarrow$(iii) in Theorem 2.2.10, we have $T_i = \frac{1}{2}(S_i + \text{Id})$ for a nonexpansive operator S_i, $i \in I$. Observe

Fig. 2.10 Basic relationships among algorithmic operators

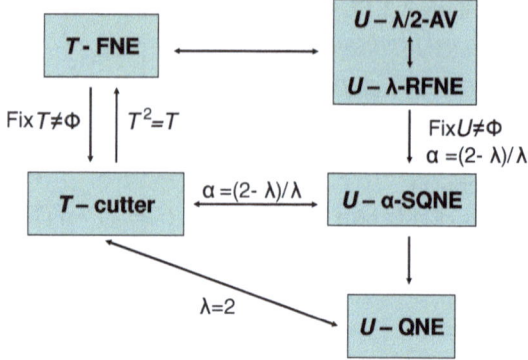

that $T = \frac{1}{2}(S + \text{Id})$ for $S := \sum_{i \in I} \omega_i S_i$. By Lemma 2.1.12 (i), the operator S is nonexpansive. The corollary follows now from the implication (iii)$\Rightarrow$(i) in Theorem 2.2.10.   □

In Fig. 2.10 we shortly present important relationships among the FNE operators, cutters, QNE operators SQNE operators and AV operators, which are proved in Sects. 2.1.3, 2.2.1 and 2.2.2. In Fig. 2.10, $T : X \rightarrow \mathcal{H}$ and $U := I_\lambda = \text{Id} +\lambda(T - \text{Id})$ is its λ-relaxation, where $\lambda \in (0, 2)$. We will extend this figure in Sect. 3.9.

2.2.3 Further Properties of the Metric Projection

The basic facts concerning firmly nonexpansive operators presented in the previous section yield further properties of the metric projection.

Theorem 2.2.21. *Let $C \subseteq \mathcal{H}$ be a nonempty closed convex subset and $P_C : \mathcal{H} \rightarrow \mathcal{H}$ be the metric projection onto C. Then the operator P_C is:*

 (i) *Idempotent, consequently* Fix $P_C = C$,
 (ii) *A cutter,*
 (iii) *Firmly nonexpansive,*
 (iv) *Monotone and nonexpansive,*
 (v) *Averaged.*

Proof. (i) The property follows directly from the definition of the metric projection.
 (ii) It follows from (i) and from the characterization of the metric projection (see Theorem 1.2.4) that $\langle z - P_C x, x - P_C x \rangle \leq 0$ for all $x \in \mathcal{H}$ and for all $z \in C = \text{Fix } P_C$, which means that P_C is a cutter.
 (iii) The property follows directly from (i), (ii) and from Theorem 2.2.5 (ii).

(iv) By Theorem 2.2.4, any firmly nonexpansive operator is monotone and nonex-
     pansive. Therefore, the property follows from (iii).
(v) By the firm nonexpansivity of P_C and by the implication (i)$\Rightarrow$(iii) in The-
     orem 2.2.10, we can write $P_C = \frac{1}{2}(S + \mathrm{Id})$ for a nonexpansive operator
     $S : X \to \mathcal{H}$. Hence, P_C is averaged.

$\square$

Definition 2.2.22. Let $C \subseteq \mathcal{H}$ be a nonempty closed convex subset. We call a
relaxation of the metric projection $P_C : \mathcal{H} \to C$ a *relaxed metric projection* onto
the subset C and we denote it by $P_{C,\lambda}$ or, shortly, by P_λ. If $\lambda < 1$, then P_λ is called
an *under-projection*. If $\lambda > 1$, then P_λ is called an *over-projection*. If $\lambda = 2$, then
P_λ is called the *reflection*.

We have

$$P_{C,\lambda} = P_\lambda = \mathrm{Id} + \lambda(P_C - \mathrm{Id}).$$

Corollary 2.2.23. Let $C \subseteq \mathcal{H}$ be a nonempty closed convex subset, $\lambda \geq 0$ and
$P_\lambda : \mathcal{H} \to \mathcal{H}$ be a relaxed metric projection. Then

 (i) P_λ is a nonexpansive operator for all $\lambda \in [0, 2]$,
(ii) Fix $P_\lambda = C$ for all $\lambda > 0$,
(iii) For all $x \in \mathcal{H}$, $z \in C$ and $\lambda \in (0, 2]$ the following inequality holds

$$\| P_\lambda x - z \|^2 \leq \| x - z \|^2 - \frac{2 - \lambda}{\lambda} \| P_\lambda x - x \|^2. \tag{2.49}$$

Consequently, P_λ is $\frac{2-\lambda}{\lambda}$-strongly quasi-nonexpansive for all $\lambda \in (0, 2]$.

Proof. Part (i) follows from the equivalence (i)$\Leftrightarrow$(ii) in Theorem 2.2.10, because
P_C is firmly nonexpansive (see Theorem 2.2.21 (iii)). Part (ii) is obvious, because
Fix $P_C = C$. Part (iii) follows now from Corollary 2.2.9. $\square$

Corollary 2.2.24. Let $C \subseteq \mathcal{H}$ be a nonempty closed convex subset and $x, y \in \mathcal{H}$.
Then

$$\| P_C x - P_C y \|^2 \leq \| x - y \|^2 - \|(P_C x - x) - (P_C y - y)\|^2 \tag{2.50}$$
$$\leq \| x - y \|^2 - (\| P_C x - x \| - \| P_C y - y \|)^2. \tag{2.51}$$

In particular,
$$\| P_C x - z \|^2 \leq \| x - z \|^2 - \| P_C x - x \|^2 \tag{2.52}$$

*for all $x \in \mathcal{H}$ and $z \in C$. Consequently, the metric projection $P_C : \mathcal{H} \to C$ is
strongly quasi-nonexpansive.*

Proof. By Theorem 2.2.21 (iii), the metric projection is firmly nonexpansive.
Therefore, inequalities (2.50) and (2.51) follow directly from the implication
(i)$\Rightarrow$(v) in Theorem 2.2.10 and from the Cauchy–Schwarz inequality. The second
part follows directly from Theorem 2.2.21 (i). $\square$

Fig. 2.11 Function
$f(x) = \|P_X(x + \alpha u) - x\|$
is nondecreasing

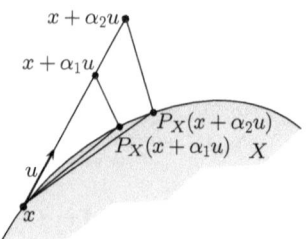

Corollary 2.2.25. *Let* $T : X \to \mathcal{H}$ *and* $\lambda \in (0, 2)$. *If* T *is a cutter, then for any* $x \in X$ *and* $z \in \mathrm{Fix}\, T$ *the following estimations hold*

$$\|P_X T_\lambda x - z\|^2 \leq \|x - z\|^2 - \lambda(2 - \lambda) \|Tx - x\|^2$$

and

$$\|P_X T_\lambda x - z\|^2 \leq \|x - z\|^2 - \frac{2 - \lambda}{\lambda} \|P_X T_\lambda x - x\|^2 . \tag{2.53}$$

Consequently, the operator $P_X T_\lambda : X \to X$ *is* $\frac{2-\lambda}{\lambda}$*-strongly quasi-nonexpansive.*

Proof. Note that P_X is a nonexpansive operator and that

$$\mathrm{Fix}\, P_X \cap \mathrm{Fix}\, T = X \cap \mathrm{Fix}\, T = \mathrm{Fix}\, T \neq \emptyset.$$

Therefore, the corollary follows from Theorem 2.1.51.   □

The following corollary will be useful in further parts of the book (see also [327, Lemma 2] and [172, Lemma 1] for related results).

Corollary 2.2.26. *Let* $x \in X$, $u \in \mathcal{H}$ *and* $0 \leq \alpha_1 < \alpha_2$. *Then the following inequality holds*

$$\|P_X(x + \alpha_2 u) - x\|^2$$
$$\geq \|P_X(x + \alpha_1 u) - x\|^2 + \|P_X(x + \alpha_2 u) - P_X(x + \alpha_1 u)\|^2 . \tag{2.54}$$

Consequently, the function $f : \mathbb{R}_+ \to \mathbb{R}_+$, $f(\alpha) := \|P_X(x + \alpha u) - x\|$ *is nondecreasing.*

Corollary 2.2.26 is illustrated in Fig. 2.11.

Proof. Inequality (2.54) is obvious for $\alpha_1 = 0$. Let now $\alpha_1 > 0$. Take $y := x + \alpha_2 u$, $z := x + \alpha_1 u$ and $\lambda = \frac{\alpha_1}{\alpha_2}$. Then we have $\lambda \in (0, 1)$ and $(x - z) = -\frac{\lambda}{1-\lambda}(y - z)$. Now inequality (2.54) can be written in the form

$$\|P_X y - x\|^2 \geq \|P_X z - x\|^2 + \|P_X y - P_X z\|^2 . \tag{2.55}$$

The characterization of the metric projection (see Theorem 1.2.4) and its monotonicity (see Theorem 2.2.21 (iv)) yield

$$\langle x - P_X z, P_X y - P_X z \rangle = \langle x - z, P_X y - P_X z \rangle + \langle z - P_X z, P_X y - P_X z \rangle$$

$$\leq -\frac{\lambda}{1-\lambda} \langle y - z, P_X y - P_X z \rangle \leq 0,$$

i.e., $\langle x - P_X z, P_X y - P_X z \rangle \leq 0$, which is equivalent to (2.55), by Lemma 1.2.5. $\square$

Let $C \subseteq \mathcal{H}$ be convex. Define the distance function $d(\cdot, C) : \mathcal{H} \to \mathbb{R}$ by $d(x, C) = \inf_{y \in C} \|x - y\|$. It follows from the continuity of the norm and from the definition
of the metric projection that

$$d(x, C) = d(x, \mathrm{cl}\, C) = \|x - P_{\mathrm{cl}\, C} x\|.$$

Therefore, we suppose without loss of generality that C is closed. It turns out that
the functions $d(\cdot, C)$ and $d^2(\cdot, C)$ are convex and differentiable.

Lemma 2.2.27. *Let $C \subseteq \mathcal{H}$ be a closed convex subset. Then the function
$f : \mathcal{H} \to \mathbb{R}$, $f(x) := \frac{1}{2} d^2(x, C)$ is differentiable and $D_f(x) = x - P_C x$ for
all $x \in \mathcal{H}$.*

Proof. (cf. [167, Proposition 2.2] and [209, Chap. IV, Example 4.1.6]) Let $x, h \in \mathcal{H}$.
It follows from the definition of the metric projection and from the properties of the
inner product that

$$f(x+h) - f(x) - \langle x - P_C x, h \rangle$$

$$= \frac{1}{2} \|x + h - P_C(x+h)\|^2 - \frac{1}{2} \|x - P_C x\|^2 - \langle x - P_C x, h \rangle$$

$$\leq \frac{1}{2} \|x + h - P_C x\|^2 - \frac{1}{2} \|x - P_C x\|^2 - \langle x - P_C x, h \rangle$$

$$= \frac{1}{2} \|h\|^2.$$

Similarly, by the definition of the metric projection, the Cauchy–Schwarz inequality
and the nonexpansivity of the metric projection, we obtain

$$f(x+h) - f(x) - \langle x - P_C x, h \rangle$$

$$= \frac{1}{2} \|x + h - P_C(x+h)\|^2 - \frac{1}{2} \|x - P_C x\|^2 - \langle x - P_C x, h \rangle$$

$$\geq \frac{1}{2} \|x + h - P_C(x+h)\|^2 - \frac{1}{2} \|x - P_C(x+h)\|^2 - \langle x - P_C x, h \rangle$$

$$= \frac{1}{2} \|h\|^2 + \langle P_C x - P_C(x+h), h \rangle$$

$$\geq \frac{1}{2} \|h\|^2 - \|P_C x - P_C(x+h)\| \cdot \|h\|$$

$$\geq \frac{1}{2} \|h\|^2 - \|h\|^2 = -\frac{1}{2} \|h\|^2.$$

Now we see that

$$-\frac{1}{2} \|h\|^2 \leq (f(x+h) - f(x) - \langle x - P_C x, h \rangle) \leq \frac{1}{2} \|h\|^2.$$

Consequently,

$$f(x+h) = f(x) + \langle x - P_C x, h \rangle + o(\|h\|).$$

Therefore, f is differentiable and $D_f(x) = x - P_C x$. $\square$

Lemma 2.2.28. *Let $C \subseteq \mathcal{H}$ be a closed convex subset. The function $h : \mathcal{H} \to \mathbb{R}$, $h(x) := d(x, C)$ is convex and differentiable for all $x \notin C$ and*

$$Dh(x) = \frac{x - P_C x}{\|x - P_C x\|}. \tag{2.56}$$

Proof. Since $h(x) = \inf_{y \in C} \|x - y\|$, the convexity of h follows from the fact that the function $p : \mathcal{H} \times \mathcal{H} \to \mathcal{H}$, $p(x, y) := \|x - y\|$ is convex (as a composition of a linear function $(x, y) \to x - y$ and a convex function $z \to \|z\|$) and from the fact that for a convex function p, the function $\inf_{y \in C} p(\cdot, y)$ is convex. Since $h = \sqrt{d^2(\cdot, C)}$, the differentiability of h as well as equality (2.56) for $x \notin C$ follow from Lemma 2.2.27 and from the formula $D(\|z\|) = \frac{z}{\|z\|}$ for $z \neq 0$. $\square$

Corollary 2.2.29. *Let $C \subseteq \mathcal{H}$ be a closed convex subset. Then the function $f : \mathcal{H} \to \mathbb{R}$, $f(x) := \frac{1}{2} d^2(x, C)$ is convex.*

Proof. The function f is convex as a composition $f = g \circ h$ of a convex function $h := d(\cdot, C)$ and of a convex and increasing function $g : \mathbb{R}^+ \to \mathbb{R}$, $g(t) := \frac{1}{2} t^2$. $\square$

2.2.4   Metric Projection onto a Closed Subspace

Let $V \subseteq \mathcal{H}$ be a closed linear subspace. Since V is convex, the metric projection P_V is well defined. The theorem below states some properties of P_V. In particular, the first part of the theorem states that the metric projection onto V is equal to the orthogonal projection onto V.

Theorem 2.2.30. *Let $V \subseteq \mathcal{H}$ be a closed subspace and $x \in \mathcal{H}$, $y \in V$. Then*

(i) $y = P_V x$ *if and only if* $\langle x - y, z \rangle = 0$ *for all* $z \in V$,
(ii) P_V *is a bounded linear operator and* $\|P_V\| = 1$,

(iii) P_V *is self-adjoint,*
(iv) $\mathrm{Id} = P_V + P_{V^\perp}$.

Proof. (i) *Necessity.* Let $y := P_V x$. By the characterization of the metric projection (see Theorem 1.2.4), $\langle x - y, z - y \rangle \le 0$ for all $z \in V$. Suppose that $\langle x - y, w - y \rangle < 0$ for some $w \in V$. Let $u := 2y - w$. Then $u \in V$ because V is a linear subspace and we have

$$\langle x - y, u - y \rangle = \langle x - y, y - w \rangle > 0.$$

This contradiction shows that $\langle x - y, z - y \rangle = 0$ for all $z \in V$. If we take $z := 0 \in V$ in the latter equality, we obtain $\langle x - y, y \rangle = 0$. Hence, $\langle x - y, z \rangle = 0$ for all $z \in V$.
Sufficiency. Let $\langle x - y, z \rangle = 0$ for all $z \in V$. Taking $z := y \in V$ we obtain in particular $\langle x - y, y \rangle = 0$. Hence, $\langle x - y, z - y \rangle = 0$ for all $z \in V$. By the characterization of the metric projection (see Theorem 1.2.4), we have $y = P_V x$.

(ii) Let $x_1, x_2 \in \mathcal{H}$, $\alpha_1, \alpha_2 \in \mathbb{R}$, $y_1 := P_V x_1$, $y_2 := P_V x_2$ and $x := \alpha_1 x_1 + \alpha_2 x_2$, $y := \alpha_1 y_1 + \alpha_2 y_2$. We show that $y = P_V x$. By (i) we have

$$\begin{aligned}
\langle x - y, z \rangle &= \langle \alpha_1 (x_1 - y_1) + \alpha_2 (x_2 - y_2), z \rangle \\
&= \alpha_1 \langle x_1 - y_1, z \rangle + \alpha_2 \langle x_2 - y_2, z \rangle \\
&= 0,
\end{aligned}$$

for all $z \in V$, i.e., $y = P_V x$. Since P_V is nonexpansive, it is bounded. Furthermore,

$$\|P_V x\| = \|P_V x - P_V 0\| \le \|x - 0\| = \|x\|$$

for all $x \in \mathcal{H}$ and $\|P_V x\| = \|x\|$ for $x \in V$. Hence, $\|P_V\| = 1$.

(iii) Let $x, u \in \mathcal{H}$. It follows from (i) that

$$\langle x, P_V u \rangle = \langle P_V x, P_V u \rangle$$

and

$$\langle u, P_V x \rangle = \langle P_V u, P_V x \rangle.$$

By the symmetry of the inner product

$$\langle P_V x, u \rangle = \langle x, P_V u \rangle,$$

i.e., P_V is self-adjoint.

(iv) Let $x \in \mathcal{H}$. By (i), we have $x - P_V x \in V^\perp$ and $x - P_V x = P_{V^\perp} x$. Since $x = P_V x + (x - P_V x)$, it holds $x = P_V x + P_{V^\perp} x$. $\square$

Corollary 2.2.31. *Let $V \subseteq \mathcal{H}$ be a closed subspace and $x \in \mathcal{H}$. Then*

$$\langle P_V x, x \rangle = \| P_V x \|^2 .$$

Proof. Since P_V is self-adjoint (see Theorem 2.2.30 (iii)), we have $\langle P_V x, u \rangle = \langle x, P_V u \rangle$ for all $u \in \mathcal{H}$. If we take $u := P_V x$ we obtain the desired property. □

Corollary 2.2.32. *A bounded linear operator is an orthogonal projection if and only if it is idempotent and self-adjoint.*

Proof. The necessity follows from Theorems 2.2.21 (i) and 2.2.30 (iii). Let now $T : \mathcal{H} \rightarrow \mathcal{H}$ be idempotent and self-adjoint. Let $V := T(\mathcal{H})$. It is clear that $V = \text{Fix } T$ and that V is a closed subspace. Now we show that $T = P_V$. Let $x \in \mathcal{H}$ and $z \in V$. Then

$$\langle Tx, z \rangle = \langle x, Tz \rangle = \langle x, z \rangle,$$

i.e., $\langle Tx - x, z \rangle = 0$. Theorem 2.2.30 (i) implies now that $T = P_V$. □

2.2.5   Metric Projection onto a Closed Affine Subspace

Let $A \subseteq \mathcal{H}$ be a closed affine subspace and $a \in A$. Then $A - a$ is a closed linear subspace. In order to show some properties of the metric projection P_A we apply Theorem 2.2.30 together with

$$P_A x = P_{A-a}(x - a) + a \tag{2.57}$$

(see Lemma 1.2.6).

Theorem 2.2.33. *Let $A \subseteq \mathcal{H}$ be a closed affine subspace and $x, u, v, w \in \mathcal{H}$, $a, y \in A$. Then*

(i) $y = P_A x$ *if and only if* $\langle x - y, z - y \rangle = 0$ *for all $z \in A$,*
(ii) $P_A u - P_A v = P_{A-a}(u - v) = P_A(u - v) - P_A 0$,
(iii) $\langle P_A u - P_A v, w \rangle = \langle u - v, P_{A-a} w \rangle = \langle u - v, P_A w - P_A 0 \rangle$,
(iv) $\langle P_A u - P_A v, u - v \rangle = \| P_A u - P_A v \|^2$,
(v) $\| u - v \|^2 = \| P_A u - P_A v \|^2 + \| (P_A u - u) - (P_A v - v) \|^2$,
(vi) P_A *is an affine operator.*

Proof. (i) Since $A - a$ is a linear subspace, $v \in A - a$ if and only if $v = z - y$, for some $z \in A$. By (2.57) and Theorem 2.2.30 (i), we have

$$y = P_A x \Leftrightarrow y - a = P_{A-a}(x - a) \Leftrightarrow \langle y - a - (x - a), z - y \rangle = 0$$

for any $z \in A$.

(ii) By (2.57) and the linearity of P_{A-a}, we have

$$
\begin{aligned}
P_A u - P_A v &= P_{A-a}(u-a) + a - (P_{A-a}(v-a) + a) \\
&= P_{A-a}(u-v) \\
&= P_{A-a}(u-v-a) - P_{A-a}(-a) \\
&= P_A(u-v) - P_A 0.
\end{aligned}
$$

(iii) Since P_{A-a} is self-adjoint, (2.57) and (ii) yield

$$
\begin{aligned}
\langle P_A u - P_A v, w \rangle &= \langle P_{A-a}(u-a) - (P_{A-a}(v-a), w \rangle \\
&= \langle u-v, P_{A-a}w \rangle \\
&= \langle u-v, P_{A-a}(w-a) - P_{A-a}(-a) \rangle \\
&= \langle u-v, P_A w - P_A 0 \rangle.
\end{aligned}
$$

(iv) Property (i) yields

$$
\langle P_A u - u, P_A u - P_A v \rangle = 0 \text{ and } \langle P_A v - v, P_A u - P_A v \rangle = 0
$$

Therefore,

$$
\langle (P_A u - u) - (P_A v - v), P_A u - P_A v \rangle = 0,
$$

i.e.,

$$
\langle P_A u - P_A v, u-v \rangle = \| P_A u - P_A v \|^2.
$$

(v) It follows from the properties of the inner product and from property (iv) that

$$
\begin{aligned}
&\| (P_A u - u) - (P_A v - v) \|^2 \\
&= \| P_A u - P_A v \|^2 + \| u-v \|^2 - 2\langle P_A u - P_A v, u-v \rangle \\
&= \| u-v \|^2 - \| P_A u - P_A v \|^2.
\end{aligned}
$$

(vi) Let $\lambda \in \mathbb{R}$. Since $A - a$ is a closed subspace, (2.57) and Theorem 2.2.30 yield

$$
\begin{aligned}
P_A((1-\lambda)u + \lambda y) &= P_{A-a}((1-\lambda)(u-a) + \lambda(y-a)) + a \\
&= (1-\lambda)(P_{A-a}(u-a) + a) + \lambda(P_{A-a}(y-a) + a) \\
&= (1-\lambda)P_A u + \lambda P_A y
\end{aligned}
$$

which completes the proof.

$\square$

2.2.6   Properties of Relaxed Firmly Nonexpansive Operators

In this section we present relationships among families of relaxed firmly nonexpansive operators, contractions, averaged operators and strongly quasi-nonexpansive operators. Furthermore, we give properties of relaxed firmly nonexpansive operators which are used in many constructions of algorithmic operators.

Theorem 2.2.34. *An α-contraction is $(1 + \alpha)$-relaxed firmly nonexpansive.*

Proof. Let $T : X \rightarrow \mathcal{H}$ be an α-contraction, i.e., $\|Tx - Ty\| \leq \alpha \|x - y\|$ for all $x, y \in X$, where $\alpha \in (0, 1)$. Let $U := \frac{2}{1+\alpha} T - \frac{1-\alpha}{1+\alpha}$ Id, or, equivalently,

$$T = \frac{1 + \alpha}{2} U + \frac{1 - \alpha}{2} \text{ Id},$$

i.e.. T is $\frac{1+\alpha}{2}$-averaged. By the convexity of the norm and the nonexpansivity of T,

$$
\begin{aligned}
\|Ux - Uy\| &= \left\| \frac{2}{1 + \alpha} (Tx - Ty) - \frac{1 - \alpha}{1 + \alpha} (x - y) \right\| \\
&\leq \frac{2}{1 + \alpha} \|Tx - Ty\| + \frac{1 - \alpha}{1 + \alpha} \|x - y\| \\
&\leq \frac{2\alpha}{1 + \alpha} \|x - y\| + \frac{1 - \alpha}{1 + \alpha} \|x - y\| \\
&= \|x - y\|,
\end{aligned}
$$

i.e., U is nonexpansive. Therefore, T is $(1 + \alpha)$-relaxed firmly nonexpansive as a $(\frac{1+\alpha}{2})$-averaged operator (see Corollary 2.2.17). $\qquad\square$

The next results show that a family of relaxed firmly nonexpansive operators is closed under convex combination and under composition.

Theorem 2.2.35. *Let $\lambda_i \in [0, 2]$ and $U_i : X \rightarrow \mathcal{H}$ be λ_i-relaxed firmly nonexpansive, $i \in I := \{1, 2, \ldots, m\}$, $U := \sum_{i=1}^{m} \omega_i U_i$ for $w = (\omega_1, \ldots, \omega_m) \in \Delta_m$. Then the operator U is λ-relaxed firmly nonexpansive, where $\lambda = \sum_{j=1}^{m} \omega_j \lambda_j$. Consequently, U is strictly relaxed firmly nonexpansive if $\lambda_i \in (0, 2)$ for some $i \in I$ and the corresponding weight $\omega_i > 0$.*

Proof. Let $U_i := \text{Id} + \lambda_i (T_i - \text{Id})$, where $T_i : X \rightarrow \mathcal{H}$ are firmly nonexpansive, $\lambda_i \in [0, 2]$, $i \in I$, and $w = (\omega_1, \ldots, \omega_m) \in \Delta_m$. It is clear that $\lambda := \sum_{j=1}^{m} \omega_j \lambda_j \in [0, 2]$. For $\lambda = 0$ the claim is obvious, because $U = \text{Id}$ in this case. Let now $\lambda \in (0, 2]$. Since

$$\sum_{i=1}^{m} \frac{\omega_i \lambda_i}{\sum_{j=1}^{m} \omega_j \lambda_j} = 1,$$

the operator

$$T := \sum_{i=1}^{m} \frac{\omega_i \lambda_i}{\sum_{j=1}^{m} \omega_j \lambda_j} T_i$$

is firmly nonexpansive as a convex combination of firmly nonexpansive operators T_i (see Corollary 2.2.20). Let $U := \sum_{i=1}^{m} \omega_i U_i$. Then we have

$$U = \sum_{i=1}^{m} \omega_i [\mathrm{Id} + \lambda_i (T_i - \mathrm{Id})]$$

$$= \mathrm{Id} + \sum_{i=1}^{m} \omega_i \lambda_i (T_i - \mathrm{Id})$$

$$= \mathrm{Id} + \left(\sum_{j=1}^{m} \omega_j \lambda_j \right) \left(\sum_{i=1}^{m} \frac{\omega_i \lambda_i}{\sum_{j=1}^{m} \omega_j \lambda_j} T_i - \sum_{i=1}^{m} \frac{\omega_i \lambda_i}{\sum_{j=1}^{m} \omega_j \lambda_j} \mathrm{Id} \right)$$

$$= \mathrm{Id} + \lambda (T - \mathrm{Id})$$

and, consequently, U is λ-relaxed firmly nonexpansive. The second part of the theorem is obvious.   □

Corollary 2.2.36. *A convex combination of averaged operators is an averaged operator.*

Proof. It suffices to apply Corollary 2.2.17 to Theorem 2.2.35.   □

Theorem 2.2.37. *Let $T, U : X \to X$ and $\lambda, \mu \in [0, 2]$. If T is λ-RFNE and U is μ-RFNE, then the composition $V := UT$ is γ-RFNE, with*

$$\gamma = \begin{cases} 0 & \text{if } \lambda = 0 \text{ and } \mu = 0 \\ 2 & \text{if } (2 - \lambda)(2 - \mu) = 0 \\ \dfrac{4(\lambda + \mu - \lambda\mu)}{4 - \lambda\mu} = \dfrac{2}{(\dfrac{\lambda}{2 - \lambda} + \dfrac{\mu}{2 - \mu})^{-1} + 1} & \text{otherwise.} \end{cases}$$

$$(2.58)$$

Proof. If $\lambda = 0$ or $\mu = 0$, then $T = \mathrm{Id}$ or $U = \mathrm{Id}$, respectively, and the claim is obvious, because the operator Id is 0-RFNE. If $\lambda = 2$ or $\mu = 2$, then T and U are nonexpansive (see Theorem 2.2.10 (ii)) and UT is nonexpansive as a composition of nonexpansive operators. Therefore, UT is 2-RFNE (see Corollary 2.2.13). Let now $\lambda, \mu \in (0, 2)$ and $x, y \in \mathcal{H}$. Denote $a_1 := Tx - x$, $a_2 := Ty - y$, $b_1 := UTx - Tx$ and $b_2 := UTy - Ty$. It is clear that

$$y - x = Ty - Tx + a_1 - a_2. \tag{2.59}$$

By Corollary 2.2.3, we have

$$\langle y - x, a_1 \rangle + \langle x - y, a_2 \rangle \geq \frac{1}{\lambda} \|a_1 - a_2\|^2$$

and

$$\langle Ty - Tx, b_1 \rangle + \langle Tx - Ty, b_2 \rangle \geq \frac{1}{\mu} \|b_1 - b_2\|^2 .$$

Therefore, the properties of the inner product, equality (2.59) and Lemma 2.1.45 yield

$$\langle y - x, UTx - x \rangle + \langle x - y, UTy - y \rangle - \frac{1}{\gamma}(\|(UTx - x) - (UTy - y)\|^2)$$

$$= \langle y - x, a_1 + b_1 \rangle + \langle x - y, a_2 + b_2 \rangle - \frac{1}{\gamma}(\|(a_1 + b_1) - (a_2 + b_2)\|^2)$$

$$= \langle y - x, a_1 \rangle + \langle x - y, a_2 \rangle + \langle Ty - Tx, b_1 \rangle + \langle Tx - Ty, b_2 \rangle$$

$$+\langle a_1 - a_2, b_1 - b_2 \rangle - \frac{1}{\gamma}(\|(a_1 + b_1) - (a_2 + b_2)\|^2)$$

$$\geq \frac{1}{\lambda} \|a_1 - a_2\|^2 + \frac{1}{\mu} \|b_1 - b_2\|^2 + \langle a_1 - a_2, b_1 - b_2 \rangle$$

$$-\frac{1}{\gamma}(\|(a_1 + b_1) - (a_2 + b_2)\|^2$$

$$= \left(\frac{1}{\lambda} - \frac{1}{\gamma}\right) \|a_1 - a_2\|^2 + \left(\frac{1}{\mu} - \frac{1}{\gamma}\right) \|b_1 - b_2\|^2 + \left(1 - \frac{2}{\gamma}\right) \langle a_1 - a_2, b_1 - b_2 \rangle$$

$$= \left\| \sqrt{\frac{1}{\lambda} - \frac{1}{\gamma}}(a_1 - a_2) - \sqrt{\frac{1}{\mu} - \frac{1}{\gamma}}(b_1 - b_2) \right\|^2 \geq 0.$$

Now it follows from Corollary 2.2.3 that UT is γ-RFNE.   □

Remark 2.2.38.   Because of Corollary 2.2.17, Theorem 2.2.37 can be stated equivalently in terms of averaged operators:
if T is α-averaged and U is β-averaged, where $\alpha, \beta \in (0, 1)$, then UT is δ-averaged, with

$$\delta := \frac{\alpha + \beta - 2\alpha\beta}{1 - \alpha\beta}. \tag{2.60}$$

This result is due to Ogura and Yamada (see [273, Theorem 3 (b)]). The fact that a composition of averaged operators $T := (1 - \alpha)\,\mathrm{Id} + \alpha R$ and $U := (1 - \beta)\,\mathrm{Id} + \beta S$ is averaged follows also from the following identity (cf. [56, Lemma 2.2 and Proposition 2.1])

$$UT = (1 - \alpha)(1 - \beta)\,\mathrm{Id} + (\alpha + \beta - \alpha\beta)[\frac{(1 - \beta)\alpha}{\alpha + \beta - \alpha\beta} R + \frac{\beta}{\alpha + \beta - \alpha\beta} ST],$$

and from the fact that the family of nonexpansive operators is closed under compositions and convex combinations (see Lemma 2.1.12). Note, however, that [273, Theorem 3 (b)] is stronger than the result mentioned above, because

$$\delta < \alpha + \beta - \alpha\beta$$

for $\alpha, \beta \in (0, 1)$ and δ given by (2.60). It follows from Corollary 2.2.17 that the result of Ogura and Yamada is equivalent to Theorem 2.2.37 with $\lambda, \mu \in (0, 2)$. Moreover, the proof of this theorem differs from the proof of [273, Theorem 3 (b)]. Note that the property of composition of relaxed cutters with a common fixed point, expressed in Theorem 2.1.46 and the property of compositions of relaxed firmly nonexpansive operators presented in Theorem 2.2.37 are similar. Therefore, it is quite natural that the proofs of both theorems are similar. But Theorem 2.1.46 is no special case of Theorem 2.2.37 because a cutter needs not to be firmly nonexpansive, even if it is nonexpansive (see Example 2.2.8).

An equivalent formulation of the following result can be found in [349, Lemma 1].

Corollary 2.2.39. *Let $T, U : \mathcal{H} \rightarrow \mathcal{H}$ be firmly nonexpansive. Then the composition $V := UT$ is $\frac{4}{3}$-relaxed firmly nonexpansive. Consequently, V_λ is firmly nonexpansive for all $\lambda \in [0, \frac{3}{4}]$ and nonexpansive for all $\lambda \in [0, \frac{3}{2}]$. If, furthermore, V has a fixed point, then V_λ is strongly quasi-nonexpansive for all $\lambda \in (0, \frac{3}{2})$.*

Proof. If we take $\lambda = \mu = 1$ in Theorem 2.2.37, we obtain that V is $\frac{4}{3}$-relaxed firmly nonexpansive. Recall that $(V_\lambda)_\mu = V_{\lambda\mu}$ (see Remark 2.1.3). Corollary 2.2.19 yields the firm nonexpansivity of V_γ for all $\gamma \in [0, \frac{3}{4}]$. By the implication (i)$\Rightarrow$(ii) in Theorem 2.2.10, V_γ is nonexpansive for all $\gamma \in [0, \frac{3}{2}]$. Now let Fix $V \neq \emptyset$ and $\gamma \in (0, \frac{3}{2})$. Then V_γ is strongly quasi-nonexpansive, by Corollary 2.2.9. $\qquad\square$

Yamada et al. also proved that, for any $\lambda > \frac{3}{2}$, there exist firmly nonexpansive operators T, U such that V_λ is not nonexpansive, where $V := UT$ (see [349, Remark 1 (b)]). This means that the constant $\frac{3}{2}$ is optimal in Corollary 2.2.39.

Remark 2.2.40. Let $T, U : \mathcal{H} \rightarrow \mathcal{H}$ be firmly nonexpansive having a common fixed point. Then it follows from Corollaries 2.2.39 and 2.2.15 that UT is $\frac{1}{2}$-strongly quasi nonexpansive. A special case of this property was proved in [152, Proposition 1] for T, U being orthogonal projections onto subspaces of $\mathcal{H}$.

Corollary 2.2.41. *Let $T : \mathcal{H} \rightarrow \mathcal{H}$ be firmly nonexpansive and $\lambda \in [0, 2]$. If V is a closed affine subspace, then the operator $U := (1 - \lambda)P_V + \lambda P_V T$ is $\frac{4}{4-\lambda}$-relaxed firmly nonexpansive.*

Proof. Let V be a closed affine subspace. By Theorem 2.2.33 (vi), the operator P_V is affine, consequently,

$$(1 - \lambda)P_V + \lambda P_V T = P_V T_\lambda.$$

Now it follows from Theorem 2.2.37 that U is $\frac{4}{4-\lambda}$-relaxed firmly nonexpansive, because the metric projection is firmly nonexpansive. $\qquad\square$

A weaker formulation of Corollary 2.2.41 can be found in [349, Lemma 2], where $T := P_C$ for a closed convex subset C.

Theorem 2.2.42. *Let $T_i : X \to X$ be λ_i-relaxed firmly nonexpansive, where $\alpha_i \in [0, 2]$, $i \in I$. Then the composition $S_m := T_m T_{m-1} \ldots T_1$ is γ_m-relaxed firmly nonexpansive, where $\gamma_m = 0$ if $\lambda_i = 0$ for all $i \in I$, $\gamma_m = 2$ if $\lambda_i = 2$ for at least one $i \in I$ and*

$$\gamma_m = \frac{2}{\left(\frac{\lambda_1}{2-\lambda_1} + \frac{\lambda_2}{2-\lambda_2} + \ldots + \frac{\lambda_m}{2-\lambda_m} \right)^{-1} + 1}, \tag{2.61}$$

otherwise. Moreover,

$$\frac{2m \min_{i \in I} \lambda_i}{(m-1) \min_{i \in I} \lambda_i + 2} \le \gamma_m \le \frac{2m \max_{i \in I} \lambda_i}{(m-1) \max_{i \in I} \lambda_i + 2}, \tag{2.62}$$

consequently, $\gamma_m < 2$ if $\lambda_i < 2$ for all $i \in I$.

Proof. Let $\lambda_i = 0$ for all $i \in I$. In this case, $S_m = \mathrm{Id}$, i.e., S_m is 0-relaxed firmly nonexpansive. Let $\lambda_i = 2$ for some $i \in I$. Then S_m is nonexpansive as a composition of nonexpansive operators, i.e., S_m is 2-RFNE (see Corollary 2.2.13).

Let now $\lambda_i \in [0, 2)$ for all $i \in I$ and $\lambda_j > 0$ for at least one $j \in I$. We prove by induction with respect to m that S_m is γ_m-RFNE, where γ_m is given by (2.61). Note that (2.61) is equivalent to

$$\frac{\gamma_m}{2 - \gamma_m} = \frac{\lambda_1}{2 - \lambda_1} + \frac{\lambda_2}{2 - \lambda_2} + \ldots + \frac{\lambda_m}{2 - \lambda_m}. \tag{2.63}$$

1^0 For $m = 2$ the above fact follows directly from Theorem 2.2.37.

2^0 Suppose that, for some $m = k$, the operator S_m is γ_m-RFNE. We prove that S_{k+1} is γ_{k+1}-RFNE. If $\lambda_{k+1} = 0$, then $T_{k+1} = \mathrm{Id}$, S_{k+1} is a composition of k operators which are relaxed firmly nonexpansive and the claim follows from the induction assumption. Let now $\lambda_{k+1} \in (0, 2)$, then we have $S_{k+1} = T_{k+1} S_k$, where T_{k+1} is λ_{k+1}-RFNE and S_k is γ_k-RFNE. It follows from Theorem 2.2.37 that S_{k+1} is γ-RFNE, where

$$\gamma = \frac{2}{\left(\frac{\gamma_k}{2-\gamma_k} + \frac{\lambda_{k+1}}{2-\lambda_{k+1}} \right)^{-1} + 1},$$

and, together with (2.63), this gives for $m = k$

$$\frac{\gamma}{2 - \gamma} = \frac{\gamma_k}{2 - \gamma_k} + \frac{\lambda_{k+1}}{2 - \lambda_{k+1}}$$

$$= \frac{\lambda_1}{2 - \lambda_1} + \frac{\lambda_2}{2 - \lambda_2} + \ldots + \frac{\lambda_k}{2 - \lambda_k} + \frac{\lambda_{k+1}}{2 - \lambda_{k+1}},$$

consequently, $\gamma = \gamma_{k+1}$. We have proved that, for any $m \in \mathbb{N}$, the operator S_m is γ_m-RFNE, where γ_m is given by (2.61).

Now we prove (2.62). By (2.63), we have

$$m \frac{\min_{i \in I} \lambda_i}{2 - \min_{i \in I} \lambda_i} \le \frac{\gamma_m}{2 - \gamma_m} \le m \frac{\max_{i \in I} \lambda_i}{2 - \max_{i \in I} \lambda_i},$$

which is equivalent to (2.62). $\square$

A part of the results presented in Theorem 2.2.42 can be found in [122, Lemma 2.2 (iii)], where it was proved that a composition of λ_i-RFNE operators T_i, where $\lambda_i \in [0, 2]$, $i \in I$, is $\frac{2m \max_{i \in I} \lambda_i}{(m-1) \max_{i \in I} \lambda_i + 2}$-SQNE.

Corollary 2.2.43. *Let $T_i : X \to X$, $i \in I$, be firmly nonexpansive. Then the operator $S_m = T_m \ldots T_1$ is γ_m-relaxed firmly nonexpansive with $\gamma_m = \frac{2m}{m+1}$. Consequently, S_m is $\frac{1}{m}$-strongly quasi-nonexpansive.*

Proof. It suffices to take $\lambda_i = 1$, $i \in I$, in (2.61). The second part of the corollary follows from Corollary 2.2.9. $\square$

Dye and Reich obtained a result which is a special case of the second part of Corollary 2.2.43 with T_i, $i \in I$, being orthogonal projections onto one-dimensional subspace of a Hilbert space (see [152, Theorem on page 109]).

Corollary 2.2.44. *Let $T_i : X \to X$ be firmly nonexpansive, $S_i := T_i \ldots T_1$, $i \in I$, and $w = (\omega_1, \ldots, \omega_m) \in \Delta_m$. Then the operator $S := \sum_{i=1}^m \omega_i S_i$ is λ-relaxed firmly nonexpansive, where*

$$\lambda = \sum_{i=1}^m \omega_i \frac{2i}{i+1}. \tag{2.64}$$

Proof. By Corollary 2.2.43, the operators S_i are γ_i-relaxed firmly nonexpansive with $\gamma_i = \frac{2i}{i+1}$. By Theorem 2.2.35, S is λ-relaxed firmly nonexpansive, where λ is given by (2.64). $\square$

The composition of firmly nonexpansive operators needs not to be firmly nonexpansive (see Exercise 2.5.10).

Definition 2.2.45. Let $T : X \to \mathcal{H}$, $\lambda \in [0, 2]$. The operator $R_\lambda : X \to \mathcal{H}$, $R_\lambda := P_X T_\lambda$ is called a *projected relaxation* of T.

The theorem below gives important properties of the projected relaxation of a firmly nonexpansive operator.

Theorem 2.2.46. *Let $T : X \to \mathcal{H}$ be firmly nonexpansive, $R_\lambda := P_X T_\lambda$, be the projected relaxation of T, where $\lambda \in (0, 2)$. Then:*

(i) *R_λ is $\frac{4}{4-\lambda}$-relaxed firmly nonexpansive.*
(ii) *Fix $R_\lambda = \text{Fix}(P_X T)$.*

(iii) *If* $\mathrm{Fix}(P_X T) \neq \emptyset$, *then the operator* R_λ *is* $\frac{2-\lambda}{2}$*-SQNE, i.e.,*

$$\|R_\lambda x - z\|^2 \leq \|x - z\|^2 - \frac{2-\lambda}{2} \|R_\lambda x - x\|^2 \qquad (2.65)$$

for all $x \in X$ *and for all* $z \in \mathrm{Fix}(P_X T)$.
(iv) *If* $\mathrm{Fix}\, T \neq \emptyset$, *then the operator* R_λ *is* $\frac{2-\lambda}{\lambda}$*-SQNE.*

Proof. (i) Since the metric projection P_X is firmly nonexpansive, it is 1-relaxed firmly nonexpansive. By Theorem 2.2.37, the operator R_λ is $\frac{4}{4-\lambda}$-RFNE.
(ii) This property follows from Corollary 1.2.10.
(iii) Since R_λ is μ-RFNE, where $\mu = \frac{4}{4-\lambda}$ (see (i)), Corollary 2.2.9 yields

$$\|R_\lambda x - z\|^2 \leq \|x - z\|^2 - \frac{2-\mu}{\mu} \|R_\lambda x - x\|^2$$

$$= \|x - z\|^2 - \frac{2-\lambda}{2} \|R_\lambda x - x\|^2.$$

(iv) The claim follows from Corollary 2.2.25.

$\square$

If $X = \mathcal{H}$, then $R_\lambda = T_\lambda$, nevertheless, estimation (2.65) is weaker than estimation (2.47). Furthermore, estimation (2.65) is weaker than estimation (2.53). Note, however, that we have supposed in Corollary 2.2.25 that the operator $T : X \to \mathcal{H}$ is a cutter, consequently $\mathrm{Fix}\, T \neq \emptyset$, while in Theorem 2.2.46 (iii) we have supposed that $\mathrm{Fix}(P_X T) \neq \emptyset$, which is weaker than the assumption $\mathrm{Fix}\, T \neq \emptyset$.

2.2.7 Fixed Points of Firmly Nonexpansive Operators

A firmly nonexpansive operator is nonexpansive (see Theorem 2.2.4), therefore, the subset of its fixed points is closed and convex (see Proposition 2.1.11). In this section we show that the subsets $\mathrm{Fix}\, T$ for FNE- and for NE-operators are intersections of half-spaces, which also yields the closedness and convexity of $\mathrm{Fix}\, T$. Equivalent formulations to the results below can be found in [185, Equalities (11.3) and (11.4)].

Theorem 2.2.47. *Let* $X \subseteq \mathcal{H}$ *be closed convex and* $T : X \to \mathcal{H}$ *be firmly nonexpansive. Then*

$$\mathrm{Fix}\, T = \bigcap_{x \in X} \{z \in X : \quad \langle Tx - x, Tx - z \rangle \leq 0\}.$$

Consequently, $\mathrm{Fix}\, T$ *is a closed convex subset.*

Proof. Since a firmly nonexpansive operator with a fixed point is a cutter (see Theorem 2.2.5), the theorem follows from Lemmas 2.1.36 and 2.1.35. □

Corollary 2.2.48. *Let $X \subseteq \mathcal{H}$ be closed and convex. The subset of fixed points of a nonexpansive operator $S : X \to \mathcal{H}$ has the form*

$$\text{Fix } S = \bigcap_{x \in X} \{z \in X : \quad 2\langle z - x, Sx - x \rangle \geq \|Sx - x\|^2\}, \tag{2.66}$$

consequently, Fix S *is a closed convex subset.*

Proof. Let $S : X \to \mathcal{H}$ be nonexpansive. By Corollary 2.2.13, we have $S = 2T - \text{Id}$ for a firmly nonexpansive operator T. It is clear that Fix $S = $ Fix T. Theorem 2.2.47 yields now

$$\text{Fix } S = \bigcap_{x \in X} \{z \in X : \quad \langle \frac{1}{2}(Sx + x) - x, \frac{1}{2}(Sx + x) - z \rangle \leq 0\}$$

which is equivalent to (2.66). □

2.3 Strongly Nonexpansive Operators

Definition 2.3.1. An operator $T : X \to \mathcal{H}$ is called *strongly nonexpansive* (SNE), if T is nonexpansive and for all sequences $\{x^k\}_{k=0}^{\infty}, \{y^k\}_{k=0}^{\infty} \subseteq X$ the following implication is true

$$\left. \begin{array}{l} (x^k - y^k) \text{ is bounded and} \\ \|x^k - y^k\| - \|Tx^k - Ty^k\| \to 0 \end{array} \right\} \implies (x^k - y^k) - (Tx^k - Ty^k) \to 0,$$

The notion of strongly nonexpansive operators in Banach spaces was proposed by Bruck and Reich in [51, Sect. 1], where also properties of these operators are proved (see also [23, Sect. 4.3]).

Remark 2.3.2. It is clear that a contraction is a strongly nonexpansive operator. Indeed, let T be a contraction, i.e., $\|Tx - Ty\| \leq \alpha \|x - y\|$ for all $x, y \in X$ and for a constant $\alpha \in (0, 1)$, and $(x^k - y^k)$ be bounded and such that $\|x^k - y^k\| - \|Tx^k - Ty^k\| \to 0$. Then we have

$$\|x^k - y^k\| - \|Tx^k - Ty^k\| \geq (1 - \alpha) \|x^k - y^k\| \to 0.$$

Consequently, $x^k - y^k \to 0$ and $Tx^k - Ty^k \to 0$, i.e., T is strongly nonexpansive.

Remark 2.3.3. (S. Reich, A private communication (2009)) Let $X \subseteq \mathcal{H}$ be compact. Then a strictly nonexpansive operator defined on X is strongly nonexpansive. Indeed, let $T : X \to \mathcal{H}$ be strictly nonexpansive, i.e.,

$$\|Tx - Ty\| < \|x - y\| \text{ or } x - y = Tx - Ty$$

for all $x, y \in X$, and X be compact. We show that T is strongly nonexpansive. Suppose that sequences $\{x^k\}_{k=0}^\infty$ and $\{y^k\}_{k=0}^\infty$ are given such that $\|x^k - y^k\| - \|Tx^k - Ty^k\| \to 0$ and that there exist subsequences $\{x^{n_k}\}_{k=0}^\infty \subseteq \{x^k\}_{k=0}^\infty$ and $\{y^{n_k}\}_{k=0}^\infty \subseteq \{y^k\}_{k=0}^\infty$ and a constant $\varepsilon > 0$ such that

$$\|(x^{n_k} - y^{n_k}) - (Tx^{n_k} - Ty^{n_k})\| \geq \varepsilon.$$

Since X is compact, we can suppose without loss of generality that $x^{n_k} \to x$ and $y^{n_k} \to y$. Since T is continuous as a nonexpansive operator, we have $Tx^{n_k} \to Tx$ and $Ty^{n_k} \to Ty$. Hence, we obtain in the limit $\|x - y\| = \|Tx - Ty\|$, which yields, due to strict nonexpansivity of T, that $x - y = Tx - Ty$. On the other hand, we have

$$\|(x - y) - (Tx - Ty)\| = \lim_k \|(x^{n_k} - y^{n_k}) - (Tx^{n_k} - Ty^{n_k})\| \geq \varepsilon,$$

a contradiction, which shows that T is strongly nonexpansive.

Theorem 2.3.4. *Let $T : X \to \mathcal{H}$ be firmly nonexpansive and $\lambda \in (0, 2)$. Then the relaxation T_λ of T is strongly nonexpansive.*

Proof. Let $\{x^k\}_{k=0}^\infty, \{y^k\}_{k=0}^\infty \subseteq X$ be such that $\|x^k - y^k\|$ is bounded and

$$\|x^k - y^k\| - \|T_\lambda x^k - T_\lambda y^k\| \to 0.$$

The firm nonexpansivity of T yields the nonexpansivity of T_λ (see Theorem 2.2.10 (ii)), consequently, the sequence $\{\|x^k - y^k\| + \|T_\lambda x^k - T_\lambda y^k\|\}_{k=0}^\infty$ is bounded. Therefore, by the obvious equality $T_\lambda x - x = \lambda(Tx - x)$ and by Corollary 2.2.15, we have

$$\|(x^k - y^k) - (T_\lambda x^k - T_\lambda y^k)\|^2$$
$$= \|(T_\lambda x^k - x^k) - (T_\lambda y^k - y^k)\|^2$$
$$\leq \frac{\lambda}{2 - \lambda} \left(\|x^k - y^k\|^2 - \|T_\lambda x^k - T_\lambda y^k\|^2 \right)$$
$$= \frac{\lambda}{2 - \lambda} \left(\|x^k - y^k\| - \|T_\lambda x^k - T_\lambda y^k\| \right) \left(\|x^k - y^k\| + \|T_\lambda x^k - T_\lambda y^k\| \right) \to 0,$$

i.e., $\|(x^k - y^k) - (T_\lambda x^k - T_\lambda y^k)\| \to 0$ and T_λ is strongly nonexpansive. $\square$

In the previous sections we have proved that the following classes of operators are closed under composition and under convex combination:

(a) The class of strictly relaxed cutters with a common fixed point (see Theorems 2.1.46 and 2.1.50),
(b) The class of strongly quasi-nonexpansive operators with a common fixed point (see Corollary 2.1.47 and Theorem 2.1.50),
(c) The class of strictly relaxed firmly nonexpansive operators (see Theorems 2.2.37 and 2.2.35)
(d) The class of averaged operators (see Remark 2.2.38 and Corollary 2.2.36).

It turns out that the class of strongly nonexpansive operators has the same properties. The first part of the theorem below was proved by Bruck and Reich in [51, Proposition 1.1] and the other one by Reich in [295, Lemma 1.3].

Theorem 2.3.5. *Let $T_1, T_2 : X \to X$ be strongly nonexpansive and T have one of the following forms:*

(i) $T := T_2 T_1$,
(ii) $T := (1 - \lambda)T_1 + \lambda T_2$, where $\lambda \in [0, 1]$.

Then T is strongly nonexpansive.

Proof. By Lemma 2.1.12, the operator T is nonexpansive. Let the sequences $\{x^k\}_{k=0}^{\infty}, \{y^k\}_{k=0}^{\infty} \subseteq X$ be such that $(x^k - y^k)$ is bounded and $\|x^k - y^k\| - \|Tx^k - Ty^k\| \to 0$.

(i) By the nonexpansivity of T_1 and T_2, we have

$$\|Tx^k - Ty^k\| = \|T_2(T_1x^k) - T_2(T_1y^k)\| \leq \|T_1x^k - T_1y^k\| \leq \|x^k - y^k\|,$$

$k \geq 0$, consequently,

$$\|x^k - y^k\| - \|T_1x^k - T_1y^k\| \to 0$$

and

$$\|T_1x^k - T_1y^k\| - \|T_2(T_1x^k) - T_2(T_1y^k)\| \to 0.$$

Since T_1 and T_2 are strongly nonexpansive, we have

$(x^k - y^k) - (Tx^k - Ty^k) =$
$(x^k - y^k) - (T_1x^k - T_1y^k) + (T_1x^k - T_1y^k) - (T_2(T_1x^k) - T_2(T_1y^k)) \to 0,$

i.e., T is strongly nonexpansive.

(ii) The assertion is clear when $\lambda = 0$ or $\lambda = 1$. Let $\lambda \in (0, 1)$. By the convexity of the norm and the nonexpansivity of T_1 and T_2, we have

Fig. 2.12 SNE operator
which is not AV

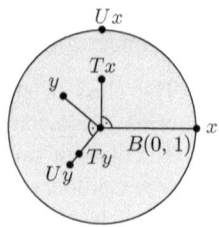

$$\|Tx^k - Ty^k\| = \|(1-\lambda)T_1x^k + \lambda T_2x^k - (1-\lambda)T_1y^k - \lambda T_2y^k\|$$
$$\leq (1-\lambda)\|T_1x^k - T_1y^k\| + \lambda\|T_2x^k - T_2y^k\|$$
$$\leq (1-\lambda)\|x^k - y^k\| + \lambda\|x^k - y^k\| = \|x^k - y^k\|,$$

consequently,

$$\|x^k - y^k\| - \|Tx^k - Ty^k\|$$
$$\geq (1-\lambda)(\|x^k - y^k\| - \|T_1x^k - T_1y^k\|) + \lambda(\|x^k - y^k\| - \|T_2x^k - T_2y^k\|).$$

Therefore,

$$\|x^k - y^k\| - \|T_1x^k - T_1y^k\| \to 0$$

and

$$\|x^k - y^k\| - \|T_2x^k - T_2y^k\| \to 0.$$

By the strong nonexpansivity of T_1 and T_2, we have now

$$(x^k - y^k) - (Tx^k - Ty^k)$$
$$= (1-\lambda)((x^k - y^k) - (T_1x^k - T_1y^k)) + \lambda((x^k - y^k) - (T_2x^k - T_2y^k)) \to 0,$$

i.e., T is strongly nonexpansive.

□

The following example shows that the class of averaged operators or, equivalently, the class of strictly relaxed firmly nonexpansive operators is a proper subclass of the class of strongly nonexpansive operators.

Example 2.3.6. Let $X := B(0,1) \subseteq \mathcal{H}$ be a unit ball, $U : \mathcal{H} \to \mathcal{H}$ be a unitary operator such that $\langle Ux, x \rangle = 0$ for all $x \in \mathcal{H}$ (e.g., $U : \mathbb{R}^2 \to \mathbb{R}^2$ is defined by $Ux := (-\xi_2, \xi_1)$ for $x = (\xi_1, \xi_2) \in \mathbb{R}^2$ with the standard inner product) and the operator $T : X \to X$ be defined by

$$Tx := \alpha(x)Ux.$$

with $\alpha(x) := 1 - \frac{1}{2}\|x\|$ (see Fig. 2.12).

It is clear that $\alpha(x)Ux = U(\alpha(x)x)$, consequently,

$$\begin{aligned}
\|Tx - Ty\| &= \|\alpha(x)Ux - \alpha(y)Uy\| \\
&= \|U(\alpha(x)x) - U(\alpha(y)y)\| \\
&= \|\alpha(x)x - \alpha(y)y\|.
\end{aligned}$$

A straightforward calculation shows that

$$\begin{aligned}
&\|x - y\|^2 - \|Tx - Ty\|^2 \\
&= \|x - y\|^2 - \|\alpha(x)x - \alpha(y)y\|^2 \\
&= -\frac{1}{4}\|x\|^4 - \frac{1}{4}\|y\|^4 + \|x\|^3 + \|y\|^3 \\
&\quad -\langle x, y\rangle(\|x\| + \|y\| - \frac{1}{2}\|x\|\cdot\|y\|) \\
&= (\|x\| - \|y\|)^2(\|x\| + \|y\|)(1 - \frac{1}{4}(\|x\| + \|y\|)) \\
&\quad +(\|x\|\cdot\|y\| - \langle x, y\rangle)(\|x\| + \|y\| - \frac{1}{2}\|x\|\cdot\|y\|).
\end{aligned}$$

We have $\|x\| + \|y\| \geq \|x\|\cdot\|y\|$, since $\|x\|, \|y\| \in [0, 1]$. This fact and the Cauchy–Schwarz inequality yield

$$\|x\| + \|y\| - \frac{1}{2}\|x\|\cdot\|y\| \geq \frac{1}{2}\|x\|\cdot\|y\| \geq \frac{1}{4}\|x\|\cdot\|y\| - \frac{1}{4}\langle x, y\rangle,$$

consequently,

$$\begin{aligned}
&\|x - y\|^2 - \|Tx - Ty\|^2 \\
&\geq (\|x\| - \|y\|)^2(\|x\| + \|y\|)(1 - \frac{1}{4}(\|x\| + \|y\|)) + \frac{1}{4}(\|x\|\cdot\|y\| - \langle x, y\rangle)^2
\end{aligned}$$

and T is nonexpansive. We apply the above inequalities to $x = x^k$ and $y = y^k$. Suppose that $x^k, y^k \in X$ and that $\|x^k - y^k\| - \|Tx^k - Ty^k\| \to 0$. Then, of course,

$$\|x^k - y^k\|^2 - \|Tx^k - Ty^k\|^2 \to 0,$$

because $\|x^k - y^k\| + \|Tx^k - Ty^k\|$ is bounded. Therefore,

$$\|x^k\| - \|y^k\| \to 0$$

(note that $1 - \frac{1}{4}(\|x^k\| + \|y^k\|) \geq \frac{1}{2}$) and

$$\|x^k\| \cdot \|y^k\| - \langle x^k, y^k \rangle \to 0.$$

Now we have

$$\|x^k - y^k\|^2 = (\|x^k\| - \|y^k\|)^2 + 2(\|x^k\| \cdot \|y^k\| - \langle x^k, y^k \rangle) \to 0,$$

i.e., $(x^k - y^k) \to 0$. Furthermore, $(Tx^k - Ty^k) \to 0$, by the nonexpansivity of T, consequently,

$$(x^k - y^k) - (Tx^k - Ty^k) \to 0,$$

i.e., T is strongly nonexpansive. Note that $z = 0$ is the unique fixed point of T. Suppose that T is α-averaged, for a constant $\alpha \in (0, 1)$. By Corollary 2.2.17, the operator T is (2α)-relaxed firmly nonexpansive. Consequently, the operator $V = T_\mu$, where $\mu = (2\alpha)^{-1} \in (\frac{1}{2}, +\infty)$ is firmly nonexpansive (see Corollary 2.2.19) and V is a cutter (see Theorem 2.2.5), i.e.,

$$-\mu\langle x, x - Tx \rangle + \mu^2 \|x - Tx\|^2$$
$$= -\mu\langle x + \mu(Tx - x), x - Tx \rangle$$
$$= \langle -T_\mu x, x - T_\mu x \rangle$$
$$= \langle 0 - Vx, x - Vx \rangle \leq 0.$$

Dividing the inequalities above by $\mu > 0$, we obtain, for all $x \neq z$,

$$\frac{1}{2} < \mu \leq \frac{\langle x, x - Tx \rangle}{\|x - Tx\|^2} = \frac{\|x\|^2}{\|x\|^2 + \alpha^2(x)\|x\|^2} = \frac{1}{1 + (1 - \frac{1}{2}\|x\|)^2}.$$

Applying the inequalities above to a sequence $\{x^k\}_{k=0}^\infty$ with $\lim_k x^k = 0$, we obtain

$$\frac{1}{2} < \mu \leq \lim_k \frac{1}{1 + (1 - \frac{1}{2}\|x^k\|)^2} = \frac{1}{2},$$

a contradiction, which proves that T is not averaged.

2.4 Generalized Relaxations of Algorithmic Operators

In the definition of a relaxation T_λ of an operator $T : X \to \mathcal{H}$ we have supposed that the relaxation parameter $\lambda \in [0, 2]$ (see Definition 2.1.2). Furthermore, the assumption $\lambda \in (0, 2)$ is necessary for the strong quasi nonexpansivity of the λ-relaxation of a firmly nonexpansive operator T with Fix $\neq \emptyset$ (see proof of

Theorem 2.1.39). However, in some applications, relaxations of operators (e.g., of firmly nonexpansive ones) with the relaxation parameter which are greater than 2 are successfully used. In general, the convergence of sequences generated by such operators is not guaranteed. It turns out that, if we allow to vary the relaxation parameter in dependence on the current point, in such a way that the relaxed operator is a cutter, then we can apply the usual convergence analysis for sequences generated by such an operator. Below we define a generalization of a relaxation of an operator, which permits us to extend the convergence results to sequences generated by the generalized relaxation.

Definition 2.4.1. Let $T : X \to \mathcal{H}$, $\lambda \in [0, 2]$ and $\sigma : X \to (0, +\infty)$. The operator $T_{\sigma,\lambda} : X \to \mathcal{H}$,

$$T_{\sigma,\lambda} x := x + \lambda \sigma(x)(Tx - x) \tag{2.67}$$

is called the *generalized relaxation* of T, the value λ is called the *relaxation parameter* and σ is called the *step size function*. If $\sigma(x) \geq 1$ for all $x \in X$, then the operator $T_{\sigma,\lambda}$ is called an *extrapolation* of T_{λ}.

Some special cases of generalized relaxations of some classes of nonexpansive operators, presented in various forms and applied in most cases to the convex feasibility problems, were studied by Gurin et al. [196, Sect. 3], Pierra [284, Sect. 1], Cegielski [62, Sect. 4.3], Kiwiel [229, Sect. 3], Bauschke [17, Sects. 7.3 and 8.3], Combettes [118, Sects. 5.4–5.8], [120, Sect. IV], Bauschke et al. [30, Sect. 3] Bauschke et al. [25] and by Cegielski and Suchocka in [76].

In this section we present properties of generalized relaxations of cutters and give conditions for a generalized relaxation to be strongly quasi-nonexpansive. These properties will be applied in one of the next chapters in order to prove the convergence of sequences generated by such operators.

Denote $T_{\sigma} = T_{\sigma,1}$.

Remark 2.4.2. Let $T : X \to \mathcal{H}$, $\lambda \in [0, 2]$ and $\sigma : X \to (0, +\infty)$.

(a) If $\sigma(x) = 1$ for all $x \in X$, then $T_{\sigma,\lambda} = T_{\lambda}$, i.e., the generalized relaxation of T is reduced to the classical relaxation of T.
(b) The values of the step size function σ for $x \in \text{Fix } T$ have no influence on the form of an operator $T_{\sigma,\lambda}$ because $T_{\sigma,\lambda} |_{\text{Fix } T} = \text{Id}$ for any step size function σ and for any $\lambda \in (0, 2]$. Therefore, we can suppose without loss of generality that $\sigma(x) = 1$ for all $x \in \text{Fix } T$.
(c) For any $x \in X$ the following equalities hold

$$T_{\sigma,\lambda} x - x = \lambda \sigma(x)(Tx - x) = \lambda(T_{\sigma} x - x), \tag{2.68}$$

i.e., $T_{\sigma,\lambda}$ is a λ-relaxation of an operator T_{σ}.
(d) For any $\lambda \neq 0$ it holds $\text{Fix } T_{\sigma,\lambda} = \text{Fix } T$ (cf. Remark 2.1.4).

The corollary below is a version of Theorem 2.1.39.

Corollary 2.4.3. *Let* $T : X \to \mathcal{H}$ *have a fixed point,* $\sigma : X \to (0, +\infty)$ *be a step size function and* $\lambda \in (0, 2)$. *Then* T_σ *is a cutter if and only if* $T_{\sigma,\lambda}$ *is* $\frac{2-\lambda}{\lambda}$-*strongly quasi-nonexpansive. In both cases*

$$\|T_{\sigma,\lambda}x - z\|^2 \leq \|x - z\|^2 - \lambda(2 - \lambda)\sigma^2(x) \|Tx - x\|^2 \tag{2.69}$$

for all $x \in X$ *and* $z \in \text{Fix } T$.

Proof. By Remark 2.4.2 (c), $T_{\sigma,\lambda}$ is the λ-relaxation of T_σ. The first part of the theorem follows now from Theorem 2.1.39. The $\frac{2-\lambda}{\lambda}$-strong quasi nonexpansivity of $T_{\sigma,\lambda}$ means

$$\|T_{\sigma,\lambda}x - z\|^2 \leq \|x - z\|^2 - \frac{2 - \lambda}{\lambda} \|T_{\sigma,\lambda}x - x\|^2 .$$

Applying now (2.68) to the inequality above we obtain (2.69).   □

Let $T : X \to \mathcal{H}$ be an operator with a fixed point. Our aim is to give sufficient conditions for the step size function $\sigma : X \to (0, +\infty)$, at which T_σ is a cutter. The following definition was proposed in [70, Definition 9.17].

Definition 2.4.4. We say that an operator $T : X \to \mathcal{H}$ with a fixed point is *oriented* if for all $x \notin \text{Fix } T$

$$\delta(x) := \inf_{z \in \text{Fix } T} \frac{\langle z - x, Tx - x \rangle}{\|Tx - x\|^2} > 0. \tag{2.70}$$

If $\delta(x) \geq \delta > 0$ for all $x \notin \text{Fix } T$, then we say that T is *strongly oriented*.

It follows from Remark 2.1.31 that $T : X \to \mathcal{H}$ is strongly oriented if and only if T is an α-relaxed cutter for some $\alpha > 0$.

Corollary 2.4.5. *Let* $T : X \to \mathcal{H}$ *be an oriented operator with* $\text{Fix } T \neq \emptyset$. *If a step size function* $\sigma : X \to (0, +\infty)$ *satisfies the inequality*

$$\sigma(x) \leq \frac{\langle z - x, Tx - x \rangle}{\|Tx - x\|^2} \tag{2.71}$$

for all $x \notin \text{Fix } T$ *and* $z \in \text{Fix } T$, *then* T_σ *is a cutter. Consequently, for any* $\lambda \in (0, 2)$, *the generalized relaxation* $T_{\sigma,\lambda}$ *of* T *is* $\frac{2-\lambda}{\lambda}$-*strongly quasi-nonexpansive.*

Proof. Let $x \notin \text{Fix } T$, $z \in \text{Fix } T$ and $\sigma : X \to (0, +\infty)$ be a step size function satisfying (2.71). The existence of σ follows from the assumption that T is oriented. Then (2.68) and inequality (2.71) yield

$$\langle z - x, T_\sigma x - x \rangle = \langle z - x, \sigma(x)(Tx - x) \rangle$$
$$\geq \|\sigma(x)(Tx - x)\|^2$$
$$= \|T_\sigma x - x\|^2 .$$

By the equivalence (a)⇔(b) in Lemma 1.2.5, we have

$$\langle z - T_\sigma x, x - T_\sigma x \rangle \le 0,$$

i.e., T_σ is a cutter. The $\frac{2-\lambda}{\lambda}$-strong quasi nonexpansivity of $T_{\sigma,\lambda}$ follows now from Corollary 2.4.3.   □

The convergence of sequences generated by generalized relaxations of an algorithmic operator U, which we present in the next chapter, requires a stronger condition than (2.71). As we will see, the convergence holds if we additionally suppose that U is strongly oriented, or, equivalently, that the step size $\sigma(x) \ge \alpha$ for all $x \in X$ and for a constant $\alpha > 0$. This leads to α-relaxed cutters (see Remark 2.1.31). It is clear that if an operator $T : X \to \mathcal{H}$ with a fixed point is an α-relaxed cutter for some $\alpha > 0$, then there exists a step size function $\sigma : X \to (0, +\infty)$ satisfying inequality (2.71), e.g., $\sigma(x) = \alpha^{-1}$ for all $x \in X$ (cf. (2.22)). In practice, however, it is important to determine a step size $\sigma(x)$ for which the difference between the right- and the left-hand side of inequality (2.71) is as small as possible for all $z \in \text{Fix } T$. Theoretically, the best possibility would be $\sigma(x) = \delta(x)$ for $x \notin \text{Fix } U$, where $\delta(x)$ is defined by (2.70), but the computation of $\delta(x)$ is, in most cases, impossible, because we usually do not know Fix T explicitly.

Having an α-relaxed cutter T we can construct its generalized relaxation $T_{\sigma,\lambda}$ with the range of the step size function σ contained in $[\alpha, +\infty)$ and satisfying assumptions of Corollary 2.4.5. The corollary below gives a collection of operators which are α-relaxed cutters.

Corollary 2.4.6. *Let $U : X \to \mathcal{H}$ have a fixed point. Then U is an α-relaxed cutter with:*

(a) $\alpha = 1$ if U is firmly nonexpansive,
(b) $\alpha = \lambda$ if U is λ-relaxed firmly nonexpansive, where $\lambda \in (0, 2]$,
(c) $\alpha = 2$ if U is nonexpansive,
(d) $\alpha = 2\nu$ if U is ν-averaged, where $\nu \in (0, 1)$,
(e) $\alpha = \frac{2}{1+\beta}$ if U is β-strongly quasi-nonexpansive, where $\beta > 0$.

Proof. (a) Let U be firmly nonexpansive. Then it follows from the first part of Theorem 2.2.5 that T is a cutter, i.e., T is a 1-relaxed cutter.
(b) Let $\lambda \in (0, 2]$ and $U := \text{Id} + \lambda(T - \text{Id})$ for a firmly nonexpansive operator T. Then, by (a), we have

$$\langle z - x, Ux - x \rangle = \lambda \langle z - x, Tx - x \rangle$$

$$\ge \lambda \|Tx - x\|^2 = \frac{1}{\lambda} \|Ux - x\|^2.$$

(c) Let U be nonexpansive. Then $U = 2T - \text{Id}$ for a firmly nonexpansive operator T (see Corollary 2.2.13) and this case is covered by (b) for $\lambda = 2$.
(d) Let $\nu \in (0, 1)$ and $U := (1 - \nu)\,\text{Id} + \nu S$ for a nonexpansive operator S. Then U is 2ν-relaxed firmly nonexpansive (see Corollary 2.2.17). The claim follows now from (b) with $\lambda = 2\nu$.

(e) Let $\beta > 0$ and U be β-strongly quasi-nonexpansive. It follows from Corollary 2.1.43 that U is a $\frac{2}{1+\beta}$-relaxed cutter.

□

In the lemma below we state some obvious properties of the generalized relaxation.

Lemma 2.4.7. *Let $T : X \rightarrow \mathcal{H}$ be an operator with a fixed point, and $\{\sigma_j\}_{j \in J} : X \rightarrow (0, +\infty)$ be a family of step size functions.*

(i) If T_{σ_j}, $j \in J$, are cutters, then $T_{\sup_{j \in J} \sigma_j}$ is a cutter.
(ii) If $\sigma_i \leq \sigma_j$ for some $i, j \in J$ and T_{σ_j} is a cutter, then T_{σ_i} is a cutter.

If T is a cutter, then there exists a step size function σ with $\sigma(x) \geq 1$ for all $x \notin \text{Fix } T$, for which T_σ is a cutter, e.g., a step size function σ defined by $\sigma(x) = \delta(x)$, where $\delta(x)$ is given by (2.70) for $x \notin \text{Fix } T$. Consequently, the generalized relaxation $T_{\sigma,\lambda}$ is strongly quasi-nonexpansive for any $\lambda \in (0, 2)$ (see Theorem 2.4.5). Note that $\sigma(x) \geq 1$, by Remark 2.1.31. The following example shows, however, that there is a cutter T such that the generalized relaxation $T_{\sigma,\lambda}$ is strongly quasi-nonexpansive for all $\lambda \in (0, 2)$ if and only if $\sigma(x) \leq 1$ for all $x \notin \text{Fix } T$.

Example 2.4.8. Let $T_{\sigma,\lambda}$ be a generalized relaxation of the metric projection $P_C : \mathcal{H} \rightarrow \mathcal{H}$, where $C \subseteq \mathcal{H}$ is a nonempty closed convex subset, i.e., $T_{\sigma,\lambda}(x) = x + \lambda\sigma(x)(P_C x - x)$ for a relaxation parameter $\lambda \in (0, 2)$ and for some step size function $\sigma : \mathcal{H} \rightarrow (0, +\infty)$. For any $x \in \mathcal{H}$ we have

$$
\begin{aligned}
\|T_{\sigma,\lambda}x - P_C x\|^2 &= \|x + \lambda\sigma(x)(P_C x - x) - P_C x\|^2 \\
&= \|x - P_C x\|^2 + \lambda^2\sigma^2(x)\|P_C x - x\|^2 - 2\lambda\sigma(x)\|P_C x - x\|^2 \\
&= \|x - P_C x\|^2 - \lambda\sigma(x)(2 - \lambda\sigma(x))\|P_C x - x\|^2,
\end{aligned}
$$

consequently,

$$\|T_{\sigma,\lambda}x - P_C x\|^2 = \|x - P_C x\|^2 - \frac{2 - \lambda\sigma(x)}{\lambda\sigma(x)}\|T_{\sigma,\lambda}x - x\|^2. \tag{2.72}$$

Let $\lambda \in (0, 2)$. Suppose that $T_{\sigma,\lambda}$ is strongly quasi-nonexpansive, i.e.,

$$\|T_{\sigma,\lambda}x - z\|^2 \leq \|x - z\|^2 - \alpha\|T_{\sigma,\lambda}(x) - x\|^2 \tag{2.73}$$

for some $\alpha > 0$, for all $x \in \mathcal{H}$ and $z \in C := \text{Fix } P_C$. Note that α can depend on λ. Let $x \notin C$ and $z = P_C x$. Then (2.72) and (2.73) yield

$$0 < \alpha \leq \frac{2 - \lambda\sigma(x)}{\lambda\sigma(x)}.$$

There exists a constant α satisfying the above inequalities for all $\lambda \in (0, 2)$ if and only if $\sigma(x) \leq 1$.

Fig. 2.13 Operators T and T_σ from Example 2.4.9

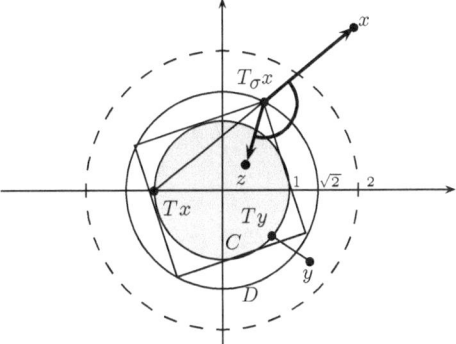

If $T : X \to \mathcal{H}$ is firmly nonexpansive with a fixed point, then T is oriented and for the function δ defined by (2.70) it holds $\delta(x) \geq 1$ for all $x \notin \operatorname{Fix} T$. Therefore, Corollary 2.4.5 applied to a firmly nonexpansive operator T with the step size $\sigma(x) := \delta(x)$ for $x \notin \operatorname{Fix} T$ is an extension of Theorem 2.2.5 (i) for generalized relaxations. Unfortunately, Theorem 2.2.5 (ii) cannot be analogously extended. The fact that $T : X \to \mathcal{H}$ is a projection and T_σ is a cutter for some step size function $\sigma : X \to (0, +\infty)$ does not yield the firm nonexpansivity of T. Even if we additionally suppose that T is nonexpansive, T needs not to be firmly nonexpansive (see Example 2.2.7). Moreover, a projection T for which T_σ is a cutter needs not to be continuous.

Example 2.4.9. Let $\mathcal{H} = \mathbb{R}^2$, $C := B(0, 1)$, $D := \operatorname{bd} B(0, \sqrt{2})$, $a = (1, 0)$. Define the operator $T : \mathbb{R}^2 \to \mathbb{R}^2$ by

$$Tx := \begin{cases} P_C x & \text{for } \|x\| \leq 2 \\ -a & \text{for } \|x\| > 2, \xi_1 \geq 0 \\ a & \text{for } \|x\| > 2, \xi_1 < 0. \end{cases}$$

It is clear that T is a projection with $\operatorname{Fix} T = C$. For $\|x\| > 2$, let Ux be the unique common point of the segment $[x, Tx]$ and the circle D. Define the function $\sigma : \mathbb{R}^2 \to \mathbb{R}$ by

$$\sigma(x) := \begin{cases} 1 & \text{if } \|x\| \leq 2 \\ \frac{\|Ux - x\|}{\|Tx - x\|} & \text{if } \|x\| > 2. \end{cases}$$

Observe that for $\|x\| > 2$ it holds $T_\sigma x = Ux$. It follows from geometrical considerations (note that the square circumscribed on the circle $\operatorname{bd} B(0, 1)$ is inscribed in the circle $\operatorname{bd} B(0, \sqrt{2})$) that for all $x \in \mathbb{R}^2$ and $z \in C = \operatorname{Fix} T$ it holds

$$\langle x - T_\sigma(x), z - T_\sigma(x) \rangle \leq 0$$

(see Fig. 2.13). Therefore, T_σ is a cutter. Note that T is not continuous, therefore, T cannot be firmly nonexpansive.

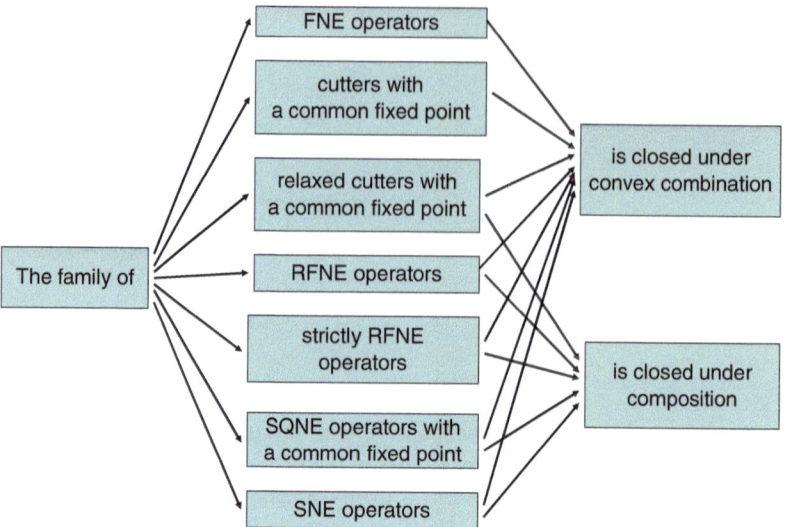

Fig. 2.14 Closedness of families of algorithmic operators

SUMMARY

In Fig. 2.14 we recall in a short form the properties of algorithmic operators which were presented in this chapter. These properties are useful in construction of projection methods. We will describe these constructions in Chaps. 4 and 5.

2.5 Exercises

Exercise 2.5.1. Show that $(T_\lambda)_\mu = T_{\lambda\mu}$ for all $\lambda, \mu \in \mathbb{R}$.

Exercise 2.5.2. Let $T : \mathbb{R} \to \mathbb{R}$,

$$Tx = \begin{cases} x^2 & \text{if } |x| \le \frac{3}{4} \\ |x| - \frac{3}{16} & \text{if } |x| > \frac{3}{4}. \end{cases}$$

Show that T is quasi-nonexpansive and continuous, but T is not a nonexpansive operator.

Exercise 2.5.3. Let $\{U_i\}_{i \in I}$ be a finite family of operators, $U_i : X \to \mathcal{H}, i \in I$. Let $w : X \to \Delta_m$ be a weight function satisfying $\omega_i(x) > 0$ for some $i(x) = \operatorname{argmax}_{i \in I} \|U_i x - x\|$ for all $x \in X$. Prove that w is appropriate with respect to the family $\{U_i\}_{i \in I}$.

Exercise 2.5.4. Prove that the assumption on the C-strict quasi nonexpansivity in Theorem 2.1.26 (i) can be weakened. In this case it suffices to suppose that all U_i are quasi-nonexpansive, $i \in I$, and at least one of them is C-strictly quasi-nonexpansive. The assumption that the weight function w is appropriate should be replaced in this case by a stronger one, namely: $w_j(x) > 0$ for all x such that $I(x) \neq \emptyset$ and for all $j \in I(x)$.

Exercise 2.5.5. Prove Corollary 2.1.29.

Exercise 2.5.6. Prove Lemma 2.1.45.

Exercise 2.5.7. Show that the operator $T : \mathbb{R}^2 \to \mathbb{R}^2$,

$$Tx := (\xi_1 \cos \varphi - \xi_2 \sin \varphi, \xi_1 \sin \varphi + \xi_2 \cos \varphi)$$

is nonexpansive and monotone for $\varphi \in (0, \pi/2)$, but T is not firmly nonexpansive.

Exercise 2.5.8. Prove Lemma 2.2.2.

Exercise 2.5.9. Show that the operator T presented in Example 2.2.7 is nonexpansive and that T is a separator of A, but T is not firmly nonexpansive.

Exercise 2.5.10. Let $\mathcal{H} = \mathbb{R}^2$, $A := \{x \in \mathbb{R}^2 : \xi_2 = 0\}$ and $B := \{x \in \mathbb{R}^2 : \xi_1 = \xi_2\}$. By Theorem 2.2.21 (iii) P_A and P_B are firmly nonexpansive. Check that $T := P_B P_A$ is not firmly nonexpansive.

Chapter 3
Convergence of Iterative Methods

3.1 Iterative Methods

Convex minimization problems are usually presented as minimization of a continuous convex function on a Hilbert space $\mathcal{H}$ or on a closed convex subset $X \subseteq$. The subset of solutions of this problem is a closed convex subset $M \subseteq X$. We recall as an example the split feasibility problem and a connected problem of minimization of a convex proximity function (see Sect. 1.3.7). The CMP is equivalent to finding a fixed point of an operator $U : X \to \mathcal{H}$ (usually nonexpansive) corresponding to the objective function. The operator U is constructed in such a way that $M = \text{Fix}\, U$. Iteration methods for solving these problems (finding an element $x^* \in M$) have the form of a recurrence

$$x^{k+1} = U x^k$$

or a more general recurrence

$$x^{k+1} = U_k x^k,$$

$k \geq 0$, where $x^0 \in X$ is arbitrary, $\{U_k\}_{k=0}^{\infty}$ is a sequence of algorithmic operators $U_k : X \to X$ such that

$$\bigcap_{k=0}^{\infty} \text{Fix}\, U_k \supseteq \text{Fix}\, U = M.$$

If $U_k = U$ for all $k \geq 0$, then we say that the iterative method is *autonomous*, otherwise, we say that the method is *nonautonomous*. In the first case we can also write $x^k = U^k x^0$. The algorithmic operator or the sequence of algorithmic operators describing the method should be constructed in such a way that any sequence $\{x^k\}_{k=0}^{\infty}$ generated by the method converges (weakly or strongly) to a solution $x^* \in M$.

A. Cegielski, *Iterative Methods for Fixed Point Problems in Hilbert Spaces*, Lecture Notes in Mathematics 2057, DOI 10.1007/978-3-642-30901-4_3, © Springer-Verlag Berlin Heidelberg 2012

3.2 Properties of the Weak Convergence

In this section we present properties of the weak convergence which will be applied in the sequel.

Lemma 3.2.1. *If* $x^k \rightharpoonup x \in \mathcal{H}$, *then* $\liminf_k \|x^k\| \geq \|x\|$.

Proof. Let $x^k \rightharpoonup x$. The lemma is clear for $x = 0$. Let now $x \neq 0$. By the Cauchy–Schwarz inequality, we have

$$\liminf_k \|x\| \cdot \|x^k\| \geq \liminf_k \langle x, x^k \rangle = \|x\|^2,$$

i.e., $\liminf_k \|x^k\| \geq \|x\|$. $\qquad\square$

Note that a sequence which is weakly convergent needs not to converge strongly.

Example 3.2.2. (cf. [267, Example 2.13]). Let $\mathcal{H} := l^2$ and $x^k = e_k = (e_{k1}, e_{k2}, \ldots)$, where

$$e_{ki} := \begin{cases} 1 \text{ for } i = k \\ 0 \text{ for } i \neq k. \end{cases}$$

Then $\|x^k\| = 1$, although $\{x^k\}_{k=0}^\infty$ does not contain a convergent subsequence, because $\|x^k - x^l\| = \sqrt{2}$ for all $k, l, k \neq l$, i.e., $\{x^k\}_{k=0}^\infty$ is not a Cauchy sequence. Note, however, that $x^k \rightharpoonup 0$.

Under some additional assumptions the weak convergence yields the strong one.

Lemma 3.2.3. *If* $x^k \rightharpoonup x \in \mathcal{H}$ *and* $\|x^k\| \to \|x\|$, *then* $x^k \to x$.

Proof. Let $x^k \rightharpoonup x$. It follows from the properties of the inner product that

$$\|x - x^k\|^2 = \|x\|^2 + \|x^k\|^2 - 2\langle x, x^k \rangle \to 0$$

as $k \to \infty$. $\qquad\square$

A Hilbert space has an important property which is expressed in the following lemma.

Lemma 3.2.4 ([279]). *If* $x^k \rightharpoonup y \in \mathcal{H}$, *then, for any* $y' \in \mathcal{H}$, $y' \neq y$, *the following inequality holds*

$$\liminf_k \|x^k - y'\| > \liminf_k \|x^k - y\|. \tag{3.1}$$

Proof. Let $x^k \rightharpoonup y$ and $y' \neq y$. Denote $\delta := \|y - y'\|^2$. Since a weakly convergent sequence is bounded, both lower limits in (3.1) are finite. The properties of the inner product yield

$$\left\| x^k - y' \right\|^2 = \left\| x^k - y + y - y' \right\|^2$$
$$= \left\| x^k - y \right\|^2 + \left\| y - y' \right\|^2 + 2\langle x^k - y, y - y' \rangle$$
$$= \left\| x^k - y \right\|^2 + \delta + 2\langle x^k - y, y - y' \rangle.$$

By assumption, $\langle x^k - y, y - y' \rangle \to 0$, consequently,

$$\liminf_k \left\| x^k - y' \right\|^2 = \liminf_k \left\| x^k - y \right\|^2 + \delta > \liminf_k \left\| x^k - y \right\|^2,$$

i.e., $\liminf_k \left\| x^k - y' \right\| > \liminf_k \left\| x^k - y \right\|$. □

Lemma 3.2.4 was proved by Zdzisław Opial in [279, Lemma 1]. The property described in this lemma is known under the name *Opial property*. A Banach space for which the property holds is called an *Opial space*. Therefore, Lemma 3.2.4 can be formulated as follows: *any Hilbert space is an Opial space.*

The following lemma, also proved by Opial in [279, Lemma 2], is a consequence of the Opial property.

Lemma 3.2.5. *Let $T : X \to \mathcal{H}$ be nonexpansive and $y \in X$ be a weak cluster point of a sequence $\{x^k\}_{k=0}^\infty$. If $\left\| Tx^k - x^k \right\| \to 0$, then $y \in \operatorname{Fix} T$.*

Proof. Let $x^{n_k} \rightharpoonup y$ for a subsequence $\{x^{n_k}\}_{k=0}^\infty$ of $\{x^k\}_{k=0}^\infty$ and $\left\| Tx^k - x^k \right\| \to 0$. Suppose that $Ty \neq y$. Then, by the triangle inequality and by Lemma 3.2.4, we have

$$\liminf_{k \to \infty} \left\| x^{n_k} - y \right\| \geq \liminf_{k \to \infty} \left\| Tx^{n_k} - Ty \right\|$$
$$= \liminf_{k \to \infty} \left\| Tx^{n_k} - x^{n_k} + x^{n_k} - Ty \right\|$$
$$\geq \liminf_{k \to \infty} (\left\| x^{n_k} - Ty \right\| - \left\| Tx^{n_k} - x^{n_k} \right\|)$$
$$= \liminf_{k \to \infty} \left\| x^{n_k} - Ty \right\|$$
$$> \liminf_{k \to \infty} \left\| x^{n_k} - y \right\|.$$

A contradiction proves that $y \in \operatorname{Fix} T$. □

Definition 3.2.6. We say that an operator $S : X \to \mathcal{H}$ is *demi-closed* at 0 if for any sequence $x^k \rightharpoonup y \in X$ with $Sx^k \to 0$ we have $Sy = 0$.

If we replace the weak convergence $x^k \rightharpoonup y$ by the strong one in Definition 3.2.6, then we obtain the definition of the closedness of S at 0. If $\mathcal{H}$ is finite dimensional, then a weakly convergent sequence is convergent. Therefore, the notions of a demi-closed operator and a closed operator coincide in a finite dimensional Hilbert space.

The property expressed in Lemma 3.2.5 is known under the name *demi-closedness principle* (see, e.g., [22, Fact 1.2]). Opial writes that the result is due to Browder [44] (see [278, Proposition 2.5]). The property says, actually, that for a nonexpansive operator T in a Hilbert space, the operator $T - \mathrm{Id}$ is demi-closed at 0.

3.3 Properties of Fejér Monotone Sequences

The notions of Fejér monotone operators, quasi-nonexpansive operators and strongly quasi-nonexpansive operators (see Definitions 2.1.15, 2.1.38) are closely related to the Fejér monotonicity of sequences which are generated by sequences of such operators.

Definition 3.3.1. Let $C \subseteq \mathcal{H}$ be nonempty. We say that a sequence $\{x^k\}_{k=0}^{\infty} \subseteq \mathcal{H}$ is:

(a) *Fejér monotone* (FM) with respect to C if

$$\left\| x^{k+1} - z \right\| \leq \left\| x^k - z \right\| \tag{3.2}$$

for all $z \in C$ and for all $k \geq 0$,
(b) *Strictly Fejér monotone* with respect to C if

$$\left\| x^{k+1} - z \right\| < \left\| x^k - z \right\|$$

for all $z \in C$ and for all $k \geq 0$ with $x^k \notin C$,
(c) *Strongly Fejér monotone* (SFM) *with respect to C if there exists a constant $\alpha > 0$ such that*

$$\left\| x^{k+1} - z \right\|^2 \leq \left\| x^k - z \right\|^2 - \alpha \left\| x^{k+1} - x^k \right\|^2,$$

for all $z \in C$ and for all $k \geq 0$.

Similarly as for the Fejér monotone operators, we can suppose without loss of generality that C is a closed convex subset in Definition 3.3.1. Let a sequence $\{x^k\}_{k=0}^{\infty} \subseteq X$ be generated by the recurrence $x^{k+1} = U_k x^k$, where $U_k : X \rightarrow X$. If the operators U_k, $k \geq 0$, are (strictly) Fejér monotone with respect to $C \subseteq X$ (see Definition 2.1.15), then $\{x^k\}_{k=0}^{\infty}$ is (strictly) Fejér monotone with respect to C. Similarly, if the operators U_k, $k \geq 0$, with a common fixed point, are α-strongly quasi-nonexpansive for some $\alpha > 0$, then $\{x^k\}_{k=0}^{\infty}$ is strongly Fejér monotone with respect to $\bigcap_{k=0}^{\infty} \mathrm{Fix}\, U_k$. It is clear that a Fejér monotone sequence is bounded, consequently it has a weak cluster point. Below, we give other important properties of Fejér monotone sequences.

Theorem 3.3.2. *Let $\{x^k\}_{k=0}^{\infty}$ be Fejér monotone with respect to a closed convex subset $C \subset \mathcal{H}$. Then:*

(i) $\{d(x^k, C)\}_{k=0}^{\infty}$ is a nonincreasing sequence, consequently, it converges.
(ii) $P_C x^k$ converges in norm to a point $z^* \in C$.
(iii) For any weak cluster point $\bar{x}$ of $\{x^k\}_{k=0}^{\infty}$ we have $P_C \bar{x} = z^*$. In particular, if $\bar{x} \in C$, then $\bar{x} = z^*$.
(iv) It holds

$$\lim_k \left\| x^k - z^* \right\| = \inf_{z \in C} \lim_k \left\| x^k - z \right\|. \tag{3.3}$$

Proof. (i) By the definition of the metric projection and the Fejér monotonicity of $\{x^k\}_{k=0}^{\infty}$, we have

$$d(x^{k+1}, C) = \left\| x^{k+1} - P_C x^{k+1} \right\| \leq \left\| x^{k+1} - P_C x^k \right\|$$
$$\leq \left\| x^k - P_C x^k \right\| = d(x^k, C).$$

(ii) (cf. [22, Theorem 2.16 (iv)]) Let $l \geq k \geq 0$. By the parallelogram law with $x = P_C x^l - x^l$ and $y = P_C x^k - x^l$, the definition of the metric projection, the convexity of C and the Fejér monotonicity of $\{x^k\}_{k=0}^{\infty}$, we have

$$\left\| P_C x^l - P_C x^k \right\|^2 = 2 \left\| P_C x^l - x^l \right\|^2 + 2 \left\| P_C x^k - x^l \right\|^2$$
$$-4 \left\| \frac{1}{2}(P_C x^l + P_C x^k) - x^l \right\|^2$$
$$\leq 2 \left\| P_C x^l - x^l \right\|^2 + 2 \left\| P_C x^k - x^l \right\|^2 - 4 \left\| P_C x^l - x^l \right\|^2$$
$$= 2(\left\| P_C x^k - x^l \right\|^2 - \left\| P_C x^l - x^l \right\|^2)$$
$$\leq 2(\left\| P_C x^k - x^k \right\|^2 - \left\| P_C x^l - x^l \right\|^2)$$
$$= 2(d(x^k, C) - d(x^l, C)).$$

By (i), the right hand side of the latter equality goes to zero as $k, l \to \infty$. Therefore, $\{P_C x^k\}_{k=0}^{\infty}$ is a Cauchy sequence, consequently, it converges in norm to a point $z^* \in C$.

(iii) Because any subsequence of a Fejér monotone sequence is Fejér monotone, we can suppose, without loss of generality, that the whole sequence converges weakly to $\bar{x} \in \mathcal{H}$. Let $x \in C$ and $\varepsilon > 0$. Denote $z^k = P_C x^k$. Let $k_0 \geq 0$ be such that

$$\max\{ \left\| z^k - x^k \right\| \cdot \left\| z^k - z^* \right\|, \left\| z^* - x \right\| \cdot \left\| z^k - z^* \right\|, \langle x^k - \bar{x}, z^* - x \rangle \} < \frac{1}{3}\varepsilon$$

for all $k \geq k_0$. Then, by the characterization of the metric projection and the Cauchy–Schwarz inequality, we have

$$\begin{aligned}
0 \geq \langle z^k - x^k, z^k - x \rangle &= \langle z^k - x^k, z^k - z^* \rangle + \langle z^k - x^k, z^* - x \rangle \\
&= \langle z^k - x^k, z^k - z^* \rangle + \langle z^k - z^*, z^* - x \rangle \\
&\quad + \langle z^* - \bar{x}, z^* - x \rangle + \langle \bar{x} - x^k, z^* - x \rangle \\
&\geq - \| z^k - x^k \| \cdot \| z^k - z^* \| - \| z^k - z^* \| \cdot \| z^* - x \| \\
&\quad + \langle z^* - \bar{x}, z^* - x \rangle + \langle \bar{x} - x^k, z^* - x \rangle \\
&\geq -\varepsilon + \langle z^* - \bar{x}, z^* - x \rangle
\end{aligned}$$

for all $k \geq k_0$, i.e., $\langle z^* - \bar{x}, z^* - x \rangle \leq \varepsilon$. Since $\varepsilon > 0$ is arbitrary, we obtain $\langle z^* - \bar{x}, z^* - x \rangle \leq 0$, i.e., $z^* = P_C \bar{x}$.

(iv) The definition of the metric projection and the continuity of the norm yield

$$\lim_k \| x^k - z \| \geq \lim_k \| x^k - P_C x^k \| = \lim_k \| x^k - z^* \|$$

for any $z \in C$. Consequently,

$$\inf_{z \in C} \lim_k \| x^k - z \| \geq \lim_k \| x^k - z^* \|. \tag{3.4}$$

But $z^* \in C$, therefore, the equality in (3.4) holds.   □

A sequence $\{x^k\}_{k=0}^\infty$ which is Fejér monotone (or even strictly Fejér monotone) with respect to a subset $C \subseteq \mathcal{H}$ needs not be weakly convergent (see Exercise 3.10.1). Furthermore, weak cluster points of a sequence which is Fejér monotone with respect to a subset $C \subseteq \mathcal{H}$ need not to have equal distance to the set C (see Exercise 3.10.2).

The following result which is due to Browder (see [47, Lemma 6]) follows immediately from Theorem 3.3.2 (iii). Nevertheless, we present below an independent simple proof of this result.

Corollary 3.3.3. *If a sequence $\{x^k\}_{k=0}^\infty \subseteq \mathcal{H}$ is Fejér monotone with respect to a nonempty subset $C \subseteq \mathcal{H}$, then $\{x^k\}_{k=0}^\infty$ has at most one weak cluster point in C. Consequently, x^k converges weakly to a point $z \in C$ if and only if all weak cluster points of $\{x^k\}_{k=0}^\infty$ belong to C.*

Proof. (cf. [22, Theorem 2.16 (ii)]). By assumption, the sequence $\{\| x^k - z \|\}_{k=0}^\infty$ is monotone, consequently, it converges for any $z \in C$. Denote

$$\alpha(z) := \lim_k \left(\| x^k - z \|^2 - \| z \|^2 \right).$$

We have

$$\lim_k \left(\| x^k \|^2 - 2 \langle x^k, z \rangle \right) = \lim_k \left(\| x^k - z \|^2 - \| z \|^2 \right) = \alpha(z).$$

Let $\{x^{m_k}\}_{k=0}^{\infty}$ and $\{x^{n_k}\}_{k=0}^{\infty}$ be two subsequences of $\{x^k\}_{k=0}^{\infty}$ such that $x^{m_k} \rightharpoonup z' \in C$ and $x^{n_k} \rightharpoonup z'' \in C$ as $k \to \infty$. Then

$$2 \lim_k \langle x^k, z' - z'' \rangle = \lim_k \left[\left(\|x^k\|^2 - 2\langle x_k, z'' \rangle \right) - \left(\|x^k\|^2 - 2\langle x_k, z' \rangle \right) \right]$$
$$= \alpha(z'') - \alpha(z').$$

But

$$\lim_k \langle x^{m_k}, z' - z'' \rangle = \langle z', z' - z'' \rangle$$

and

$$\lim_k \langle x^{n_k}, z' - z'' \rangle = \langle z'', z' - z'' \rangle.$$

Therefore, $\langle z', z' - z'' \rangle = \langle z'', z' - z'' \rangle$, i.e., $\|z' - z''\| = 0$ and $z' = z''$. $\square$

The following lemma gives a sufficient condition for the strong convergence of Fejér monotone sequences and was proved in [22, Theorem 2.16 (v)] (see also [283, Theorem 1.1] for a stronger result).

Lemma 3.3.4. *Let $\{x^k\}_{k=0}^{\infty} \subseteq \mathcal{H}$ be Fejér monotone with respect to a nonempty subset $C \subseteq \mathcal{H}$. If at least one cluster point x^* of $\{x^k\}_{k=0}^{\infty}$ belongs to C, then $x^k \to x^*$.*

Proof. Since a Fejér monotone sequence is bounded, $\{x^k\}_{k=0}^{\infty}$ has a weak cluster point x^*. Suppose that a subsequence $\{x^{n_k}\}_{k=0}^{\infty} \subseteq \{x^k\}_{k=0}^{\infty}$ converges to $x^* \in C$. We prove that the entire sequence $\{x^k\}_{k=0}^{\infty}$ converges to x^*. Suppose that $x' \in \mathcal{H}$, $x' \neq x^*$, is a cluster point of $\{x^k\}_{k=0}^{\infty}$ and that a subsequence $\{x^{m_k}\}_{k=0}^{\infty}$ converges to x'. Let $\varepsilon := \frac{1}{2} \|x' - x^*\| > 0$, k_0 be such that $\|x^{m_k} - x'\| < \varepsilon$ and $\|x^{n_k} - x^*\| < \varepsilon$, for all $k \geq k_0$ and let $m_k > n_{k_0}$. By the triangle inequality and by the Fejér monotonicity of $\{x^k\}_{k=0}^{\infty}$ with respect to C, we obtain

$$2\varepsilon = \|x' - x^*\| \leq \|x' - x^{m_k}\| + \|x^{m_k} - x^*\| < 2\varepsilon,$$

a contradiction. Therefore, $x^k \to x^*$. $\square$

3.4 Asymptotically Regular Operators

Definition 3.4.1. An operator $U : X \to X$ is called *asymptotically regular* (AR) if

$$\lim_{k \to \infty} \|U^{k+1}x - U^k x\| = 0$$

for all $x \in X$.

Asymptotically regular operators were studied, e.g., by Browder and Petryshyn [48], Opial [279], Baillon, Bruck and Reich [14], Bruck and Reich [51] and by Bauschke [19].

Remark 3.4.2. It is clear that any projection is asymptotically regular. In particular, the metric projection onto a nonempty closed convex subset $C \subseteq \mathcal{H}$ is asymptotically regular.

Since the notion of asymptotically regular operators plays an important role in iterative methods for finding fixed points of operators, below we give several sufficient conditions for an operator to be asymptotically regular.

Theorem 3.4.3. *Let $U : X \to X$ be an operator with a fixed point. If U is strongly quasi-nonexpansive, then U is asymptotically regular.*

Proof. Let U be strongly quasi-nonexpansive, $x \in X$ and $z \in \text{Fix}\, U$. For $x^k = U^k x$ and for some constant $\alpha > 0$, we have

$$\left\| x^{k+1} - z \right\|^2 = \left\| U x^k - z \right\|^2 \leq \left\| x^k - z \right\|^2 - \alpha \left\| U x^k - x^k \right\|^2.$$

Consequently, the sequence $\{ \left\| x^k - z \right\| \}_{k=0}^{\infty}$ is monotone and, therefore, it converges. By setting $k \to \infty$ in the above inequality, we obtain in the limit

$$\left\| U^{k+1} x - U^k x \right\|^2 = \left\| U x^k - x^k \right\|^2 \to 0,$$

i.e., U is asymptotically regular. $\qquad\qquad\qquad\qquad\qquad\qquad\qquad\qquad\qquad\square$

Corollary 3.4.4. *Let $U : X \to \mathcal{H}$ be an operator with a fixed point. If U is a cutter, then its projected relaxation $P_X U_\lambda : X \to X$ is asymptotically regular for any $\lambda \in (0, 2)$.*

Proof. Let U be a cutter and $\lambda \in (0, 2)$. It follows from Theorem 2.1.39 that U_λ is $\frac{2-\lambda}{\lambda}$-strongly quasi-nonexpansive. Let $x \in X$ and $z \in \text{Fix}\, U$. Note that $z = P_X z$. For $x^k = (P_X U_\lambda)^k x$ we have

$$\left\| x^{k+1} - z \right\|^2 = \left\| P_X U_\lambda x^k - z \right\|^2 = \left\| P_X U_\lambda x^k - P_X z \right\|^2$$

$$\leq \left\| U_\lambda x^k - z \right\|^2 \leq \left\| x^k - z \right\|^2 - \frac{2-\lambda}{\lambda} \left\| U_\lambda x^k - x^k \right\|^2.$$

Similarly as in the proof of Theorem 3.4.3, we obtain $\lim_k \left\| U_\lambda x^k - x^k \right\| = 0$. Since P_X is nonexpansive and $x^k \in X$, we have

$$\left\| P_X U_\lambda x^k - x^k \right\| = \left\| P_X U_\lambda x^k - P_X x^k \right\| \leq \left\| U_\lambda x^k - x^k \right\|,$$

i.e., $\lim_k \left\| P_X U_\lambda x^k - x^k \right\| = 0$ and $P_X U_\lambda$ is asymptotically regular. $\qquad\square$

Corollary 3.4.5. *Let* $T : X \to \mathcal{H}$ *be a firmly nonexpansive operator with* $\mathrm{Fix}(P_X T) \neq \emptyset$. *Then, for any* $\lambda \in (0, 2)$, *the projected relaxation* $R_\lambda = P_X T_\lambda$ *of the operator* T *is asymptotically regular.*

Proof. Let $\lambda \in (0, 2)$ and $x \in X$. By Theorem 2.2.46 (iii), the operator R_λ is strongly quasi-nonexpansive. The asymptotic regularity of R_λ follows now from Theorem 3.4.3. □

There is an essential difference between Corollaries 3.4.4 and 3.4.5. Since $\mathrm{Fix}\, T \subseteq X$, it is clear that $\mathrm{Fix}\, T \cap \mathrm{Fix}\, P_X = \mathrm{Fix}\, T \neq \emptyset$ in Corollary 3.4.4. Note that the assumption $\mathrm{Fix}(P_X T) \neq \emptyset$ in Corollary 3.4.5 is weaker than the assumption $\mathrm{Fix}\, T \neq \emptyset$ in Corollary 3.4.4. On the other hand, for an operator T with a fixed point, the assumption that T is firmly nonexpansive in Corollary 3.4.5 is stronger than the assumption that T is a cutter in Corollary 3.4.4.

Corollary 3.4.6. *Let* $T : \mathcal{H} \to \mathcal{H}$ *be a firmly nonexpansive operator with a fixed point. Then, for any* $\lambda \in (0, 2)$, *the relaxation* T_λ *of the operator* T *is asymptotically regular.*

Proof. It suffices to take $X = \mathcal{H}$ in Corollary 3.4.5. □

Corollary 3.4.7. *Let* $T_i : X \to \mathcal{H}$, $i \in I$, *be firmly nonexpansive and* $T :=$ $\sum_{i=1}^{m} \omega_i T_i$, *where* $w = (\omega_1, \ldots, \omega_m) \in \Delta_m$. *If* $\mathrm{Fix}(P_X T) \neq \emptyset$, *then, for any* $\lambda \in (0, 2)$, *the operator* R_λ *defined by*

$$R_\lambda := P_X T_\lambda = P_X (\mathrm{Id} + \lambda \sum_{i=1}^{m} \omega_i (T_i - \mathrm{Id}))$$

is asymptotically regular.

Proof. The operator $T := \sum_{i=1}^{m} \omega_i T_i$ is firmly nonexpansive (see Corollary 2.2.20). Therefore, the asymptotic regularity of R_λ follows from Corollary 3.4.5. □

Corollary 3.4.8. *Let* $T_i : X \to X$, $i \in I$, *be strictly relaxed firmly nonexpansive and the composition* $T := T_1 T_2 \ldots T_m$ *have a fixed point. Then* T *is asymptotically regular.*

Proof. By Theorem 2.2.42, the operator T is strictly relaxed firmly nonexpansive. Therefore, the asymptotic regularity of T follows from Corollary 3.4.6. □

The following result is due to Bruck and Reich (see [51, Corollary 1.1]).

Theorem 3.4.9. *Let* $T : X \to X$ *be a strongly nonexpansive operator with a fixed point. Then* T *is asymptotically regular.*

Proof. The operator T is quasi-nonexpansive, because a strongly nonexpansive operator is nonexpansive and a nonexpansive operator with a fixed point is quasi-nonexpansive (see Lemma 2.1.20). Let $x^0 = x \in X$, $x^k = T^k x$ and $z^* \in \mathrm{Fix}\, T$ be such that

$$\lim_k \left\| x^k - z^* \right\| = \inf_{z \in \operatorname{Fix} T} \lim_k \left\| x^k - z \right\|.$$

The existence and uniqueness of z^* follows from Theorem 3.3.2. Since the sequence $\{\| x^k - z^* \|\}_{k=0}^\infty$ is decreasing, it converges and the sequence $\{x^k\}_{k=0}^\infty$ is bounded. We have

$$\left\| x^k - z^* \right\| - \left\| T x^k - z^* \right\| = \left\| x^k - z^* \right\| - \left\| x^{k+1} - z^* \right\| \to 0.$$

The equality above, the strong nonexpansivity of T and the equality $T z^* = z^*$ yield now

$$\left\| T x^k - x^k \right\| = \left\| (T x^k - T z^*) - (x^k - z^*) \right\| \to 0.$$

Therefore, T is asymptotically regular. $\square$

3.5   Opial's Theorem and Its Consequences

Below, we give a theorem, which is useful in proving the convergence of sequences generated by iterative methods for convex optimization problems. The first part of the theorem below was proved by Opial in [279, Theorem 1].

Theorem 3.5.1 (Opial, 1967). *Let $X \subseteq \mathcal{H}$ be a nonempty closed convex subset of a Hilbert space $\mathcal{H}$ and $U : X \to X$ be a nonexpansive and asymptotically regular operator with a fixed point. Then, for any $x \in X$, the sequence $\{U^k x\}_{k=0}^\infty$ converges weakly to a point $z^* \in \operatorname{Fix} U$. Furthermore, z^* is the unique fixed point of U satisfying equality (3.3).*

Proof. (cf. [279, Theorem 1]) Let $x \in X$, $x^k := U^k x$ and $z \in \operatorname{Fix} U$. By the nonexpansivity of U, we have

$$\left\| x^{k+1} - z \right\| = \left\| U^{k+1} x - z \right\| = \left\| U^{k+1} x - U z \right\| \le \left\| U^k x - z \right\| = \left\| x^k - z \right\|.$$

Therefore, the sequence $\{x^k\}_{k=0}^\infty$ is bounded, consequently, it contains a subsequence $\{x^{n_k}\}_{k=0}^\infty$ which is weakly convergent to a point $y \in X$. Since U is asymptotically regular,

$$\lim_k \left\| U^{k+1} x - U^k x \right\| = \lim_k \left\| U x^k - x^k \right\| = 0.$$

Now, it follows from the demi-closedness principle (Lemma 3.2.5) that $y \in \operatorname{Fix} U$. Theorem 3.3.2 yields that $y = y^*$, where y^* is the unique fixed point of U satisfying equality (3.3). $\square$

If $X \subseteq \mathcal{H}$ is a subset of a finite dimensional Hilbert space, then theorem holds by weaker assumptions. In this case it suffices to suppose the closedness of $U - \operatorname{Id}$ at 0 and quasi nonexpansivity of U instead of its nonexpansivity. The following theorem holds.

Theorem 3.5.2. *Let* $X \subseteq \mathcal{H}$ *be a nonempty closed convex subset of a finite dimensional Hilbert space* $\mathcal{H}$ *and* $U : X \to X$ *be an operator with a fixed point and such that* $U - \text{Id}$ *is closed at* 0. *If* U *is quasi-nonexpansive and asymptotically regular, then, for an arbitrary* $x \in X$, *the sequence* $\{U^k x\}_{k=0}^{\infty}$ *converges to a point* $z \in \text{Fix}\, U$.

Proof. Let $x \in X$, $x^k := U^k x$ and $z \in \text{Fix}\, U$. By the quasi nonexpansivity of U, we have

$$\left\| x^{k+1} - z \right\| = \left\| U x^k - z \right\| \leq \left\| x^k - z \right\|,$$

consequently, $\{x^k\}_{k=0}^{\infty}$ is bounded. Let $z^* \in X$ be an arbitrary cluster point of $\{x^k\}_{k=0}^{\infty}$ and $\{x^{n_k}\}_{k=0}^{\infty}$ be a subsequence which converges to z^*. Since U is asymptotically regular,

$$\left\| U x^{n_k} - x^{n_k} \right\| = \left\| U^{n_k+1} x - U^{n_k} x \right\| \to 0.$$

By the closedness of $U - \text{Id}$ at 0, we have $U z^* = z^*$, i.e., $z^* \in \text{Fix}\, U$. By Lemma 3.3.3, the whole sequence $\{x^k\}_{k=0}^{\infty}$ converges to z^*. $\quad\square$

Corollary 3.5.3. *Let* $\lambda \in (0, 2)$ *and* $T : \mathcal{H} \to \mathcal{H}$ *be a firmly nonexpansive operator with a fixed point. Then, for any* $x \in \mathcal{H}$, *the sequence* $\{T_\lambda^k x\}_{k=0}^{\infty}$ *converges weakly to a point* $z^* \in \text{Fix}\, T$.

Proof. Let $x \in \mathcal{H}$ and $\lambda \in (0, 2)$. The operator T_λ is nonexpansive (see Theorem 2.2.10) and strongly quasi-nonexpansive (see Corollary 2.2.9). Consequently, T_λ is asymptotically regular (see Theorem 3.4.3). By Opial's theorem, the sequence $\{T_\lambda^k x\}_{k=0}^{\infty}$ converges weakly to a point $z^* \in \text{Fix}\, T$. $\quad\square$

By the equivalence of averaged operators and strictly relaxed firmly nonexpansive operators (see Corollary 2.2.17), Corollary 3.5.3 is equivalent to the following result, known in the literature as the *Krasnosel'skiĭ–Mann theorem* (see, e.g., [56, Theorem 2.1], [361] and [57, Theorem 5.16]). Actually, Krasnosel'skiĭ proved the strong convergence for compact operators (see [238, Theorem 1]).

Theorem 3.5.4. *Let* $X \subseteq \mathcal{H}$ *be a nonempty closed convex subset and* $U : X \to X$ *be an averaged operator with* $\text{Fix}\, U \neq \emptyset$. *Then, for an arbitrary* $x \in X$, *the sequence* $\{U^k x\}_{k=0}^{\infty}$ *converges weakly to a point* $z \in \text{Fix}\, U$.

The following result follows from Corollary 3.5.3.

Corollary 3.5.5. *Let* $T : X \to \mathcal{H}$ *be firmly nonexpansive and such that* $\text{Fix}(P_X T) \neq \emptyset$ *and* $R_\lambda = P_X T_\lambda$ *be a projected relaxation of* T, *where* $\lambda \in (0, 2)$. *Then, for an arbitrary* $x \in X$, *the sequence* $\{R_\lambda^k x\}_{k=0}^{\infty}$ *converges weakly to a point* $z^* \in \text{Fix}(P_X T)$.

Proof. The operator R_λ is relaxed firmly nonexpansive (see Theorem 2.2.46 (i)). Therefore, the corollary follows from Corollary 3.5.3. $\quad\square$

3.6   Generalization of Opial's Theorem

Applying Opial's theorem we can prove the convergence of sequences generated by iterating a nonexpansive and asymptotically regular operator. Below we present a generalization of Opial's theorem which can be applied to sequences generated by a sequence of operators. First, we propose a definition of an asymptotically regular sequence of operators, which extends Definition 3.4.1. Let $\{x^k\}_{k=0}^{\infty} \subseteq X$ be a sequence generated by the recurrence

$$x^{k+1} = U_k x^k, \tag{3.5}$$

where $x^0 \in X$ is arbitrary and $U_k : X \to X, k \geq 0$.

Definition 3.6.1. Let $X \subseteq \mathcal{H}$ be a nonempty closed convex subset. We say that a sequence of operators $U_k : X \to X$ is *asymptotically regular*, if for any $x \in X$

$$\lim_k \|U_k U_{k-1} \ldots U_0 x - U_{k-1} \ldots U_0 x\| = 0$$

or, equivalently,

$$\lim_k \|U_k x^k - x^k\| = 0,$$

where the sequence $\{x^k\}_{k=0}^{\infty}$ is generated by recurrence (3.5) with $x^0 = x$.

It is clear that an operator $U : X \to X$ is asymptotically regular, if a constant sequence of operators $U_k = U$ is asymptotically regular. Therefore, Definition 3.6.1 extends Definition 3.4.1 to a sequence of operators.

A weaker version of the first part of the following theorem appeared in [66, Theorem 1].

Theorem 3.6.2. *Let $X \subseteq \mathcal{H}$ be a nonempty closed convex subset, $S : X \to \mathcal{H}$ be an operator with a fixed point and such that $S - \text{Id}$ is demi-closed at 0. Let $\{U_k\}_{k=0}^{\infty}$ be an asymptotically regular sequence of quasi-nonexpansive operators $U_k : X \to X$ such that $\bigcap_{k=0}^{\infty} \text{Fix} U_k \supseteq \text{Fix } S$. Let the sequence $\{x^k\}_{k=0}^{\infty}$ be generated by recurrence (3.5) with an arbitrary $x^0 \in X$.*

(i) If the sequence of operators $\{U_k\}_{k=0}^{\infty}$ has the property

$$\lim_k \|U_k x^k - x^k\| = 0 \implies \lim_k \|S x^k - x^k\| = 0, \tag{3.6}$$

then $\{x^k\}_{k=0}^{\infty}$ converges weakly to a point $z^* \in \text{Fix } S$.

(ii) If $\mathcal{H}$ is finite dimensional and the sequence of operators $\{U_k\}_{k=0}^{\infty}$ has the property

$$\lim_k \|U_k x^k - x^k\| = 0 \implies \liminf_k \|S x^k - x^k\| = 0, \tag{3.7}$$

then $\{x^k\}_{k=0}^{\infty}$ converges to a point $z^* \in \text{Fix} S$.

Opial's Theorem is, actually, a corollary of Theorem 3.6.2 (i). Indeed, suppose that the assumptions of Theorem 3.5.1 are satisfied. Then U is quasi-nonexpansive (see Lemma 2.1.20) and $U - \mathrm{Id}$ is demi-closed at 0 (see Lemma 3.2.5). We see that all assumptions of Theorem 3.6.2 are satisfied for a constant sequence of operators $U_k = U$, $k \geq 0$. Therefore, for an arbitrary $x \in C$, the sequence $\{U^k x\}_{k=0}^{\infty}$ converges weakly to a point $x^* \in \mathrm{Fix}\, U$.

Proof of Theorem 3.6.2. (cf. [70, Theorem 9.9]) Let $x \in X$, $z \in \mathrm{Fix}\, S$ and the sequence $\{x^k\}_{k=0}^{\infty}$ be generated by recurrence (3.5). Since U_k is quasi-nonexpansive and $\mathrm{Fix}\, U_k \supseteq \mathrm{Fix}\, S$, we have

$$\left\| x^{k+1} - z \right\| = \left\| U_k x^k - z \right\| \leq \left\| x^k - z \right\|,$$

$k \geq 0$, Therefore, x^k is Fejér monotone with respect to $\mathrm{Fix}\, S$, consequently, x^k is bounded.

(i) Suppose that condition (3.6) is satisfied. By the asymptotic regularity of the sequence $\{U_k\}_{k=0}^{\infty}$, we have $\left\| U_k x^k - x^k \right\| \to 0$, consequently, $\left\| S x^k - x^k \right\| \to 0$. Let $x^* \in X$ be a weak cluster point of $\{x^k\}_{k=0}^{\infty}$ and $\{x^{n_k}\}_{k=0}^{\infty} \subseteq \{x^k\}_{k=0}^{\infty}$ be a subsequence which converges weakly to x^*. Then, of course, $\left\| S x^{n_k} - x^{n_k} \right\| \to 0$ and $x^* \in \mathrm{Fix}\, S$, by the demi-closedness of $S - \mathrm{Id}$ at 0. Since x^* is an arbitrary weak cluster point of $\{x^k\}_{k=0}^{\infty}$ and x^k is Fejér monotone with respect to $\mathrm{Fix}\, S$, the weak convergence of the whole sequence $\{x^k\}_{k=0}^{\infty}$ to x^* follows from Lemma 3.3.3.

(ii) Let $\mathcal{H}$ be finite dimensional and suppose that condition (3.7) is satisfied. By the asymptotic regularity of the sequence $\{U_k\}_{k=0}^{\infty}$, we have $\left\| U_k x^k - x^k \right\| \to 0$, consequently, $\lim_k \left\| S x^{n_k} - x^{n_k} \right\| = 0$ for a subsequence $\{x^{n_k}\}_{k=0}^{\infty} \subseteq \{x^k\}_{k=0}^{\infty}$. Due to the boundedness of x^{n_k}, there is a subsequence $\{x^{m_{n_k}}\}_{k=0}^{\infty} \subseteq \{x^{n_k}\}_{k=0}^{\infty}$ which converges to a point $x^* \in X$. Since $S - \mathrm{Id}$ is demi-closed at 0, we have $x^* \in \mathrm{Fix}\, S$. The convergence of the whole sequence $\{x^k\}_{k=0}^{\infty}$ to x^* follows now from Lemma 3.3.4. $\qquad\square$

Remark 3.6.3. We do not suppose that the operators U_k are nonexpansive in Theorem 3.6.2. It allows application of this theorem to a sequence of relaxed cutters or, equivalently, to a sequence of strongly quasi-nonexpansive operators. We only suppose that the sequence of operators $\{U_k\}_{k=0}^{\infty}$ is asymptotically regular and that condition (3.6) (or (3.7) in the finite dimensional case) is satisfied for an operator $S : X \to \mathcal{H}$ such that $S - \mathrm{Id}$ is demi-closed at 0, in order to ensure the claims of Theorem 3.6.2. Condition (3.6) is satisfied if, e.g., the inequality

$$\| U_k x - x \| \geq \alpha \| S x - x \| \tag{3.8}$$

holds for all $x \in X$, for all k and for some $\alpha > 0$. Condition (3.7) is satisfied if, e.g., inequality (3.8) holds for all $x \in X$, for infinitely many k and for some $\alpha > 0$.

Remark 3.6.4. It follows from the proof that Theorem 3.6.2 remains true if we replace the assumption that $\{U_k\}_{k=0}^{\infty}$ is asymptotically regular and the assumption (3.6) in case (i) or (3.7) in case (ii) by a weaker assumption $\lim_{k\to\infty}\|Sx^k - x^k\| = 0$ in case (i) or $\liminf_{k\to\infty}\|Sx^k - x^k\| = 0$ in case (ii), respectively. The formulation presented in Theorem 3.6.2 is preferred, because in applications, the operators U_k are often relaxed cutters with relaxation parameters guaranteeing the asymptotic regularity of $\{U_k\}_{k=0}^{\infty}$. Furthermore, various practical algorithms which apply relaxed cutters have properties which yield (3.6), (3.7) or some related conditions.

3.7   Opial-Type Theorems for Cutters

In this section we present modifications and applications of Theorem 3.6.2 for sequences of firmly nonexpansive operators, sequences of cutters and for generalized relaxations of cutters. In all these three cases we adopt Opial's theorem or its generalization (Theorem 3.6.2). Consider the following recurrence

$$x^{k+1} = P_X(x^k + \lambda_k(T_k x^k - x^k)), \qquad (3.9)$$

where $x^0 \in X$, $\lambda_k \in [0, 2]$ and $T_k : X \to \mathcal{H}$. A stronger version of the following result was presented in [70, Proposition 9.12].

Corollary 3.7.1. *Let $S : X \to \mathcal{H}$ be an operator with a fixed point and such that $S - \mathrm{Id}$ is demi-closed at 0, $x^0 \in X$ and the sequence $\{x^k\}_{k=0}^{\infty} \subseteq X$ be generated by recurrence (3.9), where $\liminf_k \lambda_k(2 - \lambda_k) > 0$ and $T_k : X \to \mathcal{H}$ is a sequence of cutters with $\bigcap_{k=0}^{\infty} \mathrm{Fix}\, T_k \supseteq \mathrm{Fix}\, S$. Then:*

(i) For all $z \in \bigcap_{k=0}^{\infty} \mathrm{Fix}\, T_k$

$$\left\| x^{k+1} - z \right\|^2 \leq \left\| x^k - z \right\|^2 - \lambda_k(2 - \lambda_k) \left\| T_k x^k - x^k \right\|^2. \qquad (3.10)$$

(ii) If the following implication holds

$$\lim_k \left\| T_k x^k - x^k \right\| = 0 \quad \Longrightarrow \quad \lim_k \left\| S x^k - x^k \right\| = 0, \qquad (3.11)$$

then x^k converges weakly to a fixed point of S.
(iii) If $\mathcal{H}$ is finite dimensional and the following implication holds

$$\lim_k \left\| T_k x^k - x^k \right\| = 0 \quad \Longrightarrow \quad \liminf_k \left\| S x^k - x^k \right\| = 0, \qquad (3.12)$$

then x^k converges to a fixed point of S.

Proof. Let $C = \bigcap_{k=0}^{\infty} \operatorname{Fix} T_k$ and $z \in C$. Denote $U_k := P_X(\operatorname{Id} + \lambda_k(T_k - \operatorname{Id}))$. Inequality (3.10) and the strong quasi nonexpansivity of U_k follow from Corollary 2.2.25. Consequently, $\|x^k - z\|$ converges as a bounded and monotone sequence and $\lim_k \lambda_k(2 - \lambda_k)\|T_k x^k - x^k\| = 0$. Since $\liminf_k \lambda_k(2 - \lambda_k) > 0$, we have $\lim_k \|T_k x^k - x^k\| = 0$. The rest part of the theorem can be proved by similar arguments as in the proof of Theorem 3.6.2. $\qquad\square$

A special case of Corollary 3.7.1 (ii) with $X = \mathcal{H}$ and $T_k = S$ for all $k \geq 0$ can be found in [253, Theorem 1]. Note that condition (3.11) holds in this case automatically.

Bauschke and Combettes proved a theorem [24, Theorem 2.9 (i)] which is related to Corollary 3.7.1 (ii). They supposed that $X = \mathcal{H}$, $\lambda_k = \lambda \in [1, 2)$ and, instead of (3.11), supposed that all cluster points of $\{x^k\}_{k=0}^{\infty}$ belong to $\bigcap_{k=0}^{\infty} \operatorname{Fix} T_k$ (see also [124, Theorem 2.6 (iii)] and [352, Theorems 2.1–2.3] for related results).

Remark 3.7.2. If $\operatorname{int} \operatorname{Fix} S \neq \emptyset$, then the range of the relaxation parameters in Corollary 3.7.1 (ii) and (iii) can be extended. By Proposition 2.1.41, for any $z \in \operatorname{int} \operatorname{Fix} S$ there is $\varepsilon > 0$ such that

$$\|x^{k+1} - z\|^2 \leq \|x^k - z\|^2 - \lambda_k(2 + \varepsilon - \lambda_k)\|T_k x^k - x^k\|^2. \tag{3.13}$$

If $\liminf_k \lambda_k > 0$ and $\limsup \lambda_k \leq 2$, then, similarly, as in in the proof of Corollary 3.7.1, we obtain $\lim_k \|T_k x^k - x^k\| = 0$, because

$$\liminf_k \lambda_k(2 + 2\varepsilon - \lambda_k) \geq \liminf_k \lambda_k(2 - \lambda_k) + 2\varepsilon \liminf_k \lambda_k > 0.$$

Therefore, if $\operatorname{int} \operatorname{Fix} T \neq \emptyset$, then the results of Corollary 3.7.1 (ii) and (iii) remain true if we suppose that $\liminf_k \lambda_k > 0$ and $\limsup \lambda_k \leq 2$ instead of $\liminf_k \lambda_k(2 - \lambda_k) > 0$. This explains the convergence of sequences generated by several projection methods which apply the extended range of relaxation parameters (see, e.g., [54, 266]).

If we iterate inequality (3.10) k-times, we obtain for an arbitrary $z \in \operatorname{Fix} S$ the following inequality

$$\|x^{k+1} - z\|^2 \leq \|x^0 - z\|^2 - \sum_{l=0}^{k} \lambda_l(2 - \lambda_l)\|T_l x^l - x^l\|^2. \tag{3.14}$$

If we know an upper approximation R of $d(x^0, \operatorname{Fix} S)$, then inequality (3.14) can be applied to the following estimation of $d(x^{k+1}, \operatorname{Fix} S)$

$$d^2(x^{k+1}, \operatorname{Fix} S) \leq R^2 - \sum_{l=0}^{k} \lambda_l(2 - \lambda_l)\|T_l x^l - x^l\|^2,$$

where $R \geq d(x^0, \operatorname{Fix} S)$.

In what follows, we will apply the following property of real sequences $\{\lambda_k\}_{k=0}^{\infty} \subseteq [0,2]$.

$$\liminf_{k} \lambda_k (2 - \lambda_k) > 0 \iff (\liminf_{k} \lambda_k > 0 \text{ and } \limsup \lambda_k < 2).$$

Corollary 3.7.3. *Let $T : X \to \mathcal{H}$ be a cutter and such that $T - \mathrm{Id}$ is demi-closed at 0 (e.g., a firmly nonexpansive operator with $\mathrm{Fix}\, T \neq \emptyset$). Further, let $\{x^k\}_{k=0}^{\infty}$ be generated by the recurrence*

$$x^{k+1} = P_X(x^k + \lambda_k(Tx^k - x^k)),$$

where $x^0 \in X$ and $\lambda_k \in [0,2]$. If $\liminf \lambda_k(2 - \lambda_k) > 0$, then x^k converges weakly to a fixed point of T.

Proof. Denote $T_k = T$, for all $k \geq 0$, and $S = T$. Then implication (3.11) is obvious. Therefore, the corollary follows from Corollary 3.7.1. $\qquad \square$

A related result to Corollary 3.7.3 was obtained by Reich [294, Theorem 2] for a uniformly convex Banach space with a Frechét differentiable norm, where it is supposed (in an equivalent form) that T is firmly nonexpansive and that $\sum_{k=0}^{\infty} \lambda_k(2 - \lambda_k) = +\infty$ instead of $\liminf \lambda_k(2 - \lambda_k) > 0$.

If we take $\lambda_k = \lambda \in (0,2)$ and $X = \mathcal{H}$ in Corollary 3.7.3, then we obtain Corollary 3.5.3.

Denote $U_k := P_X(\mathrm{Id} + \mu_k(S_k - \mathrm{Id}))$ for an operator $S_k : X \to \mathcal{H}$ and for $\mu_k \in (0,2], k \geq 0$.

Corollary 3.7.4. *Let $U : X \to \mathcal{H}$ be an operator with a fixed point and such that $U - \mathrm{Id}$ is demi-closed at 0, $x^0 \in X$ and a sequence $\{x^k\}_{k=0}^{\infty}$ be generated by the recurrence*

$$x^{k+1} = P_X(x^k + \mu_k(S_k x^k - x^k)), \tag{3.15}$$

where $\mu_k \in (0,2]$, $\limsup_k \mu_k < 2$, and $\{S_k\}_{k=0}^{\infty}$ is a sequence of firmly nonexpansive operators $S_k : X \to \mathcal{H}$ such that $\bigcap_{k=0}^{\infty} \mathrm{Fix}\,(P_X S_k) \supseteq \mathrm{Fix}\, U, k \geq 0$.

(i) If the following implication holds

$$\lim_{k} \|U_k x^k - x^k\| = 0 \implies \lim_{k} \|U x^k - x^k\| = 0, \tag{3.16}$$

then x^k converges weakly to a fixed point of U.

(ii) If $\mathcal{H}$ is finite dimensional and the following implication holds

$$\lim_{k} \|U_k x^k - x^k\| = 0 \implies \liminf_{k} \|U x^k - x^k\| = 0, \tag{3.17}$$

then x^k converges to a fixed point of U.

Fig. 3.1 Operators T_k
and U_k

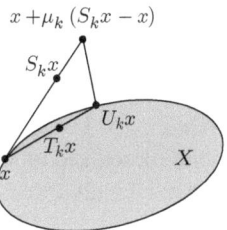

*(iii) Let $S : X \to \mathcal{H}$ be firmly nonexpansive. If $S_k = S$ and $\mu_k \in [\varepsilon, 2 - \varepsilon]$ for
all $k \geq 0$ and some $\varepsilon > 0$, then implication (3.16) holds for $U := P_X S_\varepsilon$.
Consequently, x^k converges weakly to a point $x^* \in \mathrm{Fix}\, U = \mathrm{Fix}\, P_X S$.*

Proof. (i) and (ii) The operator $U_k : X \to X$ is $\frac{4}{4-\mu_k}$-relaxed firmly nonexpansive
and $\mathrm{Fix}\, U_k = \mathrm{Fix}\,(P_X S_k) \neq \emptyset$, $k \geq 0$ (see Theorem 2.2.46). Let $T_k := (U_k)_{\frac{4-\mu_k}{4}}$
(see Fig. 3.1). Then T_k is firmly nonexpansive (see Corollary 2.2.19), T_k is a
cutter (see Theorem 2.2.5 (i)), $\mathrm{Fix}\, T_k = \mathrm{Fix}\, U_k \neq \emptyset$ (see Remark 2.1.4) and
$U_k = (T_k)_{\frac{4}{4-\mu_k}}$ (see Remark 2.1.3). It is clear that recurrence (3.15) is a special
case of (3.9) with $X = \mathcal{H}$ and $\lambda_k = \frac{4}{4-\mu_k}$. Note that $T_{k,\lambda_k} : X \to X$ and that

$$\left\| U_k x^k - x^k \right\| = \frac{4}{4 - \mu_k} \left\| T_k x^k - x^k \right\|,$$

i.e., condition (3.16) is equivalent to (3.11) with $S := U$ and condition (3.17) is
equivalent to (3.12) with $S := U$.

Setting $\lambda_k = \frac{4}{4-\mu_k}$ we easily obtain $\liminf_k \lambda_k (2 - \lambda_k) > 0$. Now the weak
convergence in (i) and the convergence in (ii) follows from Corollary 3.7.1 (i) and
(ii), respectively.

(iii) Suppose that $S_k = S$ in (3.15) and $\mu_k \in [\varepsilon, 2 - \varepsilon]$ for all $k \geq 0$ and for
some $\varepsilon \in (0, 1)$. Let $U := P_X S_\varepsilon = P_X (\mathrm{Id} + \varepsilon(S - \mathrm{Id}))$. It is clear that $\mathrm{Fix}\, U = \mathrm{Fix}\, P_X S$ (see Corollary 1.2.10). Furthermore, U is nonexpansive as a composition
of nonexpansive operators P_X and S_ε. Consequently, $U - \mathrm{Id}$ is demi-closed at 0 (see
Lemma 3.2.5). It follows from Corollary 2.2.26 that

$$\left\| U_k x^k - x^k \right\| = \left\| P_X (x^k + \mu_k (S x^k - x^k)) - x^k \right\|$$
$$\geq \left\| P_X (x^k + \varepsilon(S x^k - x^k)) - x^k \right\| = \left\| U x^k - x^k \right\|,$$

i.e., implication (3.16) is satisfied. Consequently, x^k converges weakly to a fixed
point of $P_X S$.   □

Note the difference in assumptions of Corollaries 3.7.1 and 3.7.4. An iteration in
recurrence (3.9) is defined by a projected relaxation of a cutter T_k with $\mathrm{Fix}\, T_k \supseteq$
$\mathrm{Fix}\, S$, while iteration (3.15) is defined by a projected relaxations of a firmly
nonexpansive operator S_k for which we do not suppose that $\mathrm{Fix}\, S_k \neq \emptyset$. We only

suppose that $\mathrm{Fix}(P_X S_k) \supseteq \mathrm{Fix}\, S \neq \emptyset$. Furthermore, conditions (3.11) and (3.16) have a different nature. In the first one, we apply the cutters T_k, while in the other one—projected relaxations of firmly nonexpansive operators S_k.

Remark 3.7.5. Similarly as in Theorem 3.6.2, we can replace the assumption on the demi-closedness of $S-\mathrm{Id}$ at 0 and assumptions (3.11) and (3.16) in Corollaries 3.7.1 and 3.7.4 with a weaker one:

- If $\lim_k \left\| T_k x^k - x^k \right\| = 0$ or $\lim_k \left\| U_k x^k - x^k \right\| = 0$, respectively, then all weak cluster points of x^k lie in $\mathrm{Fix}\, S$ or in $\mathrm{Fix}\, U$, respectively.

In these cases the claims remain true (cf. Remark 3.6.4). Conditions of a similar form were proposed by Schott [303, Theorem 3] and by Bauschke and Borwein [22, Definition 3.7] for slightly different recurrences.

Now we present an Opial-type theorem for generalized relaxations of algorithmic operators (see Sect. 2.4 for definitions).

Theorem 3.7.6. *Let $U : X \rightarrow \mathcal{H}$ be a strongly oriented operator with a fixed point and such that $U - \mathrm{Id}$ is demi-closed at 0, and the sequence $\{x^k\}_{k=0}^{\infty} \subseteq X$ be generated by the recurrence*

$$x^{k+1} = P_X U_{\sigma,\lambda_k}(x^k), \tag{3.18}$$

where $x^0 \in X$, $\liminf_k \lambda_k(2-\lambda_k) > 0$ and the step size function $\sigma : X \rightarrow (0, +\infty)$ satisfies the following condition

$$\alpha \leq \sigma(x) \leq \frac{\langle z - x, Ux - x \rangle}{\left\| Ux - x \right\|^2} \tag{3.19}$$

for all $x \in X$, for all $z \in \mathrm{Fix}\, U$ and for some $\alpha > 0$. Then

$$\left\| x^{k+1} - z \right\|^2 \leq \left\| x^k - z \right\|^2 - \lambda_k(2 - \lambda_k)\sigma^2(x^k) \left\| Ux^k - x^k \right\|^2 \tag{3.20}$$

and x^k converges weakly to a fixed point of U.

Proof. Let $z \in \mathrm{Fix}\, U$, $\varepsilon > 0$ and $k_0 \geq 0$ be such that $\lambda_k \in [\varepsilon, 2 - \varepsilon]$ for all $k \geq k_0$. It follows from the second inequality in (3.19) that U_σ is a cutter and that U_{σ,λ_k} is $\frac{2-\lambda_k}{\lambda_k}$-strongly quasi-nonexpansive (see Corollary 2.4.5). This, together with the nonexpansivity of the metric projection and with the first inequality in (3.19), gives

$$\begin{aligned}
\left\| x^{k+1} - z \right\|^2 &= \left\| P_X U_{\sigma,\lambda_k} x^k - z \right\|^2 \\
&= \left\| P_X U_{\sigma,\lambda_k} x^k - P_X z \right\|^2 \\
&\leq \left\| U_{\sigma,\lambda_k} x^k - z \right\|^2
\end{aligned}$$

$$\leq \left\| x^k - z \right\|^2 - \frac{2 - \lambda_k}{\lambda_k} \left\| U_{\sigma,\lambda_k} x^k - x^k \right\|^2$$

$$= \left\| x^k - z \right\|^2 - \lambda_k (2 - \lambda_k) \sigma^2(x^k) \left\| U x^k - x^k \right\|^2$$

$$\leq \left\| x^k - z \right\|^2 - \varepsilon^2 \alpha^2 \left\| U x^k - x^k \right\|^2$$

for all $k \geq k_0$. Therefore, x^k is Fejér monotone with respect to Fix U, consequently x^k is bounded and $\left\| U x^k - x^k \right\| \to 0$. Let $x^* \in X$ be a weak cluster point of $\{x^k\}_{k=0}^{\infty}$ and $\{x^{n_k}\}_{k=0}^{\infty} \subseteq \{x^k\}_{k=0}^{\infty}$ be a subsequence which converges weakly to x^*. By the demi-closedness of $U - \mathrm{Id}$, we have $x^* \in \mathrm{Fix}\, U$. This, together with the Fejér monotonicity of $\{x^k\}_{k=0}^{\infty}$ with respect to Fix U, gives the weak convergence of x^k to x^* (see Lemma 3.3.3).   □

If we iterate inequality (3.20) k-times, we obtain

$$\left\| x^{k+1} - z \right\|^2 \leq \left\| x^0 - z \right\|^2 - \sum_{l=0}^{k} \lambda_l (2 - \lambda_l) \sigma^2(x^l) \left\| U x^l - x^l \right\|^2 \qquad (3.21)$$

for any $z \in \mathrm{Fix}\, U$. If we know an upper approximation R of $d(x^0, \mathrm{Fix}\, U)$, then inequality (3.21) can be applied to the following estimation of $d(x^{k+1}, \mathrm{Fix}\, U)$

$$d^2(x^{k+1}, \mathrm{Fix}\, U) \leq R^2 - \sum_{l=0}^{k} \lambda_l (2 - \lambda_l) \sigma^2(x^l) \left\| U x^l - x^l \right\|^2, \qquad (3.22)$$

where $R \geq d(x^0, \mathrm{Fix}\, U)$.

3.8   Strong Convergence of Fejér Monotone Sequences

Theorem 3.6.2 as well as other results presented in the last two sections, yield only the weak convergence of a sequence $\{x^k\}_{k=0}^{\infty}$ generated by a sequence of asymptotically regular and quasi-nonexpansive operators to a fixed point of a nonexpansive operator. If $\mathcal{H}$ is finite dimensional, the weak convergence is equivalent to the strong convergence. Under some assumptions on the structure of the convex minimization problem we can obtain, however, the strong convergence in any Hilbert space.

The following theorem is due to Bauschke and Borwein (see [22, Theorem 2.16 (iii)]).

Theorem 3.8.1. *Let $\{x^k\}_{k=0}^{\infty} \subseteq \mathcal{H}$ be Fejér monotone with respect to a subset $C \subseteq \mathcal{H}$. If $\mathrm{int}\, C \neq \emptyset$, then $\{x^k\}_{k=0}^{\infty}$ converges strongly.*

Proof. Let $z \in \operatorname{int} C$ and $\varepsilon > 0$ be such that $B(z, \varepsilon) \subseteq C$. Denote $d^k = \frac{\varepsilon}{\|x^k - x^{k+1}\|}(x^k - x^{k+1})$. Then, obviously, we have $z^k = z + d^k \in C$. By the Fejér monotonicity of $\{x^k\}_{k=0}^{\infty}$ with respect to C,

$$\|x^{k+1} - z\|^2 + \|d^k\|^2 - 2\langle x^{k+1} - z, d^k \rangle$$

$$= \|x^{k+1} - z - d^k\|^2$$

$$= \|x^{k+1} - z^k\|^2 \leq \|x^k - z^k\|^2 = \|x^k - z - d^k\|^2$$

$$= \|x^k - z\|^2 + \|d^k\|^2 - 2\langle x^k - z, d^k \rangle,$$

i.e.,

$$\|x^{k+1} - z\|^2 \leq \|x^k - z\|^2 - 2\langle x^k - x^{k+1}, d^k \rangle$$

$$= \|x^k - z\|^2 - 2\varepsilon \|x^k - x^{k+1}\|$$

and

$$\|x^k - x^{k+1}\| \leq \frac{1}{2\varepsilon} \left(\|x^k - z\|^2 - \|x^{k+1} - z\|^2 \right).$$

Consequently,

$$\|x^k - x^{k+m}\| \leq \sum_{l=0}^{m-1} \|x^{k+l} - x^{k+l+1}\| \leq \frac{1}{2\varepsilon} \left(\|x^k - z\|^2 - \|x^{k+m} - z\|^2 \right)$$

(3.23)

for all $m \geq 1$. By the Fejér monotonicity of $\{x^k\}_{k=0}^{\infty}$ with respect to C, $\|x^k - z\|$ converges and inequalities (3.23) yield that $\{x^k\}_{k=0}^{\infty}$ is a Cauchy sequence, therefore it converges strongly.   □

Before we formulate our next result, we prove some auxiliary lemmas.

Lemma 3.8.2. *Let $C := \bigcap_{i \in J} H_-(a_i, \beta_i)$, where $a_i \in \mathcal{H}$ and $\beta_i \in \mathbb{R}$, $i \in J$, and $\{x^k\}_{k=0}^{\infty}$ be Fejér monotone with respect to C. Then $x^k \in x^0 + \operatorname{Lin}\{a_i, i \in J\}$ for all $k \geq 0$.*

Proof. Denote $V := \operatorname{Lin}\{a_i, i \in J\}$ and fix $k \geq 0$. Since $x^{k+1} = x^0 + \sum_{l=0}^{k}(x^{l+1} - x^l)$, it suffices to prove that $x^{l+1} - x^l \in V$ for all $l \in \{0, 1, \dots, k\}$. We leave it to the reader to check that the Fejér monotonicity of $\{x^k\}_{k=0}^{\infty}$ with respect to C yields that

$$\left\langle w - \frac{1}{2}\left(x^{l+1} - x^l\right), x^{l+1} - x^l \right\rangle \geq 0 \tag{3.24}$$

for any $w \in C$ (see equivalence (2.11)). Let $l \in \{0, 1, \dots, k\}$, $z \in C$, and $u^l \in V$ and $v^l \in V^{\perp}$ be such that $x^{l+1} - x^l = u^l + v^l$. Suppose that $x^{l+1} - x^l \notin V$. Then $v^l \neq 0$. For any $\alpha \in \mathbb{R}$ and $i \in J$, we have

$$\langle a_i, z + \alpha v^l \rangle = \langle a_i, z \rangle + \alpha \langle a_i, v^l \rangle = \langle a_i, z \rangle \leq \beta_i,$$

i.e., $z + \alpha v^l \in C$. Then, for

$$\alpha_l < \frac{\frac{1}{2}\left\|u^l\right\|^2 - \langle z, x^{l+1} - x^l \rangle}{\left\|v^l\right\|^2} + \frac{1}{2}$$

we obtain

$$\langle z + \alpha_l v^l - \frac{1}{2}(x^{l+1} - x^l), x^{l+1} - x^l \rangle$$

$$= \langle z, x^{l+1} - x^l \rangle + \langle \alpha_l v^l - \frac{1}{2}(x^{l+1} - x^l), x^{l+1} - x^l \rangle$$

$$= \langle z, x^{l+1} - x^l \rangle + \langle \alpha_l v^l - \frac{1}{2}(u^l + v^l), u^l + v^l \rangle$$

$$= \langle z, x^{l+1} - x^l \rangle + (\alpha_l - \frac{1}{2})\left\|v^l\right\|^2 - \frac{1}{2}\left\|u^l\right\|^2 < 0,$$

a contradiction with (3.24). Therefore, $x^{l+1} - x^l \in V$ which completes the proof.

□

Lemma 3.8.3. *Let X be a finite dimensional affine subspace of $\mathcal{H}$ and $\{x^k\}_{k=0}^\infty \subseteq X$. If x^k converges weakly to a point $z \in X$, then x^k converges strongly to z.*

Proof. Let $x^k \rightharpoonup z \in X$. Denote $y^k := x^k - x^0$. It is clear that $y^k \in V := X - x^0$, $y^k \rightharpoonup z - x^0$ and V is a finite dimensional subspace of $\mathcal{H}$. Therefore, $y^k \to z - x^0$ and $\lim_k x^k = \lim_k y^k + x^0 = z$. □

A special case of the following Theorem was proved by Gurin et al. (see [196, Theorem 1 (d)]).

Theorem 3.8.4. *Let $\{x^k\}_{k=0}^\infty \subseteq \mathcal{H}$ be Fejér monotone with respect to a subset C being the intersection of finitely many half-spaces. If $\{x^k\}_{k=0}^\infty$ converges weakly, then it converges strongly.*

Proof. Let $C := \bigcap_{i \in I} H_-(a_i, \beta_i)$, where $a_i \in \mathcal{H}$ and $\beta_i \in \mathbb{R}$, $i \in I := \{1, 2, \ldots, m\}$. Denote $X = x^0 + \text{Lin}\{a_i, i \in I\}$. Then X is a finite dimensional affine subspace. By Lemma 3.8.2, $\{x^k\}_{k=0}^\infty \subseteq X$. Suppose that $\{x^k\}_{k=0}^\infty$ converges weakly. Then the strong convergence of $\{x^k\}_{k=0}^\infty$ to a point of X follows from Lemma 3.8.3. □

Corollary 3.8.5. *Suppose that a sequence $\{x^k\}_{k=0}^\infty$ generated by recurrence (3.5) satisfies the assumptions of Theorem 3.6.2. Furthermore, suppose that one of the following conditions is satisfied:*

(i) X is a finite dimensional affine subspace of $\mathcal{H}$,
(ii) The subset $F := \bigcap_{k=0}^\infty \text{Fix } U_k$ has a nonempty interior,
(iii) The subset $F := \bigcap_{k=0}^\infty \text{Fix } U_k$ is an intersection of finitely many half-spaces.

Then for an arbitrary $x^0 \in X$ the sequence $\{x^k\}_{k=0}^\infty$ generated by recurrence (3.5) converges strongly to a fixed point of S.

Proof. The assumptions of Theorem 3.6.2 guarantee the weak convergence of $\{x^k\}_{k=0}^\infty$ to a fixed point of S. The strong convergence in case (i) follows from Lemma 3.8.3, in case (ii)—from Theorem 3.8.1 and in case (iii)—follows from Theorem 3.8.4.   □

Sufficient conditions for the strong convergence of Fejér monotone sequences, in particular of sequences generated by several projection methods for the convex feasibility problems, were presented in [196], [283, Theorem 1.1], [16, 308], [22, Sect. 5], [255] and in [256]. The strong convergence also holds under additional assumptions on the operators which generate the sequences $\{x^k\}_{k=0}^\infty$. For details see [36, Chaps. 3 and 4], [14], [114, Chap. 6], [272] and [295, Theorem 1.7]. The papers [32, 213, 258] contain examples of projection methods for the convex feasibility problems, which generate sequences converging weakly but not strongly. In the paper [24], a method is described which transforms algorithms of type (3.9) generating weakly convergent sequences to sequences which are strongly convergent. The method is based on the idea of Haugazeau [202].

3.9 Relationships Among Algorithmic Operators

In this section we recall relationships among nonexpansive operators, firmly nonexpansive operators, relaxed firmly nonexpansive operators, cutters, quasi-nonexpansive operators, strongly quasi-nonexpansive operators, strongly nonexpansive and asymptotically regular operators which were presented in Chap. 2. These relationships are presented in the form of a diagram. In Fig. 3.2, $T :$ $X \rightarrow \mathcal{H}$ and $U := T_\lambda = \mathrm{Id} + \lambda(T - \mathrm{Id})$ is a λ-relaxation of T, where $\lambda \in (0, 2)$. The figure shows an important role of firmly nonexpansive operators, cutters and strongly nonexpansive operators. It follows from Corollary 2.2.9, from Theorem 3.4.3 and from Opial's Theorem that for $\lambda \in (0, 2)$, sequences generated by the relaxation T_λ of a firmly nonexpansive operator T converge weakly to its fixed point, if Fix $T \neq \emptyset$. We also see that the weak convergence is guaranteed if U is nonexpansive, Fix $T \neq \emptyset$ and at least one of the conditions below is satisfied for $U = T_\lambda$:

(i) T is a cutter (Theorems 2.1.39 and 3.4.3),
(ii) U is strongly quasi-nonexpansive (Theorem 3.4.3),
(iii) U is strongly nonexpansive (Theorem 3.4.9).

Note that a relaxed firmly nonexpansive operator is nonexpansive (see implication (i)⇒(ii) in Theorem 2.2.10). A strongly nonexpansive operator is nonexpansive (see Definition 2.3.1). In cases (i) and (ii) we should additionally suppose that $U := T_\lambda$ is nonexpansive if we want to apply Opial's Theorem. Note, however,

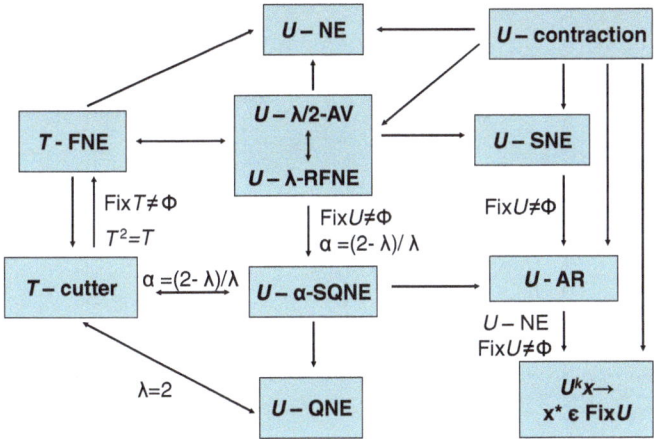

Fig. 3.2 Relationships among algorithmic operators

that the generalization of Opial's Theorem (Theorem 3.6.2) applied to a sequence of relaxations U_k of operators T_k, which satisfy one of the conditions (i)–(ii) above, does not require the nonexpansivity of U_k. In this case it suffices to suppose that implication (3.6) holds for a nonexpansive operator $S : X \rightarrow \mathcal{H}$ in order to guarantee the weak convergence of sequences generated by the sequence of operators U_k. In one of the next chapters we will present examples which show how to construct sequences of such operators U_k for a given nonexpansive operator S, which guarantee implication (3.6). The basic tools in the proofs of convergence of sequences generated by such sequences of operators are the generalizations of Opial's Theorem presented in Sects. 3.6 and 3.7.

3.10 Exercises

Exercise 3.10.1. Let $\mathcal{H} := \mathbb{R}^2$, $C := [0, 1] \times \{0\}$, $x^k := (1 + \frac{1}{k}, (-1)^k)$. Check that $\{x^k\}_{k=0}^{\infty}$ is strictly Fejér monotone with respect to C, but $\{x^k\}_{k=0}^{\infty}$ does not converge.

Exercise 3.10.2. Let $\mathcal{H} := l_2$ and $x^k := e_k$ (see Example 3.2.2). Show that the sequence $\{y^k\}_{k=0}^{\infty}$ with $y^{2k+1} = x^k$ and $y^{2k} = x^1$, $k \geq 0$, is Fejér monotone with respect to $\{0\}$ and has two weak cluster points: 0 and x^1 (the latter one is even a strong cluster point of y^k).

Chapter 4
Algorithmic Projection Operators

As we have mentioned in Chap. 2, the algorithms (or methods) for solving convex optimization problems are defined by algorithmic operators. Usually, particular iterations of these algorithms have the form

$$x^+ = Ux,$$

where x is the current approximation of a solution and x^+ is a next approximation (also called an actualization or an update). If an operator U describing this actualization belongs to the class of strongly quasi-nonexpansive operators (or, equivalently, to the class of relaxed cutters), then we call them *algorithmic projection operators*. The name projection operator can be explained by the following property of a cutter $T : \mathcal{H} \to \mathcal{H}$: For any $x \notin \text{Fix } T$, Tx is the metric projection of x onto a hyperplane

$$H(x - Tx, \langle Tx, x - Tx \rangle) = \{y \in \mathcal{H} : \langle y - Tx, x - Tx \rangle = 0\}$$

which separates the point x from the subset Fix T.

In this chapter we give examples of algorithmic projection operators and we show their properties. These properties are, in most cases, corollaries of general properties of operators presented in Chap. 2. Since the metric projection plays an important role in the construction of algorithmic projection operators, in Sect. 4.1 we give the formulas for the metric projection onto simple closed convex subsets.

4.1 Examples of Metric Projections

4.1.1 Metric Projection onto a Hyperplane

A hyperplane in a Hilbert space $\mathcal{H}$ has the form

$$H(a, \beta) := \{z \in \mathcal{H} : \langle a, z \rangle = \beta\},$$

A. Cegielski, *Iterative Methods for Fixed Point Problems in Hilbert Spaces*,
Lecture Notes in Mathematics 2057, DOI 10.1007/978-3-642-30901-4_4,
© Springer-Verlag Berlin Heidelberg 2012

Fig. 4.1 Metric projection
onto a hyperplane

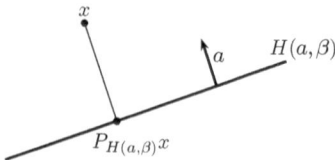

where $a \in \mathcal{H}$, $a \neq 0$ and $\beta \in \mathbb{R}$. It is clear that $H(a, \beta)$ is a closed convex subset.
We show that

$$P_{H(a,\beta)}x = x - \frac{\langle a, x \rangle - \beta}{\|a\|^2}a. \tag{4.1}$$

Let $y := x - \|a\|^{-2}(\langle a, x \rangle - \beta)a$. A straightforward computation shows that
$\langle a, y \rangle = \beta$, i.e., $y \in H(a, \beta)$. Furthermore, for any $z \in H(a, \beta)$, we have

$$\langle x - y, z - y \rangle = \frac{\langle a, x \rangle - \beta}{\|a\|^2}(\langle a, z \rangle - \langle a, y \rangle) = 0.$$

Now, the characterization of the metric projection (see Theorem 1.2.4) yields $y = P_{H(a,\beta)}x$ (Fig. 4.1). We can also write

$$P_{H(a,\beta)}x = x - \frac{\rho(x)}{\|a\|^2}a,$$

where $\rho(x) = \langle a, x \rangle - \beta$ is the residuum of the equality $\langle a, x \rangle = \beta$ at the point
$x \in \mathcal{H}$.

Example 4.1.1. In a Euclidean space ($\mathbb{R}^n$ with the standard inner product) we have

$$H(a, \beta) = \{x \in \mathbb{R}^n : a^{\top}x = \beta\},$$

where $a \in \mathbb{R}^n$ and $\beta \in \mathbb{R}$. In this case equality (4.1) can be written in the form

$$P_{H(a,\beta)}x = x - \frac{a^{\top}x - \beta}{\|a\|^2}a$$

or

$$P_{H(a,\beta)}x = x - \frac{\rho(x)}{\|a\|^2}a,$$

where $\rho(x) = a^{\top}x - \beta$.

Example 4.1.2. Denote by $P_C^G x$ the metric projection of $x \in \mathbb{R}^n$ onto a closed
convex subset C of $\mathbb{R}^n$ equipped with the inner product $\langle \cdot, \cdot \rangle_G$ induced by a positive
definite matrix G of type $n \times n$, defined by $\langle x, y \rangle_G := x^{\top}Gy$. We call $P_C^G x$ an
oblique projection. We calculate the oblique projection $P_{H(a,\beta)}^G x$ for $x \in \mathbb{R}^n$, where

Fig. 4.2 Oblique projection onto a hyperplane

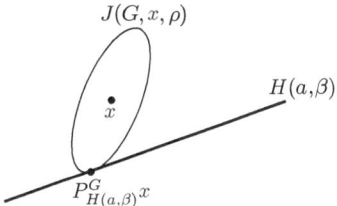

$a \in \mathbb{R}^n$, $a \neq 0$ and $\beta \in \mathbb{R}$. It follows from the definition of the metric projection that $P^G(x)$ is a solution of the following differentiable minimization problem

$$\begin{aligned}
\text{minimize } & f(z) = \tfrac{1}{2}\|z - x\|_G^2 \\
\text{subject to } & \langle a, z \rangle = \beta \\
& z \in \mathbb{R}^n
\end{aligned} \tag{4.2}$$

(see Fig. 4.2). By the properties of the metric projection, this problem has a unique solution. Since the minimization problem (4.2) is convex, the solution $z = P^G_{H(a,\beta)}x$ can be derived from the Karush–Kuhn–Tucker conditions (see Theorem 1.3.7). The Lagrange function $L : \mathbb{R}^n \times \mathbb{R} \to \mathbb{R}$ has the form

$$L(z, \lambda) = \frac{1}{2}\|z - x\|_G^2 + \lambda(\langle a, z \rangle - \beta)$$

and the KKT-point (z, λ) is the unique solution of the KKT-system

$$G(z - x) + \lambda a = 0$$

$$\langle a, z \rangle = \beta.$$

Consequently,

$$\langle G^{-1}a, G(z - x) + \lambda a \rangle = 0$$

and

$$\lambda = -\frac{\langle a, z - x \rangle}{\langle a, G^{-1}a \rangle} = \frac{\langle a, x \rangle - \beta}{\|a\|_{G^{-1}}^2}.$$

If we put the computed Lagrange multiplier λ into the first equation of the KKT-system, we obtain

$$P^G(x) = x - \frac{\langle a, x \rangle - \beta}{\|a\|_{G^{-1}}^2} G^{-1}a. \tag{4.3}$$

Example 4.1.3. In $\mathcal{H} = L_2([\alpha, \beta])$ with the inner product defined by

$$\langle f, g \rangle := \int_\alpha^\beta f(x)g(x)dx$$

Fig. 4.3 Metric projection
onto an affine subspace

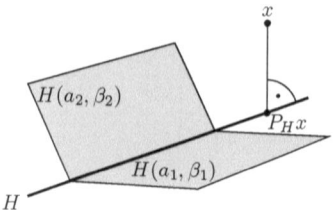

we have

$$H(g, \gamma) = \{h \in L_2([\alpha, \beta]) : \int_\alpha^\beta g(x)h(x)dx = \gamma\},$$

where $g \in L_2([\alpha, \beta])$, $g \neq 0$, and $\gamma \in \mathbb{R}$. By applying equality (4.1), we obtain

$$P_{H(g,\gamma)} f = f - \frac{\int_\alpha^\beta g(x)f(x)dx - \gamma}{\int_\alpha^\beta g^2(x)dx} g.$$

4.1.2 Metric Projection onto a Finite Dimensional Affine Subspace

Let H be an intersection of a finite number of hyperplanes in $\mathbb{R}^n$, i.e., $H = \bigcap_{i=1}^m H(a_i, \beta_i)$, where $a_i \in \mathbb{R}^n$, $a_i \neq 0$ and $\beta_i \in \mathbb{R}$, $i = 1, 2, \ldots, m$. Suppose that $H \neq \emptyset$. Obviously, H is an affine subspace in $\mathbb{R}^n$. It is clear that H is closed and convex as an intersection of closed convex subsets. Let $A = [a_1, \ldots, a_m]^\top$ be a matrix, with rows a_i and $b = (\beta_1, \ldots, \beta_m)$. Suppose that $\mathbb{R}^n$ is equipped with a standard inner product $\langle x, y \rangle := x^\top y$ and with the norm $\|\cdot\| := \sqrt{\langle x, x \rangle}$. Then $H = \{y \in \mathbb{R}^n : Ay = b\}$. Let $x \in \mathbb{R}^n$ be arbitrary. Note that

$$H - x = \{y \in \mathbb{R}^n : A(y + x) = b\}.$$

We have $P_{H-x}0 = A^+(b - Ax)$, where A^+ denotes the Moore–Penrose pseudoinverse of A (see Example 1.3.11). Lemma 1.2.6 yields

$$P_H x = x - A^+(Ax - b). \tag{4.4}$$

If A has full row rank, then $A^+ = A^\top (AA^\top)^{-1}$ and

$$P_H x = x - A^\top (AA^\top)^{-1}(Ax - b) \tag{4.5}$$

(Fig. 4.3). If $m = 1$, we have $A = a^\top \in \mathbb{R}^n$ and $b = \beta \in \mathbb{R}$. In this case we have $AA^\top = a^\top a = \|a\|^2$ and (4.5) obtains the form (4.1).

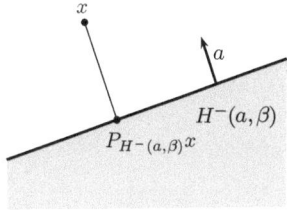

Fig. 4.4 Metric projection
onto a half-space

4.1.3 Metric Projection onto a Half-Space

A half-space in a Hilbert space $\mathcal{H}$ has the form

$$H_-(a, \beta) := \{z \in \mathcal{H} : \langle a, z \rangle \leq \beta\},$$

where $a \in \mathcal{H}$, $a \neq 0$ and $\beta \in \mathbb{R}$. It is clear that $H_-(a, \beta)$ is closed and convex. We show that

$$P_{H_-(a,\beta)}x = \begin{cases} x - \frac{\langle a,x \rangle - \beta}{\|a\|^2}a & \text{if } \langle a, x \rangle > \beta \\ x & \text{if } \langle a, x \rangle \leq \beta. \end{cases} \tag{4.6}$$

Equality (4.6) is clear if $x \in H_-(a, \beta)$, i.e., $\langle a, x \rangle \leq \beta$. Let now $\langle a, x \rangle > \beta$ and $y := x - \|a\|^{-2}(\langle a, x \rangle - \beta)a$. We easily see that $\langle a, y \rangle = \beta$, i.e., $y \in H(a, \beta) \subseteq H_-(a, \beta)$. Furthermore, for an arbitrary $z \in H_-(a, \beta)$, we have

$$\langle x - y, z - y \rangle = \frac{\langle a, x \rangle - \beta}{\|a\|^2}(\langle a, z \rangle - \langle a, y \rangle) \leq 0.$$

Now, it follows from the characterization of the metric projection (see Theorem 1.2.4) that $y = P_{H_-(a,\beta)}x$. We can also write

$$P_{H_-(a,\beta)}x = x - \frac{(\langle a, x \rangle - \beta)_+}{\|a\|^2}a \tag{4.7}$$

(Fig. 4.4) or

$$P_{H_-(a,\beta)}x = x - \frac{\rho_+(x)}{\|a\|^2}a,$$

where $\rho_+(x) = (\langle a, x \rangle - \beta)_+$.

4.1.4 Metric Projection onto a Band

A *band* in a Hilbert space $\mathcal{H}$ has the form

$$C := \{x \in \mathcal{H} : \beta_1 \leq \langle a, x \rangle \leq \beta_2\},$$

Fig. 4.5 Metric projection
onto a band

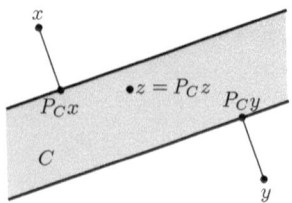

where $a \in \mathcal{H}$, $a \neq 0$, $\beta_1, \beta_2 \in \mathbb{R}$ and $\beta_1 < \beta_2$. Obviously, C is closed and
convex as an intersection of closed convex subsets $H_-(a, \beta_2)$ and $H_-(-a, -\beta_1)$.
By equality (4.6), we easily obtain

$$
P_C x = \begin{cases}
x - \frac{\langle a, x \rangle - \beta_2}{\|a\|^2} a & \text{if } \langle a, x \rangle > \beta_2 \\
x & \text{if } \beta_1 \leq \langle a, x \rangle \leq \beta_2 \\
x - \frac{\langle a, x \rangle - \beta_1}{\|a\|^2} a & \text{if } \langle a, x \rangle < \beta_1
\end{cases}
$$

(see Fig. 4.5).

4.1.5 Metric Projection onto the Orthant

Let $\mathbb{R}^n_+$ be the nonnegative orthant in the Euclidean space $\mathbb{R}^n$. Obviously, $\mathbb{R}^n_+$ is
closed and convex. We show that $P_{\mathbb{R}^n_+}(x) = x_+$. Let $y := x_+$. It is clear that $y \in \mathbb{R}^n_+$. Let $z \in \mathbb{R}^n_+$ be arbitrary. Since $x = x_+ - x_-$, $\langle x_-, z \rangle \geq 0$ and $\langle x_+, x_- \rangle = 0$,
we have

$$
\begin{aligned}
\langle x - y, z - y \rangle &= \langle x - x_+, z - x_+ \rangle \\
&= \langle -x_-, z - x_+ \rangle \\
&= -\langle x_-, z \rangle + \langle x_-, x_+ \rangle \leq 0.
\end{aligned}
$$

The characterization of the metric projection (see Theorem 1.2.4) yields

$$
P_{\mathbb{R}^n_+} x = x_+
$$

(see Fig. 4.6).

In a similar way we can show that

$$
P_{\mathbb{R}^n_-} x = -x_-. \tag{4.8}
$$

We obtain similar results in $\mathcal{H} = L_2([\alpha, \beta])$, where $-\infty \leq \alpha < \beta \leq +\infty$,
equipped with the inner product $\langle f, g \rangle := \int_\alpha^\beta f(x)g(x)dx$, for the subsets $L_2^+ :=$

Fig. 4.6 Metric projection onto $\mathbb{R}^n_+$

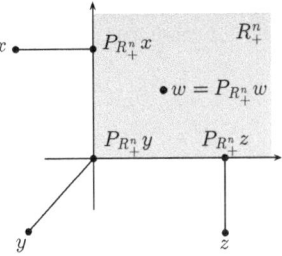

$\{f \in L_2([\alpha, \beta]) : f \geq 0\}$ and $L_2^- := \{f \in L_2 : f \leq 0\}$. Obviously, both subsets are closed and convex. By applying the same arguments as above, we obtain

$$P_{L_2^+} f = f_+$$

and

$$P_{L_2^-} f = -f_-.$$

Example 4.1.4. Let $Q := \{u \in \mathbb{R}^m : u \leq b\}$, where $b \in \mathbb{R}^m$. We have $Q = \mathbb{R}^m_- + b$ or, equivalently, $\mathbb{R}^m_- = Q - b$. Applying equalities (1.12) and (4.8), we obtain

$$\begin{aligned}
P_Q(y) &= P_{\mathbb{R}^m_-}(y - b) + b \\
&= -(y - b)_- + b \\
&= y - (y - b)_+.
\end{aligned}$$

If we take $y := Ax$ and $r(x) := Ax - b$, we obtain

$$(P_Q - \mathrm{Id})Ax = -(Ax - b)_+ = -r_+(x). \tag{4.9}$$

This equality will be used in the sequel.

4.1.6 Metric Projection onto Box Constraints

The box constraints in $\mathbb{R}^n$ have the form $a \leq z \leq b$, where $a, b \in \mathbb{R}^n$, $a \leq b$. Let $C := \{z \in \mathbb{R}^n : a \leq z \leq b\}$. Obviously, the subset C is closed and convex as the Cartesian product of closed intervals,

$$C = [\alpha_1, \beta_1] \times \ldots \times [\alpha_n, \beta_n]. \tag{4.10}$$

In a one-dimensional case the metric projection of a point $\xi \in \mathbb{R}$ onto the interval $[\alpha, \beta]$ has the form

$$P_{[\alpha, \beta]}\xi = \mathrm{median}\{\xi, \beta, \alpha\} := \max\{\min\{\xi, \beta\}, \alpha\}.$$

Fig. 4.7 Metric projection
onto box constraints

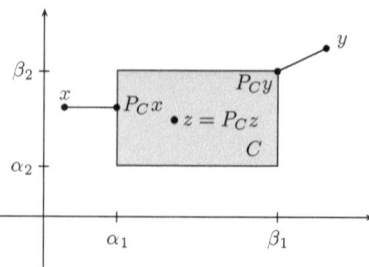

Fig. 4.8 Metric projection
onto $C = \{f \in L_2 : g \leq f \leq h\}$

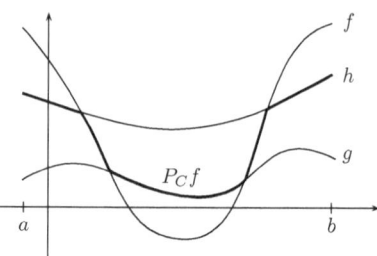

It follows from the above equality and and from Lemma 1.2.8 that

$$P_C x = \max\{\min\{x, b\}, a\}$$

(see Fig. 4.7). This formula can also be used when $\alpha_j = -\infty$ or $\beta_j = +\infty$ for some j, $j = 1, 2, \ldots, n$.

We obtain a similar formula in $\mathcal{H} = L_2 := L_2([a, b])$ for the metric projection of $f \in L_2$ onto the subset $C := \{f \in L_2 : g \leq f \leq h\}$, where $g, h \in L_2$ and $g \leq h$. Obviously, C is closed and convex. It follows from the characterization of the metric projection (see Theorem 1.2.4) that

$$P_C f = \max\{\min\{f, h\}, g\},$$

i.e.,

$$(P_C f)(x) = \begin{cases} g(x) & \text{if } f(x) < g(x) \\ f(x) & \text{if } g(x) \leq f(x) \leq h(x) \\ h(x) & \text{if } f(x) > h(x) \end{cases}$$

(see Fig. 4.8).

The formula above can be extended to the case $C = \{f \in L_2 : g \leq f \leq h$ on $S\}$, where $S \subseteq [a, b]$ is a measurable subset and $g \leq h$ on S. We leave it to the reader to check that

$$(P_C f)(x) = \begin{cases} g(x) & \text{if } f(x) < g(x) \text{ and } x \in S \\ f(x) & \text{if } g(x) \leq f(x) \leq h(x) \text{ or } x \notin S \\ h(x) & \text{if } f(x) > h(x) \text{ and } x \in S. \end{cases}$$

Fig. 4.9 Metric projection
onto a ball

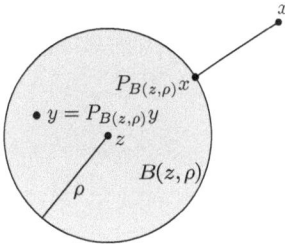

If we take $g = h$ on S, then $C = \{f \in L_2 : f(x) = g(x) \text{ for } x \in S\}$ and we obtain, in particular

$$(P_C f)(x) = \begin{cases} g(x) \text{ if } x \in S \\ f(x) \text{ if } x \notin S. \end{cases}$$

The subset C defined by (4.10) is a special case of a polytope. The metric projection onto the latter subset can be evaluated by several methods, e.g., by a finite procedure called an active set method (see [339] for details).

4.1.7 Metric Projection onto a Ball

Obviously, a ball $B(z, \rho) \subseteq \mathcal{H}$, where $z \in \mathcal{H}$ and $\rho > 0$, is a nonempty closed convex subset. It follows easily from the characterization of the metric projection (see Theorem 1.2.4) and from the Cauchy–Schwarz inequality that

$$P_{B(z,\rho)}(x) = \begin{cases} x & \text{if } \|x - z\| \le \rho \\ z + \frac{\rho}{\|x-z\|}(x - z) & \text{if } \|x - z\| > \rho \end{cases}$$

(see Fig. 4.9). The details are left to the reader.

4.1.8 Metric Projection onto an Ellipsoid

An *ellipsoid* in $\mathbb{R}^n$ has the form

$$C = J(D, z, \rho) := \{y \in \mathbb{R}^n : (y - z)^\top D(y - z) \le \rho\},$$

where D is a positive definite matrix, $z \in \mathbb{R}^n$ and $\rho > 0$. An ellipsoid is a closed convex subset as a sublevel set of a convex function

$$f(x) = \frac{1}{2}(x - z)^\top D(x - z)$$

Fig. 4.10 Metric projection
onto an ellipsoid

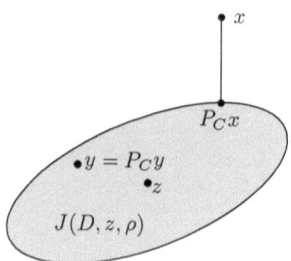

(note that the Hessian $\nabla^2 f = D$ is positive definite). We present a method for calculating the metric projection of a point $x \in \mathbb{R}^n$ onto the ellipsoid C (cf. [316, Sect. 3.4] and [228]). A different method was also presented in [128].

It follows from the definition of the metric projection that $y = P_C x$ if and only if y is a solution of the following convex minimization problem

$$
\begin{aligned}
&\text{minimize} \quad f(y) = \tfrac{1}{2} \|y - x\|^2 \\
&\text{subject to} \quad (y - z)^\top D(y - z) \le \rho \\
&\qquad\qquad y \in \mathbb{R}^n
\end{aligned} \tag{4.11}
$$

(see Fig. 4.10). This problem has a unique solution, because the objective is strongly convex. The same also follows from Theorem 1.2.3. All assumptions of the Karush–Kuhn–Tucker theorem are satisfied. The Slater constraints qualification for problem (4.11) is satisfied, because int $J(D, z, \rho) \neq \emptyset$. Therefore, $y \in \mathbb{R}^n$ is a solution of this problem if and only if there exists $\lambda \in \mathbb{R}_+$ such that (y, λ) is a KKT-point (see Theorems 1.3.6 and 1.3.7). Define the Lagrange function $L : \mathbb{R}^n \times \mathbb{R} \to \mathbb{R}$ by the equality

$$
L(y, \lambda) = \frac{1}{2} \|y - x\|^2 + \lambda[(y - z)^\top D(y - z) - \rho].
$$

The Karush–Kuhn–Tucker system has the form

$$
\begin{aligned}
y - x + 2\lambda D(y - z) &= 0 \\
(y - z)^\top D(y - z) &\le \rho \\
\lambda &\ge 0 \\
\lambda[(y - z)^\top D(y - z) - \rho] &= 0.
\end{aligned} \tag{4.12}
$$

If $(x - z)^\top D(x - z) \le \rho$, then $x \in J(D, z, \rho)$ and $(y, \lambda) = (x, 0)$ is the only KKT-point, i.e., $P_C x = x$. If $(x - z)^\top D(x - z) > \rho$, then $x \notin J(D, z, \rho)$, i.e., $y \neq x$. It follows from the first and from the third equality in (4.12) that $\lambda > 0$, consequently,

$$
(y - z)^\top D(y - z) - \rho = 0, \tag{4.13}
$$

by the complementary condition (fourth equality in (4.12)). The first equality in the KKT-system gives

$$y = (\mathrm{Id} + 2\lambda D)^{-1}(x + 2\lambda Dz)$$

(note that $\mathrm{Id} + 2\lambda D$ is positive definite, because D is positive definite). Setting it in equality (4.13), we obtain

$$[(x + 2\lambda Dz)^\top (\mathrm{Id} + 2\lambda D)^{-1} - z^\top] D[(\mathrm{Id} + 2\lambda D)^{-1}(x + 2\lambda Dz) - z] = \rho. \quad (4.14)$$

By the existence and uniqueness of the metric projection $P_C x$, equation (4.14) has a unique solution λ. Summarizing, we obtain

$$P_C x = \begin{cases} x & \text{if } (x - z)^\top D(x - z) \le \rho \\ (\mathrm{Id} + 2\lambda D)^{-1}(x + 2\lambda Dz) & \text{if } (x - z)^\top D(x - z) > \rho, \end{cases}$$

where $\lambda > 0$ is the unique solution of equality (4.14).

If $D = W = \mathrm{diag}\, w$ for $w \in \mathbb{R}^n_{++}$, $z = 0$ and $\rho = 1$, we have $C = J(W, 0, 1) = \{y \in \mathbb{R}^n : y^\top W y \le 1\}$ and (4.14) obtains the form

$$x^\top (\mathrm{Id} + 2\lambda W)^{-1} W (\mathrm{Id} + 2\lambda W)^{-1} x = 1. \quad (4.15)$$

Since

$$(\mathrm{Id} + 2\lambda W)^{-1} = \begin{bmatrix} (1 + 2\lambda\omega_1)^{-1} & 0 & \cdots & 0 \\ 0 & (1 + 2\lambda\omega_2)^{-1} & \cdots & 0 \\ \cdots & \cdots & \cdots & \cdots \\ 0 & 0 & \cdots & (1 + 2\lambda\omega_n)^{-1} \end{bmatrix},$$

(4.15) can be written in the form

$$\sum_{j=1}^n \frac{\omega_j \xi_j^2}{(1 + 2\lambda\omega_j)^2} = 1 \quad (4.16)$$

(note that this equation is considered only in the case $x \notin J(W, 0, 1)$, i.e., $\sum_{j=1}^n \omega_j \xi_j^2 > 1$). As mentioned before, this equation has a unique solution $\lambda > 0$, which can be computed, e.g., by the Newton method with the starting point $\lambda_0 = 0$. Note that the function $h : \mathbb{R}_+ \to \mathbb{R}$, $h(\lambda) = \sum_{j=1}^n \frac{\omega_j \xi_j^2}{(1 + 2\lambda\omega_j)^2}$ is convex ($h''(\lambda) > 0$ for all $\lambda \ge 0$), $h(0) > 1$ and $\lim_{\lambda \to +\infty} h(\lambda) = 0$. Therefore, the Newton method generates an increasing sequence λ_k which converges to a unique solution λ^* of (4.16). The details can be found in [316, Theorem 3.4-2].

Fig. 4.11 Ice-cream cone
and its polar

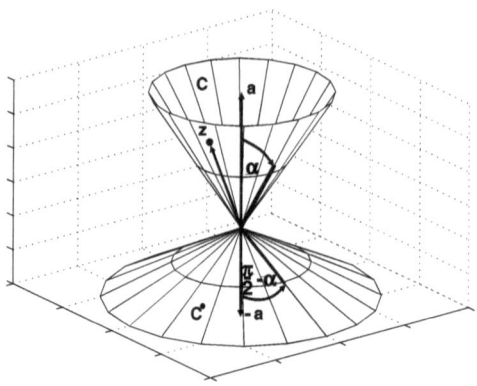

4.1.9 Metric Projection onto an Ice Cream Cone

Let $a \in \mathcal{H}$ $a \neq 0$ and $\alpha \in [0, \frac{\pi}{2}]$. A subset

$$C(a, \alpha) := \{z \in \mathcal{H} : \sphericalangle(z, a) \leq \alpha\} \cup \{0\}$$

is called an *ice cream cone* with axis a and angle α. Without loss of generality we
suppose that $\|a\| = 1$. Then the ice cream cone has the form

$$C(a, \alpha) = \{z \in \mathcal{H} : \langle a, z \rangle \geq \gamma \|z\|\},$$

where $\gamma = \cos \alpha \in [0, 1]$. This subset is closed and convex as a sublevel set $S(f, 0)$
of a continuous convex function $f : \mathcal{H} \to \mathbb{R}$, $f(x) := \gamma \|x\| - \langle a, x \rangle$. It is clear
that $C(a, \alpha)$ is a cone. The polar cone to an ice cream cone is again an ice cream
cone and has the form

$$(C(a, \alpha))^* = C(-a, \frac{\pi}{2} - \alpha) \tag{4.17}$$

(see, e.g., [186, Sect. 3.1] or [187]) (Fig. 4.11).

We will give a formula for $P_{C(a,\alpha)}x$, where $x \in \mathcal{H}$ and $\alpha \in (0, \frac{\pi}{2})$. If $x \in$
$C(a, \alpha)$, then, obviously, $P_{C(a,\alpha)}x = x$. Let $x \in C(-a, \frac{\pi}{2} - \alpha)$. Then it follows
from equality (4.17) and from the characterization of the metric projection (see
Theorem 1.2.4) that $P_{C(a,\alpha)}x = 0$. Now suppose that $x \in C(a, \alpha)' \cap C(-a, \frac{\pi}{2} - \alpha)'$.
Let $u := \lambda a$ with $\lambda \in \mathbb{R}$, be such that

$$\frac{\langle a, u - x \rangle}{\|u - x\|} = \sin \alpha \tag{4.18}$$

Fig. 4.12 Metric projection
onto an ice cream cone

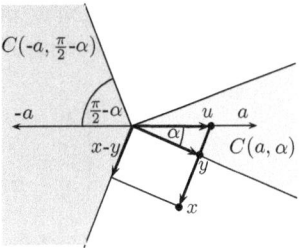

or, equivalently,

$$\frac{\sqrt{\|x\|^2 - \langle a, x \rangle^2}}{\|u - x\|} = \cos \alpha. \tag{4.19}$$

Obviously, $x \neq u$, because $\alpha > 0$, and $\|x\|^2 - \langle a, x \rangle^2 \geq 0$, by the Cauchy–Schwarz inequality. Let $y := x + \mu(u - x)$ with $\mu \in \mathbb{R}$, be such that $\langle y, x - u \rangle = 0$ (see Fig. 4.12).

A simple calculation shows that

$$\lambda = \langle a, x \rangle + \sqrt{\|x\|^2 - \langle a, x \rangle^2} \tan \alpha \tag{4.20}$$

and

$$\mu = \frac{\langle x, x - u \rangle}{\|u - x\|^2} = \frac{\|x\|^2 - \lambda \langle a, x \rangle}{\|\lambda a - x\|^2}. \tag{4.21}$$

Consequently,

$$\langle y, a \rangle = \langle x + \mu(u - x), a \rangle = \frac{\lambda(\|x\|^2 - \langle a, x \rangle^2)}{\|\lambda a - x\|^2} \tag{4.22}$$

and

$$\|y\| = \|x + \mu(u - x)\| = \frac{\lambda \sqrt{\|x\|^2 - \langle a, x \rangle^2}}{\|\lambda a - x\|}. \tag{4.23}$$

By (4.19), (4.22) and (4.23), we obtain $\langle y, a \rangle = \|y\| \cos \alpha$, i.e., $y \in C(a, \alpha)$. Furthermore, (4.17) and (4.18) yield $x - y = \mu(x - u) \in C^*(a, \alpha)$, and we have

$$\langle x - y, y \rangle = \mu \langle x - u, y \rangle = 0,$$

consequently,

$$\langle x - y, z - y \rangle = \langle x - y, z \rangle \leq 0$$

for all $z \in C(a, \alpha)$. Now, the characterization of the metric projection yields that $y = P_{C(a,\alpha)}x$. Subsuming all cases, we obtain

$$P_{C(a,\alpha)}x = \begin{cases} x & \text{if } \langle a, x \rangle \geq \|x\| \cos \alpha \\ 0 & \text{if } \langle a, x \rangle \leq -\|x\| \sin \alpha \\ x + \mu(\lambda a - x) & \text{if } -\|x\| \sin \alpha < \langle a, x \rangle < \|x\| \cos \alpha, \end{cases}$$

where λ and μ are given by (4.20) and (4.21). Equivalent formulas can also be found in [17, Theorem 3.3.6] or in [316, Sect. 3.5].

4.2 Cutters

Recall that an operator $T : \mathcal{H} \to \mathcal{H}$ with a fixed point is a cutter if

$$\langle Tx - x, Tx - z \rangle \leq 0$$

for all $x \in \mathcal{H}$ and for all $z \in \text{Fix } T$ (see Definition 2.1.30).

4.2.1 Characterization of Cutters

It follows from Lemma 2.1.36 that, for T being a cutter, Fix T is closed and convex, consequently, Fix T is an intersection of half-spaces, i.e., Fix $T = \bigcap_{i \in J} H_-(a_i, \beta_i)$, where $a_i \in \mathcal{H}$, $a_i \neq 0$, $\beta_i \in \mathbb{R}$, $i \in J$. This fact is applied in the theorem below which gives a necessary and sufficient condition for T to be a cutter. Denote $V := \text{Lin}\{a_i, i \in J\}$.

Theorem 4.2.1. *Let* $T : \mathcal{H} \to \mathcal{H}$. *For any* $y \in \mathcal{H}$ *the following statements are equivalent:*

(i) T *is a cutter.*
(ii) *For all* $x \in \mathcal{H}$ *and for all* $w \in (V + y) \cap \text{Fix } T$, *it holds*

$$\langle Tx - x, Tx - w \rangle \leq 0 \tag{4.24}$$

and

$$Tx - x \in V. \tag{4.25}$$

Proof. Let $z \in \text{Fix } T$, $v^\perp \in V^\perp$ and $\alpha \in \mathbb{R}$. We prove that $z + \alpha v^\perp \in \text{Fix } T$. Note that $\langle a_i, v^\perp \rangle = 0$ and $\langle a_i, z \rangle \leq \beta_i$, $i \in J$, Therefore, we have

$$\langle a_i, z + \alpha v^\perp \rangle = \langle a_i, z \rangle + \alpha \langle a_i, v^\perp \rangle \leq \beta_i$$

for all $i \in J$, i.e., $z + \alpha v^\perp \in \bigcap_{i \in J} H_-(a_i, \beta_i) = \text{Fix } T$.

Fig. 4.13 Cutter T with
affine Fix T

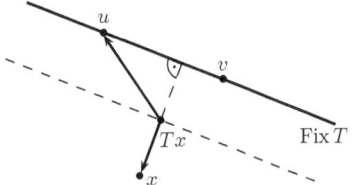

(ii)$\Rightarrow$(i) Let $x, y \in \mathcal{H}$ and $z \in$ Fix T. Suppose that conditions (4.24) and (4.25) are satisfied for all $w \in (V + y) \cap$ Fix T. Let $\bar{y}, \bar{z} \in V$ and $y^{\perp}, z^{\perp} \in V^{\perp}$ be such that $y = \bar{y} + y^{\perp}$ and $z = \bar{z} + z^{\perp}$. It follows from the first part of the proof that $\bar{z} \in$ Fix T and that $\bar{z} + y^{\perp} \in$ Fix T. It is clear that $V + y = V + y^{\perp}$, consequently, $\bar{z} + y^{\perp} \in V + y$. By (4.24) and (4.25), we have

$$\langle Tx - x, Tx - z \rangle = \langle Tx - x, Tx - (\bar{z} + y^{\perp}) \rangle - \langle Tx - x, z^{\perp} - y^{\perp} \rangle \leq 0$$

and T is a cutter.

(i)$\Rightarrow$(ii) Suppose that T is a cutter. Let $x \in \mathcal{H}$ and $w \in (V + y) \cap$ Fix T. Inequality (4.24) is obvious. Suppose that (4.25) does not hold. Then $Tx - x = \bar{v} + v^{\perp}$ for some $\bar{v} \in V$ and $v^{\perp} \in V^{\perp}$, $v^{\perp} \neq 0$. It follows from the first part of the proof that $w + \alpha v^{\perp} \in$ Fix T for any $\alpha \in \mathbb{R}$. For $\alpha < \langle Tx - x, Tx - w \rangle / \|v^{\perp}\|^2$ we obtain

$$0 \geq \langle Tx - x, Tx - (w + \alpha v^{\perp}) \rangle = \langle Tx - x, Tx - w \rangle - \alpha \langle \bar{v} + v^{\perp}, v^{\perp} \rangle$$
$$= \langle Tx - x, Tx - w \rangle - \alpha \|v^{\perp}\|^2 > 0,$$

a contradiction which shows that $Tx - x \in V$. $\square$

Corollary 4.2.2. *Let* $T : \mathcal{H} \to \mathcal{H}$ *be a cutter. If* Fix T *is a polyhedral subset, i.e.,* Fix $T = \bigcap_{i \in I} H_-(a_i, \beta_i)$, *where* $a_i \in \mathcal{H}$, $a_i \neq 0$ *and* $\beta_i \in \mathbb{R}$, $i \in I :=$ $\{1, 2, \ldots, m\}$, *then* $Tx \in x + \mathrm{Lin}\{a_i : i \in I\}$.

4.2.2   Cutters with Subsets of Fixed Points Being Affine Subspaces

In this section we apply Theorem 4.2.1 to an operator $T : \mathcal{H} \to \mathcal{H}$ with the subset of fixed points being a nonempty and affine subspace (Fig. 4.13).

Corollary 4.2.3. *Let* $T : \mathcal{H} \to \mathcal{H}$ *be an operator with* Fix T *being a nonempty closed and affine subspace. The operator* T *is a cutter if and only if*

$$\langle Tx - x, Tx - u \rangle \leq 0 \tag{4.26}$$

for all $x \in \mathcal{H}$ and for some $u \in \text{Fix } T$ and

$$\langle Tx - x, v - u \rangle = 0 \tag{4.27}$$

for all $x \in \mathcal{H}$ and for all $v, u \in \text{Fix } T$.

Proof. Since $\text{Fix } T$ is a closed affine subspace, it has the form $\text{Fix } T = W + u$, where W is a closed linear subspace and $u \in \text{Fix } T$. Denote $V := W^{\perp}$. It is clear that $(V + u) \cap (W + u) = \{u\}$. The corollary follows now from Theorem 4.2.1. □

Corollary 4.2.3 can also be proved directly.

2nd Proof of Corollary 4.2.3. The sufficiency of the conditions and the necessity of (4.26) is obvious. Suppose that T is a cutter and that

$$\langle Tx - x, v - u \rangle \neq 0 \tag{4.28}$$

for some $x \in \mathcal{H}$ and for some $v, u \in \text{Fix } T$. By the symmetry of (4.28) with respect to u and v, it suffices to consider only the case $\langle Tx - x, v - u \rangle > 0$. Let $w = u + t(v - u)$ for $t \in \mathbb{R}$. Of course, $w \in \text{Fix } T$, because $\text{Fix } T$ is an affine subspace. For

$$t < \frac{\langle Tx - x, Tx - u \rangle}{\langle Tx - x, v - u \rangle}$$

it holds

$$\langle Tx - x, Tx - w \rangle > 0.$$

We have obtained a contradiction with the assumption that T is a cutter, which yields the necessity of (4.27). □

4.2.3 Subgradient Projection

Definition 4.2.4. Let $f : \mathcal{H} \to \mathbb{R}$ be a continuous convex function. Let $g_f(x) \in \partial f(x)$ be a subgradient of f at x, $x \in \mathcal{H}$ (note that f is subdifferentiable, i e., for any $x \in \mathcal{H}$, there exists $g_f(x)$). Let $\alpha \in \mathbb{R}$. The operator $P_{f,\alpha} : \mathcal{H} \to \mathcal{H}$ defined by

$$P_{f,\alpha}x := \begin{cases} x - \frac{(f(x)-\alpha)_+}{\|g_f(x)\|^2} g_f(x) & \text{if } g_f(x) \neq 0 \\ x & \text{if } g_f(x) = 0 \end{cases} \tag{4.29}$$

is called a *subgradient projection relative to f by a level α*. The operator $P_f := P_{f,0}$ is called a *subgradient projection relative to f* or, shortly, a *subgradient projection*.

If $g_f(x) \neq 0$, then

$$P_{f,\alpha}x = P_{S(\bar{f}_x, \alpha)}(x), \tag{4.30}$$

Fig. 4.14 Subgradient projection

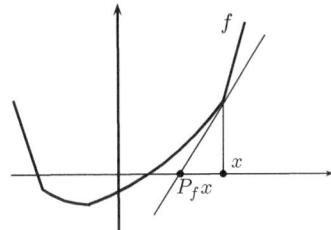

Fig. 4.15 $P_{f,\alpha}x = P_{S(\bar{f}_x,\alpha)}x$

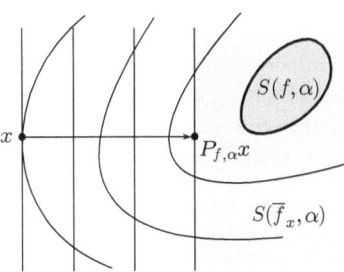

where $\bar{f}_x = \langle g_f(x), \cdot - x \rangle + f(x)$ denotes a linearization of f at x. Indeed, for $g_f(x) \neq 0$ the sublevel set $S(\bar{f}_x, \alpha)$ is a half-space

$$S(\bar{f}_x, \alpha) = \{y \in \mathcal{H} : \langle g_f(x), y - x \rangle + f(x) \leq \alpha\}$$

and, by (4.7),

$$P_{S(\bar{f}_x,\alpha)}x = x - \frac{(f(x) - \alpha)_+}{\|g_f(x)\|^2} g_f(x) = P_{f,\alpha}x.$$

Equality (4.30) is illustrated in Figs. 4.14 and 4.15.

If $\alpha \geq \inf_{x \in \mathcal{H}} f(x)$ it holds that $\mathrm{Argmin}_{x \in \mathcal{H}} f(x) \subseteq S(f, \alpha)$ and we can write

$$P_{f,\alpha}x = \begin{cases} x - \frac{f(x)-\alpha}{\|g_f(x)\|^2} g_f(x) & \text{if } x \notin S(f, \alpha) \\ x & \text{if } x \in S(f, \alpha), \end{cases} \tag{4.31}$$

because for $x \notin S(f, \alpha)$ we have $(f(x) - \alpha)_+ = f(x) - \alpha$ and $g_f(x) \neq 0$ (see Theorem 1.3.2), and for $x \in S(f, \alpha)$ we have $(f(x) - \alpha)_+ = 0$. Obviously, $P_{f,\alpha}x$ depends on the choice of $g_f(x) \in \partial f(x)$, $x \in \mathcal{H}$. Nevertheless, the properties of the subgradient projection presented below hold for any choice of $g_f(x) \in \partial f(x)$, $x \in \mathcal{H}$. Note that $P_{f,\alpha}$ is well defined even if $S(f, \alpha) = \emptyset$, i.e., if $\alpha < \inf_{x \in \mathcal{H}} f(x)$ or if $\alpha = \inf_{x \in \mathcal{H}} f(x)$ and the function f does not attain its minimum.

Lemma 4.2.5. *Let $P_{f,\alpha} : \mathcal{H} \to \mathcal{H}$ be a subgradient projection relative to a convex subdifferentiable function f by a level $\alpha \in \mathbb{R}$. If $S(f, \alpha) \neq \emptyset$, then*

$$\text{Fix } P_{f,\alpha} = S(f,\alpha).$$

Proof. The inclusion $S(f,\alpha) \subseteq \text{Fix } P_{f,\alpha}$ is obvious. Suppose that $x \notin S(f,\alpha) \neq \emptyset$. Then $f(x) > \alpha \geq \inf_{x\in\mathcal{H}} f(x)$, consequently, $g_f(x) \neq 0$ (see Theorem 1.3.2) and

$$\frac{f(x) - \alpha}{\left\|g_f(x)\right\|^2} g_f(x) \neq 0,$$

i.e., $x \notin \text{Fix } P_{f,\alpha}$. Therefore, $\text{Fix } P_{f,\alpha} = S(f,\alpha)$.   □

Let $\alpha \in \mathbb{R}$. It is clear that $S(f,\alpha) = S(f - \alpha, 0)$. Furthermore, f is a convex and subdifferentiable function if and only if $f - \alpha$ is convex and subdifferentiable. Moreover, for any $x \in \mathcal{H}$ we have $\partial(f - \alpha)(x) = \partial f(x)$. Therefore, we can restrict our further analysis of $P_{f,\alpha}$ to the case $\alpha = 0$.

Corollary 4.2.6. *Let f be convex and subdifferentiable and $S(f,0) \neq \emptyset$. A subgradient projection P_f is a cutter.*

Proof. By Lemma 4.2.5, we have $\text{Fix } P_f = S(f,0)$. Let $x \notin S(f,0)$ and $z \in S(f,0)$, i.e., $f(z) \leq 0 < f(x)$. By the definition of a subgradient of a convex function, we have

$$\langle P_f x - x, P_f x - z \rangle = \left\| P_f x - x \right\|^2 + \langle P_f x - x, x - z \rangle$$

$$= \left(\frac{f(x)}{\left\|g_f(x)\right\|} \right)^2 - \frac{f(x)}{\left\|g_f(x)\right\|^2} \langle g_f(x), x - z \rangle$$

$$\leq \left(\frac{f(x)}{\left\|g_f(x)\right\|} \right)^2 - \frac{f(x)}{\left\|g_f(x)\right\|^2} (f(x) - f(z))$$

$$\leq 0$$

which completes the proof.   □

Theorem 4.2.7. *Let $f : \mathcal{H} \to \mathbb{R}$ be a convex function which is globally Lipschitz continuous on bounded subsets (this holds if, e.g., $\mathcal{H} = \mathbb{R}^n$). Then the operator $P_f - \text{Id}$ is demi-closed at 0.*

Proof. Let $x^k \rightharpoonup z \in \mathcal{H}$ and $\left\| P_f x^k - x^k \right\| \to 0$. Then we have

$$\frac{(f(x^k))_+}{\left\|g_f(x^k)\right\|} = \left\| P_f x^k - x^k \right\| \to 0. \tag{4.32}$$

Note that $\{x^k\}_{k=0}^{\infty}$ is bounded as a weakly convergent sequence. By the global Lipschitz continuity of f on bounded subsets, there exists a constant $\kappa > 0$ such that

$$f'(x^k, g_f(x^k)) \leq \kappa \left\| g_f(x^k) \right\|$$

(see Theorem 1.1.50). By Theorem 1.1.58,

$$\left\| g_f(x^k) \right\|^2 \le \sup_{g \in \partial f(x^k)} \langle g, g_f(x^k) \rangle = f'(x^k, g_f(x^k)) \le \kappa \left\| g_f(x^k) \right\|,$$

i.e., $\left\| g_f(x^k) \right\| \le \kappa$ and the sequence $\{g_f(x^k)\}_{k=0}^{\infty}$ is bounded. Now (4.32) yields $(f(x^k))_+ \to 0$. Recall that a continuous and convex function is weakly lower semi-continuous (see Theorem 1.1.51). Therefore,

$$f(z) \le \liminf_k f(x^k) \le \liminf_k (f(x^k))_+ = 0$$

and $P_f z = z$.   □

4.3 Alternating Projection

4.3.1 Basic Properties

Definition 4.3.1. Let $A, B \subseteq \mathcal{H}$ be a nonempty closed convex subset. We call an operator $T := P_A P_B : \mathcal{H} \to A$ an *alternating projection* (Fig. 4.16).

The alternating projection was introduced by John von Neumann [271], who studied the convergence of sequences generated by this operator for closed convex subspaces $A, B \subseteq \mathcal{H}$. The alternating projection or its modifications were studied by Aronszajn [11], Gurin et al. [196], Deutsch [138, 139, 141], Bauschke and Borwein [20, 21], Combettes [117], Bauschke et al. [30], Hundal [213], Kopecká and Reich [235, 236], Bauschke et al. [25], Scolnik et al. [306], Cegielski and Dylewski [74] and by Cegielski and Suchocka [75, 76].

First we give properties of an alternating projection $T = P_A P_B$ in the case $A \cap B \ne \emptyset$. It follows from Theorem 2.1.26 (ii) that in this case we have Fix $T = A \cap B$, because the metric projection is strictly quasi-nonexpansive (see Corollary 2.2.24).

Lemma 4.3.2. *Let $A, B \subseteq \mathcal{H}$ be nonempty closed convex subsets and $A \cap B \ne \emptyset$. Then for all $x \in A$ and $z \in A \cap B$ the following inequalities hold*

$$\langle z - Tx, x - Tx \rangle \le \langle P_B x - Tx, x - P_B x \rangle \le -\|Tx - P_B x\|^2,$$

where $T := P_A P_B$ is an alternating projection. Consequently, the operator $T \mid_A$ is a cutter and its projected relaxation $P_A T_\lambda$ is asymptotically regular for all $\lambda \in (0, 2)$.

Proof. Let $x \in A$ and $z \in \text{Fix } T = A \cap B$. It follows from the characterization of the metric projections $P_A(P_B x)$ and $P_B x$ and from the equivalence (a)⇔(b) in Lemma 1.2.5 that

$$\langle z - Tx, x - Tx \rangle = \langle z - Tx, x - P_B x \rangle + \langle z - Tx, P_B x - Tx \rangle$$

$$\le \langle z - Tx, x - P_B x \rangle$$

Fig. 4.16 Alternating projection

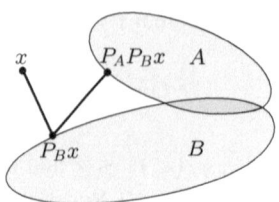

$$= \langle z - P_B x, x - P_B x \rangle + \langle P_B x - T x, x - P_B x \rangle$$
$$\leq \langle P_B x - T x, x - P_B x \rangle$$
$$\leq - \| T x - P_B x \|^2 \leq 0,$$

i.e., the operator $T \mid_A$ is a cutter. The asymptotic regularity of of the projected relaxation $P_A T_\lambda$ for $\lambda \in (0, 2)$ follows now from Corollary 3.4.4.   □

If $A \cap B \neq \emptyset$, then Lemma 4.3.2 and Theorem 2.1.39 imply the strong quasi nonexpansivity of the relaxation T_λ of the alternating projection $T := P_A P_B$ for any $\lambda \in (0, 2)$. Note that Lemma 4.3.2 is not true without the assumption $A \cap B \neq \emptyset$. In the case $A \cap B = \emptyset$ the asymptotic regularity of the projected relaxation $P_A T_\lambda$ holds only for a narrower range of the relaxation parameter λ.

Corollary 4.3.3. *Let $A, B \subseteq \mathcal{H}$ be nonempty closed convex subsets. Then the alternating projection $T := P_A P_B$ is $\frac{4}{3}$-relaxed firmly nonexpansive. Furthermore, for any $\mu \in [0, \frac{3}{4}]$, the relaxation T_μ of T is firmly nonexpansive. Consequently, for any $\lambda \in (0, \frac{3}{2})$, the operator T_λ is relaxed firmly nonexpansive and T_λ is asymptotically regular whenever* Fix $T \neq \emptyset$.

Proof. By Corollary 2.2.39, the operator T is $\frac{4}{3}$-relaxed firmly nonexpansive as a composition of FNE operators P_A and P_B. Moreover, by Corollary 2.2.19, its relaxation T_μ is firmly nonexpansive for any $\mu \in [0, \frac{3}{4}]$. Suppose that Fix $T \neq \emptyset$. Then Corollary 3.4.6 yields the asymptotic regularity of the operator T_λ for any $\lambda \in (0, \frac{3}{2})$.   □

If we take $\lambda = 1$ in Corollary 4.3.3, then we obtain that the alternating projection $T := P_A P_B$ is relaxed firmly nonexpansive and asymptotically regular, whenever Fix $T \neq \emptyset$.

4.3.2 Fixed Points of the Alternating Projection

Recall that the distance function $d(\cdot, C) : \mathcal{H} \to \mathbb{R}$, where $C \subseteq \mathcal{H}$, is defined by $d(x, C) := \inf_{z \in C} \| x - z \|$. Denote $d(A, B) := \inf_{x \in A, y \in B} \| x - y \|$, where A, $B \subseteq \mathcal{H}$.

If $A \cap B \neq \emptyset$, then Fix $P_A P_B = A \cap B$ (see Theorem 2.1.26 (ii)). In a general case, the alternating projection does not need to have fixed points (e.g., for $A := \{(x, y) \in \mathbb{R}^2 : y = 0\}$ and $B := \{(x, y) \in \mathbb{R}^2 : y \geq e^x\}$ we have Fix $P_A P_B = \emptyset$). The theorem below gives relationships between fixed points of operators $P_A P_B$ and $P_B P_A$ and minimizers of the distance functions $d(\cdot, B) : A \to \mathbb{R}$ and $d(\cdot, A) : B \to \mathbb{R}$. The theorem presents a part of the results of [21, Lemma 2.2].

Theorem 4.3.4. *Let $A, B \subseteq \mathcal{H}$ be nonempty closed convex subsets, $x^* \in A$ and $y^* \in B$. The following conditions are equivalent:*

(i) $x^* \in \mathrm{Fix}\, P_A P_B$ and $y^* = P_B x^*$,
(ii) $y^* \in \mathrm{Fix}\, P_B P_A$ and $x^* = P_A y^*$,
(iii) $\|x^* - y^*\| = d(x^*, B) \leq d(x, B)$ for all $x \in A$,
(iv) $\|x^* - y^*\| = d(y^*, A) \leq d(y, A)$ for all $y \in B$.
(v) $\|x^* - y^*\| = d(A, B)$.

Proof. The inequality in condition (iii) denotes that the function $f := \frac{1}{2} d^2(\cdot, B) : A \to \mathbb{R}$ attains its minimum at $x^* \in A$. The function f is convex (see Corollary 2.2.29) and differentiable and $Df = \mathrm{Id} - P_B$ (see Lemma 2.2.27). Now, it follows from the necessary and sufficient optimality condition for the convex differentiable minimization (see Theorem 1.3.4) and from Lemma 1.2.9 that

$$
\begin{aligned}
x^* &\in \operatorname*{Argmin}_{x \in A} f(x) \\
&\Longleftrightarrow P_B x^* - x^* = -Df(x^*) \in N_A(x^*) \\
&\Longleftrightarrow x^* = P_A P_B x^* \\
&\Longleftrightarrow x^* \in \mathrm{Fix}\, P_A P_B.
\end{aligned}
\tag{4.33}
$$

(i)$\Leftrightarrow$(ii) Let $P_A P_B x^* = x^*$ and $y^* = P_B x^*$. Then $P_A y^* = P_A P_B x^* = x^*$ and, consequently, $P_B P_A y^* = P_B x^* = y^*$. The converse implication can be shown in a similar way.

(i)$\Leftrightarrow$(iii) The equivalence of the first part of (i) and of the inequality in (iii) was already proved in (4.33). If $y^* = P_B x^*$, then the equality in (iii) follows from the definition of the metric projection P_B. The converse implication follows from the uniqueness of the metric projection. In a similar way, one can show the equivalence (ii)$\Leftrightarrow$(iv).

(i)$\Leftrightarrow$(v) The function $h : A \times B \to \mathbb{R}$ defined by $h(x, y) := \frac{1}{2}\|x - y\|^2$ is convex as a composition of a linear function $(x, y) \mapsto x - y$ and of a convex function $u \mapsto \frac{1}{2}\|u\|^2$. Moreover, h is differentiable and

$$
Dh(x, y) = (x - y, y - x)
$$

for all $x \in A$ and $y \in B$. We leave it to the reader to check that $N_{A \times B}(x, y) = N_A(x) \times N_B(y)$. Now, using similar arguments as in the first part of the proof, we

Fig. 4.17 $\mathrm{Fix}(P_A P_B)$,
$\mathrm{Fix}(P_B P_A)$ and minimizers
of $d(\cdot, A)$ and of $d(\cdot, B)$

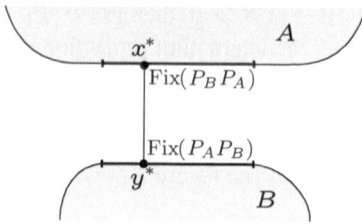

obtain

$$(x^*, y^*) \in \underset{x \in A, y \in B}{\mathrm{Argmin}}\, h(x, y)$$

$$\Longleftrightarrow (y^* - x^*, x^* - y^*) = -Dh(x^*, y^*) \in N_{A \times B}(x^*, y^*)$$

$$\Longleftrightarrow y^* - x^* \in N_A(x^*) \text{ and } x^* - y^* \in N_B(y^*)$$

$$\Longleftrightarrow x^* = P_A y^* \text{ and } y^* = P_B x^*$$

$$\Longleftrightarrow x^* = P_A P_B x^* \text{ and } y^* = P_B x^*,$$

i.e., conditions (i) and (v) are equivalent.   □

Theorem 4.3.4 is illustrated in Fig. 4.17.
Part (ii) of the following corollary was proved by Cheney and Goldstein (see [112, Theorem 2]).

Corollary 4.3.5. *Let $A, B \subseteq \mathcal{H}$ be nonempty closed convex subsets. Then:*

(i) $\mathrm{Fix}\, P_A P_B \neq \emptyset \Longleftrightarrow \mathrm{Fix}\, P_B P_A \neq \emptyset$,
(ii) $\mathrm{Fix}\, P_A P_B = \mathrm{Argmin}_{x \in A} d(x, B)$,
(iii) $\mathrm{Fix}\, P_B P_A = \mathrm{Argmin}_{y \in B} d(y, B)$,
(iv) $\mathrm{Fix}\, P_A P_B \times \mathrm{Fix}\, P_B P_A = \mathrm{Argmin}_{(x,y) \in A \times B} \|x - y\|$,
(v) If $A \cap B \neq \emptyset$, then $\mathrm{Fix}\, P_A P_B = A \cap B$.

The corollary below gives a sufficient condition for the nonemptiness of the subset $\mathrm{Fix}\, P_A P_B$.

Corollary 4.3.6. *Let $A, B \subseteq \mathcal{H}$ be nonempty closed convex subsets. If at least one of the subsets A and B is bounded, then $\mathrm{Fix}\, P_A P_B \neq \emptyset$.*

Proof. Suppose that A is bounded. It is clear that $P_A P_B : A \to A$ is nonexpansive. It follows from the Browder–Göhde–Kirk theorem that $\mathrm{Fix}\, P_A P_B \neq \emptyset$. Suppose now that B is bounded. Then, $\mathrm{Fix}\, P_B P_A \neq \emptyset$ and the corollary follows from Corollary 4.3.5.   □

If we suppose that $\mathcal{H} = \mathbb{R}^n$ and $A, B \subseteq \mathbb{R}^n$ are polytopes, then Corollary 4.3.6 holds without any assumption on the boundedness of A or B (see [112, Theorem 5]).

4.3.3 Alternating Projection for a Closed Affine Subspace

The alternating projection $P_A P_B$ does not need not to be firmly nonexpansive even if A an B are closed subspaces of $\mathcal{H}$ (see Exercise 2.5.9). Nevertheless, if we suppose that A is a closed affine subspace and if we restrict the operator $P_A P_B$ to the subset A, then $P_A P_B$ is firmly nonexpansive. This fact is formulated in the following theorem which is an extension of an important result of Combettes [117, Proposition 3].

Theorem 4.3.7. *Let $A \subseteq \mathcal{H}$ be a closed affine subspace and $B \subseteq \mathcal{H}$ be a closed convex subset. Then, for all $x, y \in A$,*

$$\langle Tx - Ty, x - y \rangle \geq \|Tx - Ty\|^2 + \|(Tx - P_B x) - (Ty - P_B y)\|^2 \quad (4.34)$$

and

$$\langle Tx - Ty, x - y \rangle \geq \|Tx - Ty\|^2 + (\|Tx - P_B x\| - \|Ty - P_B y\|)^2. \quad (4.35)$$

Consequently, the alternating projection

$$T := P_A P_B : A \to A,$$

is firmly nonexpansive.

Proof. Let $a \in A$. Applying the property

$$\langle P_A u - P_A v, w \rangle = \langle u - v, P_{A-a} w \rangle$$

for $u, v, w \in \mathcal{H}$ (see Theorem 2.2.33 (iii)), the fact $x - y \in A - a$ for $x, y \in A$, the firm nonexpansivity of P_B and Corollary 2.2.24, we obtain

$$
\begin{aligned}
& \langle Tx - Ty, x - y \rangle \\
&= \langle P_A P_B x - P_A P_B y, x - y \rangle \\
&= \langle P_B x - P_B y, P_{A-a}(x - y) \rangle \\
&= \langle P_B x - P_B y, x - y \rangle \\
&\geq \|P_B x - P_B y\|^2 \\
&\geq \|P_A P_B x - P_A P_B y\|^2 + \|(P_A P_B x - P_B x) - (P_A P_B y - P_B y)\|^2 \\
&= \|Tx - Ty\|^2 + \|(Tx - P_B x) - (Ty - P_B y)\|^2 \\
&\geq \|Tx - Ty\|^2 + (\|Tx - P_B x\| - \|Ty - P_B y\|)^2 \\
&\geq \|Tx - Ty\|^2
\end{aligned}
$$

for all $x, y \in A$. Consequently, T is firmly nonexpansive. $\qquad\square$

Corollary 4.3.8. *Let $A \subseteq \mathcal{H}$ be a closed affine subspace and $B \subseteq \mathcal{H}$ a closed convex subset. Further, let $T_\lambda : A \to A$ be a relaxation of the alternating projection $T := P_A P_B$, where $\lambda \in (0, 2)$. Then*

$$\|T_\lambda x - T_\lambda y\|^2 \le \|x - y\|^2 - \frac{2 - \lambda}{\lambda} \|(T_\lambda x - x) - (T_\lambda y - y)\|^2 \qquad (4.36)$$

for arbitrary $x, y \in A$. If, furthermore, $\mathrm{Fix}\, T \ne \emptyset$, then, for any $x \in A$ and $z \in \mathrm{Fix}\, T$ it holds

$$\|T_\lambda x - z\|^2 \le \|x - z\|^2 - \frac{2 - \lambda}{\lambda} \|T_\lambda x - x\|^2,$$

i.e., the operator T_λ is $\frac{2-\lambda}{\lambda}$-strongly quasi-nonexpansive and asymptotically regular.

Proof. By Theorem 4.3.7, the operator $T := P_A P_B$ is firmly nonexpansive. Therefore, the corollary follows from Corollary 2.2.15 and Theorem 3.4.3. □

4.3.4 Generalized Relaxation of the Alternating Projection

Let $A, B \subseteq \mathcal{H}$ be nonempty closed and convex and $T := P_A P_B$ be the alternating projection with a fixed point. Consider a generalized relaxation $T_{\sigma,\lambda} : A \to \mathcal{H}$ of T, given by the formula

$$T_{\sigma,\lambda}(x) = x + \lambda \sigma(x)(Tx - x), \qquad (4.37)$$

where the relaxation parameter $\lambda \in [0, 2]$ and $\sigma : A \to (0, +\infty)$ is a step size function. It is clear that if $\sigma(x) = 1$ for all $x \in A$, then $T_{\sigma,\lambda}$ coincides with the classical relaxation T_λ of the alternating projection $T := P_A P_B$. Corollaries 2.4.3 and 2.4.5 suggest that a step size function σ with values larger than 1, for which the operator T_σ is a cutter, leads to an acceleration (at least local) of the corresponding iterative procedures of the form

$$x^{k+1} = P_A T_{\sigma,\lambda_k}(x^k).$$

Therefore, the ability to construct such step sizes is of big interest. In the remaining part of this section we will present various step sizes for which T_σ is a cutter. Recall that $\mathrm{Fix}\, T_{\sigma,\lambda} = \mathrm{Fix}\, T$ for all $\lambda > 0$ (see Remark 2.4.2 (d)). Denote

$$\delta := d(A, B) = \inf_{x \in A, y \in B} \|x - y\|$$

and

$$\bar{\delta}(x) := \|Tx - P_B x\|,$$

where $x \in A$. If $A \cap B \neq \emptyset$, then, of course, $\delta = 0$. Suppose that we know an upper approximation $\tilde{\delta}(x) \in [\delta, \bar{\delta}(x)]$ of δ, for any $x \in A$. Recall that we still suppose that Fix $T \neq \emptyset$.

Lemma 4.3.9. *Let $x \in A$ be such that $Tx \notin$ Fix T. Then the vectors $x - P_B x$ and $Tx - P_B x$ are linearly independent.*

Proof. The assumption yields that $x - P_B x$ and $Tx - P_B x$ are nonzero vectors. Suppose that $x - P_B x$ and $Tx - P_B x$ are linearly dependent, i.e.,

$$Tx - P_B x = \gamma(x - P_B x) \tag{4.38}$$

for some γ. By the characterization of the metric projection $P_A(P_B x)$, we have

$$\|Tx - P_B x\|^2 = \gamma \langle x - P_B x, Tx - P_B x \rangle \geq \gamma \|Tx - P_B x\|^2 > 0. \tag{4.39}$$

Therefore, $\gamma > 0$. By Lemma 1.2.9, we have $x - P_B x \in N_B(P_B x)$, consequently, $\gamma(x - P_B x) \in N_B(P_B x)$ and, again by Lemma 1.2.9,

$$P_B(P_B x + \gamma(x - P_B x)) = P_B x. \tag{4.40}$$

Now, by (4.38) and (4.40), we obtain

$$T^2 x = P_A P_B Tx = P_A P_B(P_B x + \gamma(x - P_B x))$$
$$= P_A P_B P_B x = P_A P_B x = Tx,$$

a contradiction with the assumption $Tx \notin$ Fix T. Therefore, the vectors $x - P_B x$ and $Tx - P_B x$ are linearly independent. □

Define the step size function $\sigma : A \to (0, +\infty)$ by the formula

$$\sigma(x) := \begin{cases} \dfrac{\|Tx - P_B x\|^2 - \tilde{\delta}(x) \|P_B x - x\| + \langle P_B x - x, Tx - x \rangle}{\|Tx - x\|^2} & \text{if } x \notin \text{Fix } T, \\ 1 & \text{if } x \in \text{Fix } T, \end{cases} \tag{4.41}$$

where $\tilde{\delta}(x)$ is a fixed element of the segment $[\delta, \|Tx - P_B x\|]$ (see Fig. 4.18). This step size was proposed in [76, equality (13)].

Lemma 4.3.10. *Let $x \in A$ and the step size $\sigma(x)$ be defined by (4.41). Then*

$$\sigma(x) \geq \frac{1}{2} \tag{4.42}$$

and the inequality is strict if $Tx \notin$ Fix T.

Proof. If $x \in$ Fix T, then $\sigma(x) = 1$. Let now $x \notin$ Fix T. Denote $a := P_B x - x$, $b := Tx - x$, $c := P_B x - Tx$ (see Fig. 4.18) and $\tilde{\delta} := \tilde{\delta}(x)$.

Fig. 4.18 Step size $\sigma(x)$
given by (4.41) with
$\tilde{\delta}(x) := \|Tx - P_B x\|$

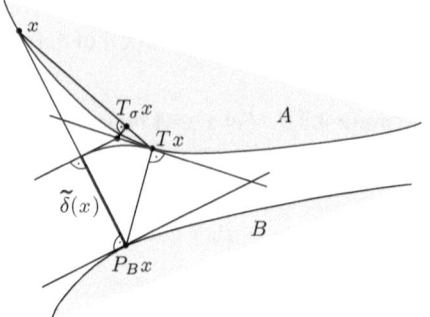

Since $b \neq 0$, $b = a - c$ and $\tilde{\delta} \leq \|c\|$, the Cauchy–Schwarz inequality yields

$$
\begin{aligned}
\sigma(x) &= \frac{\|c\|^2 - \tilde{\delta}\,\|a\| + \langle a, b \rangle}{\|b\|^2} \\
&\geq \frac{\|c\|^2 - \|a\| \cdot \|c\| + \langle a, b \rangle}{\|b\|^2} \\
&= \frac{\|c\|^2 - \|a\| \cdot \|c\| + \langle a, a - c \rangle}{\|a - c\|^2} \\
&= \frac{\|a\|^2 + \|c\|^2 - \|a\| \cdot \|c\| - \langle a, c \rangle}{\|a\|^2 + \|c\|^2 - 2\langle a, c \rangle} \\
&= \frac{(\|a\| - \|c\|)^2 + \|a\| \cdot \|c\| - \langle a, c \rangle}{(\|a\| - \|c\|)^2 + 2(\|a\| \cdot \|c\| - \langle a, c \rangle)}.
\end{aligned}
$$

Let $\beta \in \mathbb{R}$. Note that the function $f : (-2\beta, +\infty) \to \mathbb{R}$, $f(\alpha) := \frac{\alpha + \beta}{\alpha + 2\beta}$ is increasing. Therefore, if we take $\alpha := (\|a\| - \|c\|)^2$ and $\beta := \|a\| \cdot \|c\| - \langle a, c \rangle$ in the above inequalities, we obtain $\alpha \geq 0 \geq -2\beta$ and

$$
\sigma(x) \geq f(\alpha) \geq f(0) = \frac{1}{2},
$$

i.e., $\sigma(x) \geq \frac{1}{2}$. If $Tx \notin \operatorname{Fix} T$, then a and c are linearly independent by Lemma 4.3.9. In this case, $\beta = \|a\| \cdot \|c\| - \langle a, c \rangle > 0$ and $\sigma(x) > \frac{1}{2}$. $\square$

Inequality (4.42) can be strengthened. One can prove that

$$
\sigma(x) \geq \frac{1}{1 + \cos \alpha(x)} \geq \frac{1}{2},
$$

where $\alpha(x)$ denotes the angle between the nonzero vectors $x - P_B x$ and $Tx - P_B x$ (for details see [76, Lemmas 3 and 4]).

Fig. 4.19 Solution of
system (4.44)–(4.45) with
$\tilde{\delta}(x) := \delta$

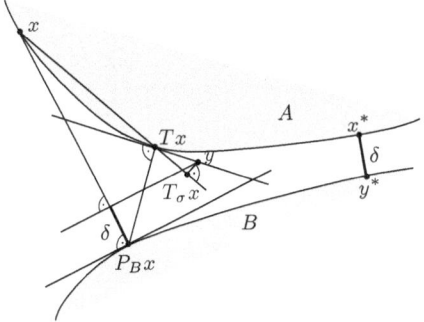

Let $x \in A$ be such that $Tx \notin \mathrm{Fix}\,T$. One can prove that the step size $\sigma(x)$ given
by (4.41) is characterized by the equality

$$\langle x + \sigma(x)(Tx - x) - y, Tx - x \rangle = 0, \tag{4.43}$$

where $y \in \mathrm{aff}(x, P_B x, Tx)$ is a unique solution of the system

$$\langle y - P_B x, x - P_B x \rangle = \tilde{\delta}(x) \| P_B x - x \| \tag{4.44}$$

$$\langle y - P_B x, Tx - P_B x \rangle = \| Tx - P_B x \|^2 \tag{4.45}$$

and $\tilde{\delta}(x) \in [\delta, \| Tx - P_B x \|]$ (see Fig. 4.19). By Lemma 4.3.9, such a solution is
defined uniquely. For details, see [76, Lemma 6].

Theorem 4.3.11. *Let $A, B \subseteq \mathcal{H}$ be nonempty closed convex subsets, $T := P_A P_B$,
$\mathrm{Fix}\,T \neq \emptyset$ and the step size function $\sigma : A \to (0, +\infty)$ be defined by (4.41). Then,
for any $x \notin \mathrm{Fix}\,T$ and $z \in \mathrm{Fix}\,T$,*

$$\sigma(x) \leq \frac{\langle z - x, Tx - x \rangle}{\| Tx - x \|^2} \tag{4.46}$$

*holds and the operator $T_\sigma : A \to \mathcal{H}$ is a cutter. Consequently, for any $\lambda \in (0, 2)$,
the operator $T_{\sigma,\lambda}$ is $\frac{2-\lambda}{\lambda}$-strongly quasi-nonexpansive and asymptotically regular.*

Proof. Let $z \in \mathrm{Fix}\,T$, $x \notin \mathrm{Fix}\,T$ and $\lambda \in (0, 2)$. It is clear that

$$\mathrm{Fix}\,T_{\sigma,\lambda} = \mathrm{Fix}\,T_\sigma = \mathrm{Fix}\,T.$$

Denote $w := P_B z$. Note that $\| z - w \| = \delta$ (see the implication (i)$\Rightarrow$(v) in
Theorem 4.3.4). By the characterization of the metric projection $P_B x$ and by the
Cauchy–Schwarz inequality, we have

$$\langle z - P_B x, P_B x - x \rangle = \langle z - w, P_B x - x \rangle + \langle w - P_B x, P_B x - x \rangle$$
$$\geq \langle z - w, P_B x - x \rangle$$
$$\geq - \|z - w\| \cdot \|P_B x - x\| = -\delta \|P_B x - x\|.$$

Therefore, if we apply again the characterization of the metric projection $P_A(P_B x)$, we obtain

$$\langle z - x, Tx - x \rangle = \langle z - P_B x, Tx - x \rangle + \langle P_B x - x, Tx - x \rangle$$
$$= \langle z - P_B x, Tx - P_B x \rangle + \langle z - P_B x, P_B x - x \rangle$$
$$+ \langle P_B x - x, Tx - x \rangle$$
$$\geq \|Tx - P_B x\|^2 - \delta \|P_B x - x\| + \langle P_B x - x, Tx - x \rangle.$$

Now, (4.46) follows from the inequality $\delta \leq \tilde{\delta}(x)$ and from the definition of the step size given by (4.41). We see that the step size $\sigma(x)$ satisfies the assumptions of Corollary 2.4.5. Therefore, the operator T_σ is a cutter and the operator $T_{\sigma,\lambda}$ is $\frac{2-\lambda}{\lambda}$-strongly quasi-nonexpansive. The asymptotic regularity of $T_{\sigma,\lambda}$ follows from Theorem 3.4.3.   □

If $A \cap B \neq \emptyset$, then, of course, $\delta = 0$. If we take $\tilde{\delta}(x) := \delta = 0$ in (4.41), then we obtain

$$\sigma(x) = \begin{cases} \dfrac{\|Tx - P_B x\|^2 + \langle P_B x - x, Tx - x \rangle}{\|Tx - x\|^2} & \text{if } x \notin A \cap B \\ 1 & \text{if } x \in A \cap B. \end{cases} \qquad (4.47)$$

We have the following result (cf. [76, Lemma 10]).

Lemma 4.3.12. *Let $A \cap B \neq \emptyset$ and $x \in A \backslash B$. Then*

$$\frac{\|Tx - P_B x\|^2 + \langle P_B x - x, Tx - x \rangle}{\|Tx - x\|^2} \geq \frac{\|P_B x - x\|^2}{\langle P_B x - x, Tx - x \rangle}, \qquad (4.48)$$

Furthermore, the equality holds in (4.48) if A is a closed affine subspace.

Proof. Denote $a := P_B x - x$, $b := Tx - x$ and $c := P_B x - Tx$. We have $a = b + c$ and $b \neq 0$, because $x \notin A \cap B = \text{Fix } T$. By the characterization of the metric projection $P_A(P_B x)$, we have

$$\langle a, b \rangle = \langle c, b \rangle + \|b\|^2 \geq \|b\|^2 > 0, \qquad (4.49)$$

i.e., both sides of inequality (4.48) are well defined. Inequality (4.48) can be written in the form

$$\frac{\|b - a\|^2 + \langle a, b \rangle}{\|b\|^2} \geq \frac{\|a\|^2}{\langle a, b \rangle}$$

Fig. 4.20 Step sizes $\sigma(x)$
given by (4.47) and by (4.52)

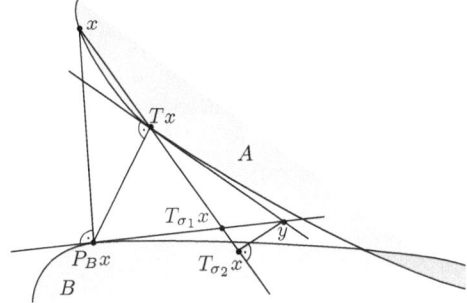

which is equivalent to

$$(\|a\|^2 - \langle a, b \rangle)(\langle a, b \rangle - \|b\|^2) \geq 0. \tag{4.50}$$

It follows from the characterization of the metric projection $P_A(P_B x)$ that

$$\langle b, a - b \rangle = \langle b, c \rangle \geq 0. \tag{4.51}$$

Therefore, the inequalities in (4.49), the Cauchy–Schwarz inequality and the nonexpansivity of the metric projection P_A yield

$$0 < \|b\|^2 \leq \langle a, b \rangle \leq \|a\| \cdot \|b\| \leq \|a\|^2,$$

consequently, (4.50) is true. Suppose now that A is a closed affine subspace. Then $\langle c, b \rangle = 0$ (see Theorem 2.2.33 (i)), consequently, $\langle a, b \rangle = \|b\|^2$ and the equality in (4.50) holds. □

Suppose that $A \cap B \neq \emptyset$. Define the step size function $\sigma : A \to (0, +\infty)$ by the formula

$$\sigma(x) = \begin{cases} \dfrac{\|P_B x - x\|^2}{\langle P_B x - x, T x - x \rangle} & \text{if } x \in A \backslash B, \\ 1 & \text{if } x \in A \cap B, \end{cases} \tag{4.52}$$

where $T := P_A P_B$ is the alternating projection. The step size defined by (4.52) was proposed by Gurin et al. in [196, Sect. 3] and was applied in an accelerated alternating projection method of the form $x^{k+1} = P_A T_\sigma x^k$. Note that Lemma 4.3.12 says that in the case $A \cap B \neq \emptyset$, the step size given by (4.47) is equal to at least the step size given by (4.52). Therefore, the estimation (2.69) for $T := P_A P_B$ with the step size $\sigma(x)$ given by (4.47) is better than the one with the step size $\sigma(x)$ given by (4.52) (see Fig. 4.20). The following lemma shows that both step sizes lead to generalized relaxations T_σ which are extrapolations of T.

Lemma 4.3.13. *Let $A \cap B \neq \emptyset$ and the step size function $\sigma : A \to (0, +\infty)$ be defined by (4.52). Then $\sigma(x) \geq 1$ for all $x \in A$ and the equality holds if and only if $P_B x \in A$.*

Proof. Let $x \in A$. If $P_B x \in A$, then $\sigma(x) = 1$. Let now $P_B x \notin A$. Then, of course, $x \notin B$. Similarly as in the proof of Lemma 4.3.12, one can prove that $\langle P_B x - x, Tx - x \rangle > 0$, i.e., the step size $\sigma(x)$ is well defined. Furthermore, the characterization of the metric projection $P_A(P_B x)$ yields

$$
\begin{aligned}
\langle P_B x - x, Tx - x \rangle &= \| P_B x - x \|^2 + \langle P_B x - x, Tx - P_B x \rangle \\
&\leq \| P_B x - x \|^2 - \| Tx - P_B x \|^2 \\
&< \| P_B x - x \|^2 .
\end{aligned}
$$

Therefore, $\sigma(x) \geq 1$ and the equality holds if and only if $P_B x \in A$. □

Corollary 4.3.14. *Let $A, B \subseteq \mathcal{H}$ be nonempty closed convex subsets, $A \cap B \neq \emptyset$ and the step size function $\sigma : A \to (0, +\infty)$ be defined by (4.52). Then the operator $T_\sigma : A \to \mathcal{H}$ is a cutter. Consequently, for any $\lambda \in (0, 2)$, the operator $T_{\sigma,\lambda}$ is $\frac{2-\lambda}{\lambda}$-strongly quasi-nonexpansive.*

Proof. Let $x \in A$ and $\lambda \in (0, 2)$. It follows from Lemma 4.3.12 that $\sigma(x)$ is not greater than the step size given by (4.47). Theorem 4.3.11 yields now that $\sigma(x)$ satisfies conditions of Corollary 2.4.5. Therefore, T_σ is a cutter and $T_{\sigma,\lambda}$ is $\frac{2-\lambda}{\lambda}$-strongly quasi-nonexpansive. □

Remark 4.3.15. Let $x \in A \backslash B$. Note that the step size $\sigma(x)$ defined by (4.52) is the unique solution of the equality

$$
\langle x + \overset{\prime}{\sigma}(x)(Tx - x) - P_B x, P_B x - x \rangle = 0
$$

(see Fig. 4.20).

Suppose that A is a closed affine subspace and that $A \cap B \neq \emptyset$. Then Theorem 2.2.33 (i) yields

$$
\begin{aligned}
\langle P_B x - x, Tx - x \rangle &= \| Tx - x \|^2 + \langle P_B x - Tx, Tx - x \rangle \\
&= \| Tx - x \|^2 ,
\end{aligned}
$$

where $x \in A$ and $T = P_A P_B$. The step size $\sigma(x)$ defined by (4.52) can be now written in the form

$$
\sigma(x) = \begin{cases} \dfrac{\| P_B x - x \|^2}{\| Tx - x \|^2} & \text{if } x \in A \backslash B, \\ 1 & \text{if } x \in A \cap B, \end{cases} \tag{4.53}
$$

(see Fig. 4.21). This step size was proposed by Bauschke et al. in [25, Corollary 4.11] and was applied in an accelerated alternating projection method of the form $x^{k+1} = T_\sigma x^k$.

Now suppose that A and B are closed subspaces of $\mathcal{H}$. In this case, the metric projections P_A, P_B are, actually, orthogonal projections (see Theorem 2.2.30 (i)) and

Fig. 4.21 Step size $\sigma(x)$ given by (4.53)

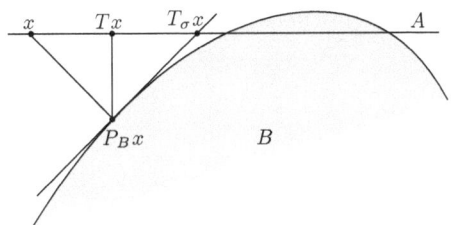

Fig. 4.22 Step size $\sigma(x)$ given by (4.54)

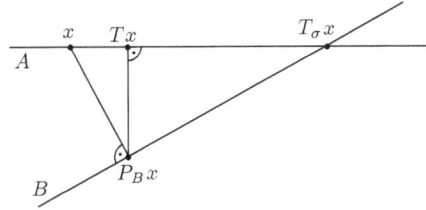

$$\|P_B x - x\|^2 = \langle P_B x, P_B x - x \rangle + \langle x, x - P_B x \rangle$$
$$= \langle x, x - P_B x \rangle$$
$$= \langle x, x - T x \rangle + \langle x, T x - P_B x \rangle$$
$$= \langle x, x - T x \rangle,$$

where $x \in A$ and $T := P_A P_B$. The step size $\sigma(x)$ defined by (4.53) can now be written in the form

$$\sigma(x) = \begin{cases} \dfrac{\langle x, x - Tx \rangle}{\|Tx - x\|^2} & \text{if } x \in A \backslash B \\ 1 & \text{if } x \in A \cap B, \end{cases} \tag{4.54}$$

(see Fig. 4.22). This step size was applied in [30, equality 3.1.2] in order to accelerate globally the alternating projection method (see [30, Theorem 3.23]).

Now we consider the case in which A is a closed affine subspace and B is a closed convex subset. We will propose a step size function $\sigma : A \to (0, +\infty)$ which leads to an extrapolation T_σ of T and we will show that the generalized relaxation $T_{\sigma,\lambda}$ which employs this step size function is strongly quasi-nonexpansive. Define the step size function $\sigma : A \to (0, +\infty)$ by the formula

$$\sigma(x) := \begin{cases} 1 + \dfrac{(\|Tx - P_B x\| - \tilde{\delta}(x))^2}{\|Tx - x\|^2} & \text{if } x \notin \text{Fix } T \\ 1 & \text{if } x \in \text{Fix } T, \end{cases} \tag{4.55}$$

where $\tilde{\delta}(x) \in [\delta, \|Tx - P_B x\|]$ is an upper approximation of $\delta := d(A, B) = \inf_{x \in A, y \in B} \|x - y\|$. A geometrical interpretation of the steps size $\sigma(x)$ is presented in Fig. 4.23, where $\sigma(x) \leq \tilde{\delta}(x) < \|Tx - P_B x\|$, $\tan \alpha = \dfrac{\|Tx - x\|}{\|Tx - P_B x\| - \tilde{\delta}(x)}$. We have $\sigma(x) = 1 + \cot^2 \alpha = \dfrac{1}{\sin^2 \alpha}$ and $\|T_\sigma x - x\| = \sigma(x) \|Tx - x\| = \dfrac{\|Tx - x\|}{\sin^2 \alpha}$.

Fig. 4.23 Step size $\sigma(x)$
given by (4.55)

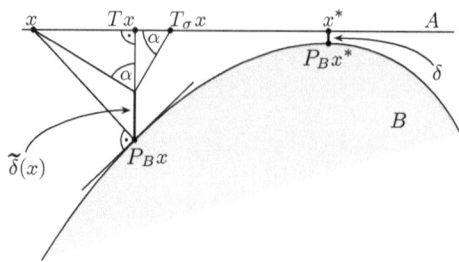

Theorem 4.3.16. *Let $A \subseteq \mathcal{H}$ be a closed affine subspace, $B \subseteq \mathcal{H}$ be a nonempty closed convex subset and the step size function $\sigma : A \rightarrow (0, +\infty)$ be defined by (4.55). Then the operator $T_\sigma : A \rightarrow A$ is a cutter. Consequently, for any $\lambda \in (0, 2)$, the operator $T_{\sigma,\lambda}$ is strongly quasi-nonexpansive and asymptotically regular.*

Proof. Let $z \in \text{Fix } T_\sigma$, $x \notin \text{Fix } T_\sigma$ and $\lambda \in (0, 2)$. Note that $\text{Fix } T_{\sigma,\lambda} = \text{Fix } T_\sigma = \text{Fix } T$. It is clear that $\sigma(x) \geq 1$. We prove that

$$\langle z - x, Tx - x \rangle \geq \sigma(x) \|Tx - x\|^2$$

The inequality is obvious for $x \in \text{Fix } T$. Let now $x \notin \text{Fix } T$. Since $\delta = \|Tz - P_B z\|$ (see equivalence (i)$\Leftrightarrow$(v) in Theorem 4.3.4), inequality (4.35) yields

$$
\begin{aligned}
\langle z - x, Tx - x \rangle &= \|Tx - x\|^2 + \langle z - Tx, Tx - x \rangle \\
&= \|Tx - x\|^2 + \langle Tz - Tx, z - x \rangle - \|Tz - Tx\|^2 \\
&\geq \|Tx - x\|^2 + (\|Tx - P_B x\| - \|Tz - P_B z\|)^2 \\
&= (1 + \frac{(\|Tx - P_B x\| - \delta)^2}{\|Tx - x\|^2}) \|Tx - x\|^2 \\
&\geq (1 + \frac{(\|Tx - P_B x\| - \tilde{\delta}(x))^2}{\|Tx - x\|^2}) \|Tx - x\|^2 \\
&= \sigma(x) \|Tx - x\|^2 .
\end{aligned}
$$

We see that the step size $\sigma(x)$ satisfies the assumptions of Theorem 2.4.5. Therefore, the operator T_σ is a cutter and the operator $T_{\sigma,\lambda}$ is $\frac{2-\lambda}{\lambda}$-strongly quasi-nonexpansive. The asymptotic regularity of $T_{\sigma,\lambda}$ follows now from Theorem 3.4.3. $\square$

4.3.5 Averaged Alternating Reflection

Let $A, B \subseteq \mathcal{H}$ be closed convex subsets. An operator $T : \mathcal{H} \rightarrow \mathcal{H}$ defined by

$$T := \frac{1}{2}(R_A R_B + \text{Id}), \tag{4.56}$$

Fig. 4.24 Averaged
alternating reflection

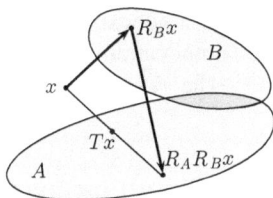

where $R_A := 2P_A - \text{Id}$ and $R_B := 2P_B - \text{Id}$ are reflection operators onto A and B, respectively, is called an *averaged alternating reflection* (AAR) (see Fig. 4.24). The properties of AAR were studied in [149, 246], [157, Sect. 4], [26, Sect. 5.D], [27, Sect. 3], [28, Sect. 3] and in [29]. AAR is a special case of Elser's difference map (for details, see [161]). It is clear that a λ-relaxation of T has the form

$$T_\lambda = (1 - \frac{\lambda}{2})\text{Id} + \frac{\lambda}{2}R_A R_B$$

More general results than the following one can be found in [28, Sect. 3].

Corollary 4.3.17. *Let $A, B \subseteq \mathcal{H}$ be closed convex, $U := R_A R_B : \mathcal{H} \to \mathcal{H}$ and $T := \frac{1}{2}(U + \text{Id})$ be the averaged alternating reflection. Then*

(i) $A \cap B \subseteq \text{Fix } T$,

(ii) If $\text{int}(A \cap B) \neq \emptyset$, then $A \cap B = \text{Fix } T$,

(iii) If $A \cap B \neq \emptyset$ and $x \in \text{Fix } T$, then $P_B x = P_A R_B x$, consequently, $P_B(\text{Fix } T) = A \cap B$,

(iv) T is firmly nonexpansive, consequently, T_λ is nonexpansive and $\frac{2-\lambda}{\lambda}$-strongly quasi-nonexpansive for any $\lambda \in (0, 2)$.

Proof. (i) The inclusion is obvious (see Remark 2.1.1).

(ii) Let $C := \text{int}(A \cap B) \neq \emptyset$. It is clear that

$$C = \text{int } A \cap \text{int } B = \text{int Fix } R_A \cap \text{int Fix } R_B \subseteq \text{Fix } R_A \cap \text{Fix } R_B.$$

By Proposition 2.1.41, R_A and R_B are int C-strictly quasi-nonexpansive. Now Theorem 2.1.26 (ii) and the facts that $\text{Fix } R_A = A$ and $\text{Fix } R_B = B$ yield $\text{Fix } T = \text{Fix } U = \text{Fix } R_A \cap \text{Fix } R_B = A \cap B$.

(iii) Let $x \in \text{Fix } T$ ($= \text{Fix } U$), $b := P_B x$, $y := R_B x = 2b - x$ and $a = P_A y$. It is clear that $y = b + \frac{1}{2}(y - x)$ and $x = R_A y = a + \frac{1}{2}(x - y)$. Consequently, $b = a$, i.e., $P_B x = P_A R_B x$. Therefore, $P_B x \in A \cap B$, i.e., $P_B(\text{Fix } T) = P_B(\text{Fix } U) \subseteq A \cap B$. The converse inclusion follows immediately from (i) and from the fact that $P_B(A \cap B) = A \cap B$.

(iv) The metric projections P_A and P_B are firmly nonexpansive (see Theorem 2.2.21 (iii)). By the implication (i)$\Rightarrow$(ii) in Theorem 2.2.10, the reflections R_A and R_B are nonexpansive. Therefore, $U := R_A R_B$ is nonexpansive. Now the implication (iii)$\Rightarrow$(i) in Theorem 2.2.10 yields the firm nonexpansivity of $T := \frac{1}{2}(U + \mathrm{Id})$. Let $\lambda \in (0, 2)$. Applying again the implication (i)$\Rightarrow$(ii) in Theorem 2.2.10, we obtain that T_λ is nonexpansive. The $\frac{2-\lambda}{\lambda}$-strong quasi nonexpansivity of T_λ follows from Corollary 2.2.9. $\square$

4.4 Simultaneous Projection

Definition 4.4.1. Let $C_i \subseteq \mathcal{H}$ be nonempty closed convex subsets, $i \in I := \{1, 2, \ldots, m\}$, and $w = (\omega_1, \ldots, \omega_m) \in \Delta_m$ be a vector of weights. The operator

$$T := \sum_{i \in I} \omega_i P_{C_i}$$

is called a *simultaneous projection*. The operator

$$\sum_{i \in I} \omega_i P_{C_i, \lambda_i},$$

where $P_{C_i, \lambda_i} := \mathrm{Id} + \lambda_i (P_{C_i} - \mathrm{Id})$, $\lambda_i \in [0, 2]$, $i \in I$, is called a *simultaneous relaxed projection* (see Fig. 4.25).

The simultaneous projection was introduced by Gianfranco Cimmino [116] and was investigated by many authors, e.g., by Auslender [12], Censor and Elfving [87], Reich [295], Pierra [284], De Pierro and Iusem [134–137,215], Butnariu and Censor [53], Combettes [117,118,120], Bauschke [17], Bauschke and Borwein [22], Censor and Zenios [108] and by Matoušková and Reich [258].

One can also consider a generalization of the simultaneous projection of the form $T := \sum_{i \in I} \omega_i P_{C_i}$, where $w : \mathcal{H} \to \Delta_m$ is a weight function (see Sect. 4.8). In this section, however, we restrict ourselves to the classical simultaneous projection with constant weights. Denote $C := \bigcap_{i \in I} C_i$. The metric projection P_{C_i} is a special case of a simultaneous projection $T := \sum_{i \in I} \omega_i P_{C_i}$, where $w = e_i$, $i \in I$.

Corollary 4.4.2. *Let $C_i \subseteq \mathcal{H}$ be nonempty closed convex subsets, $i \in I$, $C \neq \emptyset$, $w \in \mathrm{ri}\, \Delta_m$ and $T := \sum_{i \in I} \omega_i P_{C_i}$. Then $\mathrm{Fix}\, T = C$.*

Proof. By the nonexpansivity of the metric projection and by the fact $\bigcap_{i \in I} \mathrm{Fix}\, P_{C_i} = \bigcap_{i \in I} C_i = C \neq \emptyset$, the claim follows from Theorem 2.1.14. $\square$

Remark 4.4.3. Let $\mu := \sum_{i \in I} \omega_i \lambda_i$. If $\mu = 0$, then it is clear that $\sum_{i \in I} \omega_i P_{C_i, \lambda_i} = \mathrm{Id}$. Now let $\mu > 0$ and $\nu_i = \frac{\omega_i \lambda_i}{\mu}$, where $w = (\omega_1, \omega_2, \ldots, \omega_m) \in \mathrm{ri}\, \Delta_m$ and $\lambda_i \in [0, 2]$, $i \in I$. Then

$$\sum_{i \in I} \omega_i P_{C_i, \lambda_i} = \sum_{i \in I} \omega_i (\mathrm{Id} + \lambda_i (P_{C_i} - \mathrm{Id}))$$

$$= \mathrm{Id} + \sum_{i \in I} \omega_i \lambda_i (P_{C_i} - \mathrm{Id})$$

Fig. 4.25 Simultaneous
projection and simultaneous
relaxed projection

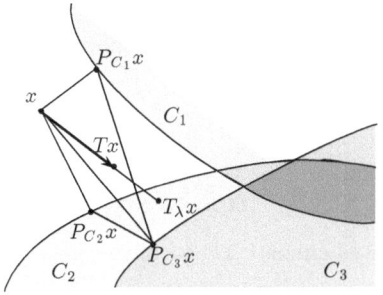

$$= \mathrm{Id} + \mu \sum_{i \in I} \frac{\omega_i \lambda_i}{\mu} (P_{C_i} - \mathrm{Id})$$

$$= \mathrm{Id} + \mu \left(\sum_{i \in I} \nu_i P_{C_i} - \mathrm{Id} \right)$$

$$= \sum_{i \in I} \nu_i (\mathrm{Id} + \mu(P_{C_i} - \mathrm{Id}))$$

$$= \sum_{i \in I} \nu_i P_{C_i, \mu}.$$

We see that in both cases ($\mu = 0$ and $\mu > 0$), the simultaneous relaxed projection $\sum_{i \in I} \omega_i P_{C_i, \lambda_i}$ can be presented as a relaxation U_μ of a simultaneous projection $U := \sum_{i \in I} \nu_i P_{C_i}$ or as a simultaneous relaxed projection $\sum_{i \in I} \nu_i P_{C_i, \mu}$ with the same relaxation parameter $\mu \in [0, 2]$ for all projections P_{C_i}. Therefore, we restrict our further analysis of simultaneous relaxed projection operators $\sum_{i \in I} \omega_i P_{C_i, \lambda_i}$ to the case $\lambda_i = \lambda \in [0, 2]$ for all $i \in I$, or, equivalently, to the relaxation T_λ of a simultaneous projection $T := \sum_{i \in I} \omega_i P_{C_i}$ with a relaxation parameter $\lambda \in [0, 2]$.

4.4.1 Simultaneous Projection as an Alternating Projection in a Product Space

Let $C_i \subseteq \mathcal{H}$ be nonempty closed convex subsets, $i \in I$, $w = (\omega_1, \omega_2, \ldots, \omega_m) \in \mathrm{ri}\, \Delta_m$ and $T := \sum_{i \in I} \omega_i P_{C_i}$ be a simultaneous projection. In this subsection we present the operator T as an alternating projection in a product Hilbert space $\mathbf{H} = \mathcal{H}^m$ with the inner product $\langle \cdot, \cdot \rangle_{\mathbf{H}} : \mathbf{H} \to \mathbb{R}$ defined by

$$\langle \mathbf{u}, \mathbf{v} \rangle_{\mathbf{H}} := \sum_{i=1}^{m} \omega_i \langle u_i, v_i \rangle,$$

where $\mathbf{u} = (u_1, u_2, \ldots, u_m) \in \mathbf{H}$, $\mathbf{v} = (v_1, v_2, \ldots, v_m) \in \mathbf{H}$, and with the norm $\|\cdot\|_{\mathbf{H}}$ induced by this inner product. Define the subsets $\mathbf{C} \subseteq \mathbf{H}$ and $\mathbf{D} \subseteq \mathbf{H}$ by

$$\mathbf{C} := C_1 \times \ldots \times C_m$$

and

$$\mathbf{D} := \{\mathbf{x} = (x_1, x_2, \ldots, x_m) \in \mathbf{H} : x_1 = x_2 = \ldots = x_m\}.$$

By Lemma 1.2.8, we have

$$P_{\mathbf{C}}\mathbf{x} = (P_{C_1}x_1, P_{C_2}x_2, \ldots, P_{C_m}x_m),$$

where $\mathbf{x} = (x_1, \ldots, x_m) \in \mathbf{H}$. One can easily show that

$$P_{\mathbf{C},\lambda}\mathbf{x} = (P_{C_1,\lambda}x_1, P_{C_2,\lambda}x_2, \ldots, P_{C_m,\lambda}x_m),$$

where $\lambda \in [0, 2]$. Now, we calculate the metric projection $P_{\mathbf{D}}\mathbf{y}$ for $\mathbf{y} = (y_1, y_2, \ldots, y_m) \in \mathbf{H}$. By the definition of the metric projection, we have

$$P_{\mathbf{D}}\mathbf{y} = \operatorname*{argmin}_{\mathbf{x} \in \mathbf{D}} \frac{1}{2} \|\mathbf{x} - \mathbf{y}\|_{\mathbf{H}}^2$$

$$= \operatorname*{argmin}_{x \in \mathcal{H}} \frac{1}{2} \sum_{i=1}^{m} \omega_i \|x - y_i\|^2.$$

Since the function $f : \mathcal{H} \to \mathbb{R}$ defined by $f(x) := \frac{1}{2} \sum_{i=1}^{m} \omega_i \|x - y_i\|^2$ is differentiable and convex, x is a minimizer of this function if and only if x is its stationary point. After differentiation we obtain the following condition

$$\sum_{i=1}^{m} \omega_i (x - y_i) = 0.$$

By the equality $\sum_{i=1}^{m} \omega_i = 1$, we obtain $x = \sum_{i=1}^{m} \omega_i y_i$, i.e.,

$$P_{\mathbf{D}}\mathbf{y} = (\sum_{i=1}^{m} \omega_i y_i, \ldots, \sum_{i=1}^{m} \omega_i y_i).$$

Subsuming, for the operator $\mathbf{T} := P_{\mathbf{D}} P_{\mathbf{C}} : \mathbf{D} \to \mathbf{D}$, we obtain

$$\mathbf{T}\mathbf{x} = (\sum_{i=1}^{m} \omega_i P_{C_i} x, \ldots, \sum_{i=1}^{m} \omega_i P_{C_i} x) = (Tx, \ldots, Tx).$$

Therefore, $\mathbf{x} \in \operatorname{Fix} P_{\mathbf{D}} P_{\mathbf{C}}$ if and only if $x \in \operatorname{Fix} T$. Furthermore, for the operator $\mathbf{T}_\lambda = \operatorname{Id} + \lambda(P_{\mathbf{D}} P_{\mathbf{C}} - \operatorname{Id})$ it holds

$$\mathbf{T}_\lambda \mathbf{x} = \mathbf{x} + \lambda (P_\mathbf{D} P_\mathbf{C} \mathbf{x} - \mathbf{x})$$

$$= \left(\sum_{i=1}^{m} \omega_i P_{C_i, \lambda} x, \ldots, \sum_{i=1}^{m} \omega_i P_{C_i, \lambda} x \right)$$

$$= (T_\lambda x, T_\lambda x, \ldots, T_\lambda x) = P_\mathbf{D} P_{\mathbf{C}, \lambda} \mathbf{x},$$

where $\mathbf{x} = (x, \ldots, x) \in \mathbf{D}$. We see that the simultaneous projection $T := \sum_{i \in I} \omega_i P_{C_i} : \mathcal{H} \to \mathcal{H}$ can be presented equivalently as an alternating projection $\mathbf{T} = P_\mathbf{D} P_\mathbf{C} : \mathbf{D} \to \mathbf{D}$ in the product space $\mathbf{H} = \mathcal{H}^m$. Note that $\mathbf{D} \subseteq \mathbf{H}$ is a closed affine subspace, consequently, $\mathbf{T}$ is firmly nonexpansive (see Theorem 4.3.7). This leads to the firm nonexpansivity of the simultaneous projection $T := \sum_{i \in I} \omega_i P_{C_i}$. This property also follows from Corollary 2.2.20.

The idea of representing the simultaneous projection as an alternating projection is due to Pierra [284, Sect. 1], where $\omega_i = \frac{1}{m}$, $i \in I$. The idea was continued by Combettes in [117, Sect. III] and [120, Sect. IV].

4.4.2 Properties of the Simultaneous Projection

Let $C_i \subseteq \mathcal{H}$ be nonempty closed convex subsets, $i \in I$, and $w = (\omega_1, \ldots, \omega_m) \in \Delta_m$ be a vector of weights. In this section we give basic properties of the simultaneous projection $T := \sum_{i \in I} \omega_i P_{C_i}$ and of the proximity function $f : \mathcal{H} \to \mathbb{R}$ defined by

$$f(x) := \frac{1}{2} \sum_{i \in I} \omega_i \| P_{C_i} x - x \|^2. \tag{4.57}$$

Corollary 4.4.4. *The simultaneous projection* $T := \sum_{i \in I} \omega_i P_{C_i}$, *where* $w \in \Delta_m$, *is firmly nonexpansive. If* $\text{Fix}\, T \neq \emptyset$, *then* T *is a cutter and, for any* $\lambda \in (0, 2)$, *its relaxation* T_λ *is* $\frac{2-\lambda}{\lambda}$-*strongly quasi-nonexpansive and asymptotically regular. Furthermore,*

$$\| T_\lambda x - z \|^2 \leq \| x - z \|^2 - \lambda (2 - \lambda) \left\| \sum_{i \in I} (\omega_i P_{C_i} x - x) \right\|^2 \tag{4.58}$$

for any $z \in \text{Fix}\, T$.

Proof. Since T is a convex combination of firmly nonexpansive operators P_{C_i}, $i \in I$, the firm nonexpansivity of T follows from Corollary 2.2.20. Let $\text{Fix}\, T \neq \emptyset$. Then T is a cutter, by Theorem 2.2.5 (i). Corollary 2.2.9 yields the $\frac{2-\lambda}{\lambda}$-strong quasi nonexpansivity of T_λ, i.e.,

$$\| T_\lambda x - z \|^2 \leq \| x - z \|^2 - \frac{2 - \lambda}{\lambda} \| T_\lambda x - x \|^2.$$

Now inequality (4.58) follows from the obvious equality $T_\lambda x - x = \lambda(Tx - x)$. The asymptotic regularity of T_λ follows from Theorem 3.4.3.   □

If $\bigcap_{i \in I} C_i \neq \emptyset$, then inequality (4.58) can be strengthened. In the following theorem we allow that the weights ω_i, $i \in I$, can depend on $x \in \mathcal{H}$.

Theorem 4.4.5. *Let $C_i \subseteq \mathcal{H}$ be nonempty closed convex subsets, $i \in I$, $C := \bigcap_{i \in I} C_i \neq \emptyset$, $w : \mathcal{H} \to \Delta_m$ be an appropriate weight function and $T := \sum_{i \in I} \omega_i P_{C_i}$. Then*

$$\|T_\lambda x - z\|^2 \leq \|x - z\|^2 - \lambda(2 - \lambda) \sum_{i \in I} \omega_i(x) \| P_{C_i} x - x \|^2, \qquad (4.59)$$

for all $x \in \mathcal{H}$, $z \in C$ and $\lambda \in [0, 2]$.

Proof. Let $x \in \mathcal{H}$, $z \in C$ and $\lambda \in [0, 2]$. We have $\bigcap_{i \in I} \operatorname{Fix} T_i = \bigcap_{i \in I} C_i$. By Corollary 2.2.24, P_{C_i} is strictly quasi-nonexpansive, $i \in I$, consequently, $\operatorname{Fix} T = \bigcap_{i \in I} C_i$ (see Theorem 2.1.26). Furthermore, P_{C_i} is a cutter (see Theorem 2.2.21 (ii)), i.e., $\langle P_{C_i} x - x, z - x \rangle \geq \| P_{C_i} x - x \|^2$ (see Remark 2.1.31). Now the convexity of the function $\|\cdot\|^2$ yields

$$\|T_\lambda x - z\|^2 = \left\| x + \lambda \sum_{i \in I} \omega_i(x)(P_{C_i} x - x) - z \right\|^2$$

$$= \|x - z\|^2 + \lambda^2 \left\| \sum_{i \in I} \omega_i(x)(P_{C_i} x - x) \right\|^2$$

$$- 2\lambda \sum_{i \in I} \omega_i(x) \langle z - x, P_{C_i} x - x \rangle$$

$$\leq \|x - z\|^2 + \lambda^2 \sum_{i \in I} \omega_i(x) \| P_{C_i} x - x \|^2$$

$$- 2\lambda \sum_{i \in I} \omega_i(x) \| P_{C_i} x - x \|^2$$

$$= \|x - z\|^2 - \lambda(2 - \lambda) \sum_{i \in I} \omega_i(x) \| P_{C_i} x - x \|^2$$

which completes the proof.   □

Let $C \neq \emptyset$. The strong convexity of the function $\|\cdot\|^2$ yields that estimation (4.59) with a constant weight function w is stronger than estimation (4.58).

Theorem 4.4.6. *Let $T := \sum_{i \in I} \omega_i P_{C_i}$ be a simultaneous projection, where $w \in \Delta_m$, and a proximity function $f : \mathcal{H} \to \mathbb{R}$ be defined by (4.57) Then*

$$\operatorname{Fix} T = \operatorname*{Argmin}_{x \in \mathcal{H}} f(x).$$

If at least one of the subsets C_i, $i \in I$, is bounded and the corresponding weight $\omega_i > 0$, then $\operatorname{Fix} T \neq \emptyset$.

Proof. The proximity function f is convex and differentiable, and

$$Df(x) = \sum_{i \in I} \omega_i (x - P_{C_i} x) = x - Tx.$$

(see Lemma 2.2.27) Therefore, the sufficient and necessary optimality condition (see Corollary 1.3.3) yields that $z \in \operatorname{Argmin}_{x \in \mathcal{H}} f(x)$ if and only if $Df(z) = z - Tz = 0$, i.e., $z \in \operatorname{Fix} T$. Now suppose that a subset C_i is bounded for some $i \in I$ and that $\omega_i > 0$. Then the function f is coercive. By Corollary 1.1.53, $\operatorname{Argmin}_{x \in \mathcal{H}} f(x) \neq \emptyset$ and $\operatorname{Fix} T \neq \emptyset$. $\qquad\square$

The proximity function $f : \mathcal{H} \to \mathbb{R}$ defined by (4.57) has the following nice property (cf. [215, Lemma 2]).

Lemma 4.4.7. *Let* $T := \sum_{i \in I} \omega_i P_{C_i}$, *where* $w \in \Delta_m$. *Then, for any* $x \in \mathcal{H}$, *it holds*

$$f(Tx) \leq f(x) - \frac{1}{2} \|Tx - x\|^2.$$

Proof. Let $x \in \mathcal{H}$. It follows from the definition of the metric projection that $\|P_{C_i} Tx - Tx\| \leq \|P_{C_i} x - Tx\|$. Therefore, the properties of the inner product yield

$$f(Tx) = \frac{1}{2} \sum_{i \in I} \omega_i \|P_{C_i} Tx - Tx\|^2 \leq \frac{1}{2} \sum_{i \in I} \omega_i \|P_{C_i} x - Tx\|^2$$

$$= \frac{1}{2} \sum_{i \in I} \omega_i \|P_{C_i} x - x\|^2 + \frac{1}{2} \sum_{i \in I} \omega_i \|Tx - x\|^2 - \sum_{i \in I} \omega_i \langle P_{C_i} x - x, Tx - x \rangle$$

$$= \frac{1}{2} \sum_{i \in I} \omega_i \|P_{C_i} x - x\|^2 + \frac{1}{2} \sum_{i \in I} \omega_i \|Tx - x\|^2 - \|Tx - x\|^2$$

$$= f(x) - \frac{1}{2} \|Tx - x\|^2$$

which completes the proof. $\qquad\square$

Let $x \in \mathcal{H}$. By the equality $Df(x) = x - Tx$, Lemma 4.4.7 can be written in the form

$$f(x - Df(x)) \leq f(x) - \|Df(x)\|^2.$$

The equality above is related to the efficiency of a minimization method defined by the recurrence $x^{k+1} = x^k - Df(x^k)$ (see [178, Definition 4.5]). It is known that in the case $\mathcal{H} = \mathbb{R}^n$ the sequence $\{x^k\}_{k=0}^{\infty}$ generated by this recurrence converges to a minimizer of f (see [178, Theorem 4.6]).

It is clear that for a simultaneous projection $T := \sum_{i \in I} \omega_i P_{C_i}$ the following inclusion is true $C := \bigcap_{i \in I} C_i \subseteq \operatorname{Fix} T$. If $C \neq \emptyset$ and $w \in \operatorname{ri} \Delta_m$, then $C = \operatorname{Fix} T$

(see Theorem 2.1.14) and, for the proximity function $f : \mathcal{H} \to \mathbb{R}$ defined by (4.57) it holds that $\min_{x \in \mathcal{H}} f(x) = 0$.

4.4.3 Simultaneous Projection for a System of Linear Equations

Suppose that $C_i \subseteq \mathcal{H}$ are hyperplanes, $C_i := \{y \in \mathcal{H} : \langle a_i, y \rangle = \beta_i\}$, where $a_i \in \mathcal{H}$, $a_i \neq 0$, $\beta_i \in \mathbb{R}$, $i \in I$. Then the convex feasibility problem reduces to the solution of a system of linear equations $\langle a_i, y \rangle = \beta_i$, $i \in I$. We have $P_{C_i} x = x + \frac{\beta_i - \langle a_i, x \rangle}{\|a_i\|^2} a_i$ (see (4.1)) and $\| P_{C_i} x - x \| = \frac{|\langle a_i, x \rangle - \beta_i|}{\|a_i\|}$. Now the simultaneous projection can be presented in the form

$$T x = x + \sum_{i \in I} \omega_i \frac{\beta_i - \langle a_i, x \rangle}{\|a_i\|^2} a_i, \tag{4.60}$$

where $w = (\omega_1, \ldots, \omega_m) \in \Delta_m$ is a vector of weights. The proximity function f defined by (4.57) has the form

$$f(x) = \frac{1}{2} \sum_{i \in I} \omega_i \frac{(\langle a_i, x \rangle - \beta_i)^2}{\|a_i\|^2}. \tag{4.61}$$

If $\|a_i\| = 1$, $i \in I$, then T has the form

$$T x = x + \sum_{i \in I} \omega_i (\beta_i - \langle a_i, x \rangle) a_i. \tag{4.62}$$

If $\mathcal{H} = \mathbb{R}^n$, the operator T and the proximity function f can be written in the following, more convenient, matrix forms

$$T x = x - A^\top D (A x - b) \tag{4.63}$$

and

$$f(x) = \frac{1}{2} (A x - b)^\top D (A x - b), \tag{4.64}$$

where $D := \operatorname{diag}(\frac{\omega_1}{\|a_1\|^2}, \ldots, \frac{\omega_m}{\|a_m\|^2})$ We can suppose without loss of generality that $w \in \operatorname{ri} \Delta_m$. In this case, the proximity function f has the form

$$f(x) = \frac{1}{2} \| A x - b \|_D^2, \tag{4.65}$$

where $\|u\|_D = (u^\top D u)^{\frac{1}{2}}$ denotes the norm of the vector $u \in \mathbb{R}^m$, induced by the positive definite matrix D. The operator T is firmly nonexpansive and its relaxation T_λ is $\frac{2-\lambda}{\lambda}$-strongly quasi-nonexpansive and asymptotically regular (see Corollary 4.4.4), and Fix $T = \operatorname{Argmin}_{x \in \mathcal{H}} f(x)$ (see Theorem 4.4.6). Furthermore, both subsets are nonempty.

Theorem 4.4.8. *Let a function* $f : \mathcal{H} \to \mathbb{R}$ *be defined by (4.61). Then* $\operatorname{Argmin}_{x \in \mathcal{H}} f(x) \neq \emptyset$.

Proof. The minimization of the proximity function $f : \mathcal{H} \to \mathbb{R}$ given by (4.61) is equivalent to the following minimization problem

$$
\begin{aligned}
&\text{minimize } h(y) := \tfrac{1}{2} \|y\|^2 \\
&\text{subject to } \eta_i = \sqrt{\omega_i} \frac{(\langle a_i, x \rangle - \beta_i)}{\|a_i\|}, \, i = 1, 2, \dots, m, \\
&(x, y) \in \mathcal{H} \times \mathbb{R}^m,
\end{aligned} \qquad (4.66)
$$

where $y = (\eta_1, \eta_2, \dots, \eta_m) \in \mathbb{R}^m$ and the product Hilbert space $\mathcal{H} \times \mathbb{R}^m$ is equipped with a scalar product $\langle\langle \cdot, \cdot \rangle\rangle$ defined by $\langle\langle (x, y), (v, z) \rangle\rangle = \langle x, v \rangle + y^\top z$, where $(x, y), (v, z) \in \mathcal{H} \times \mathbb{R}^m$. Note that h is continuous, convex and coercive. Furthermore, the subset

$$
X := \{(x, z) \in \mathcal{H} \times \mathbb{R}^m : \zeta_i = \sqrt{\omega_i} \frac{(\langle a_i, x \rangle - \beta_i)}{\|a_i\|}, i = 1, 2, \dots, m\}
$$

is closed and convex as an intersection of hyperplanes. Consequently, the function h has a minimizer (x^*, y^*) on X (see Corollary 1.1.53), i.e., $x^* \in \operatorname{Argmin}_{x \in \mathcal{H}} f(x)$.
□

De Pierro and Iusem proved a property which yields Theorem 4.4.8 in the case $\mathcal{H} = \mathbb{R}^n$ (see [137, Proposition 10]).

4.4.4 Simultaneous Projection for the Linear Feasibility Problem

Suppose that $C_i \subseteq \mathcal{H}$ are closed half-spaces, i.e., $C_i := \{y \in \mathcal{H} : \langle a_i, y \rangle \leq \beta_i\}$, where $a_i \in \mathcal{H}$, $a_i \neq 0$, $\beta_i \in \mathbb{R}$, $i \in I$. Then the convex feasibility problem reduces to the solution of a system of linear inequalities $\langle a_i, y \rangle \leq \beta_i$, $i \in I$, $P_{C_i} x = x - \frac{(\langle a_i, x \rangle - \beta_i)_+}{\|a_i\|^2} a_i$ (see (4.7)) and $\|P_{C_i} x - x\| = \frac{(\langle a_i, x \rangle - \beta_i)_+}{\|a_i\|}$. Similarly as in the linear case (see Sect. 4.4.3), the simultaneous projection can be presented in the form

$$
T x = x - \sum_{i \in I} \omega_i \frac{(\langle a_i, x \rangle - \beta_i)_+}{\|a_i\|^2} a_i, \qquad (4.67)
$$

where $w = (\omega_1, \dots, \omega_m) \in \Delta_m$ is a vector of weights. The proximity function f defined by (4.57) has the form

$$f(x) = \frac{1}{2} \sum_{i \in I} \omega_i \frac{[(\langle a_i, x \rangle - \beta_i)_+]^2}{\|a_i\|^2} \tag{4.68}$$

(cf. (1.34)). If $\|a_i\| = 1$, $i \in I$, then we have

$$Tx = x - \sum_{i \in I} \omega_i (\langle a_i, x \rangle - \beta_i)_+ a_i. \tag{4.69}$$

If $\mathcal{H} = \mathbb{R}^n$, the operator T and the corresponding proximity function f can be written in the following more convenient matrix form

$$Tx = x - A^\top D(Ax - b)_+ \tag{4.70}$$

and

$$f(x) = \frac{1}{2}(Ax - b)_+^\top D(Ax - b)_+, \tag{4.71}$$

where $D := \mathrm{diag}(\frac{\omega_1}{\|a_1\|^2}, \ldots, \frac{\omega_m}{\|a_m\|^2})$. We can suppose without loss of generality that $w \in \mathrm{ri}\, \Delta_m$. In this case, the proximity function f has the form

$$f(x) = \frac{1}{2} \|(Ax - b)_+\|_D^2. \tag{4.72}$$

The operator T is firmly nonexpansive and its relaxation T_λ is $\frac{2-\lambda}{\lambda}$-strongly quasi-nonexpansive and asymptotically regular (see Corollary 4.4.4), and Fix $T = \mathrm{Argmin}_{x \in \mathcal{H}} f(x)$ (see Theorem 4.4.6). Furthermore, both subsets are nonempty.

Theorem 4.4.9. *Let a function* $f : \mathcal{H} \to \mathbb{R}$ *be defined by (4.68). Then* $\mathrm{Argmin}_{x \in \mathbb{R}^n} f(x) \neq \emptyset$.

Proof. The minimization of the proximity function $f : \mathcal{H} \to \mathbb{R}$ given by (4.68) is equivalent to the following minimization problem

$$\begin{aligned}
&\text{minimize } h(y) := \tfrac{1}{2} \|y\|^2 \\
&\text{subject to } \eta_i \geq \sqrt{\omega_i} \tfrac{(\langle a_i, x \rangle - \beta_i)}{\|a_i\|}, i = 1, 2, \ldots, m, \\
&\qquad\qquad \eta_i \geq 0, i = 1, 2, \ldots, m, \\
&\qquad\qquad (x, y) \in \mathcal{H} \times \mathbb{R}^m.
\end{aligned}$$

The remaining part of the proof is similar to the proof of Theorem 4.4.8 with X replaced by

$$X := \{(x, z) \in \mathcal{H} \times \mathbb{R}^m : \zeta_i \geq \sqrt{\omega_i} \frac{(\langle a_i, x \rangle - \beta_i)}{\|a_i\|}, \zeta_i \geq 0, i = 1, 2, \ldots, m\}.$$

Note that X is closed and convex as an intersection of closed half-spaces.   □

De Pierro and Iusem proved a property which yields Theorem 4.4.9 in case $\mathcal{H} = \mathbb{R}^n$ (see [137, Proposition 13]).

4.5 Cyclic Projection

Let $C_i \subseteq \mathcal{H}$ be nonempty closed convex subsets, $i \in I := \{1, 2, \ldots, m\}$.

Definition 4.5.1. The operator $T : \mathcal{H} \to \mathcal{H}$ defined by

$$T := P_{C_m} P_{C_{m-1}} \ldots P_{C_1} \tag{4.73}$$

is called a *cyclic projection* (see Fig. 4.26).

The cyclic projection was introduced by Stefan Kaczmarz for hyperplanes in $\mathbb{R}^n$ (see [223]). The properties of the cyclic projection were investigated by many authors, e.g., by Halperin [198], Bregman [42], Gurin et al. [196], Gordon et al. [189], Tanabe [321], Herman [204], McCormick [259], Censor [78], Censor et al. [86], Babenko [13], De Pierro and Iusem [137], Dax [129–132], Dye et al. [151], Bauschke and Borwein [22], Bauschke et al. [23], Popa [285], and by Deutsch and Hundal [142, 143]. In this section we present the basic properties of the cyclic projection defined by (4.73).

Corollary 4.5.2. *If $\bigcap_{i \in I} C_i \neq \emptyset$, then Fix $T = \bigcap_{i \in I} C_i$.*

Proof. Let $\bigcap_{i \in I} C_i \neq \emptyset$. Since Fix $P_{C_i} = C_i$ and the metric projection is strictly quasi-nonexpansive, the claim follows from Theorem 2.1.26. □

Without the assumption $\bigcap_{i \in I} C_i \neq \emptyset$ the existence of a fixed point of a cyclic projection T is not guaranteed in general. One can prove, however, that Fix $T \neq \emptyset$ if all $C_i \subseteq \mathbb{R}^n$ are half-spaces, $i \in I$ (see [137, Lemma 1 and Proposition 13]).

Corollary 4.5.3. *The cyclic projection $T := P_{C_m} P_{C_{m-1}} \ldots P_{C_1}$ is $\frac{2m}{m+1}$-relaxed firmly nonexpansive. Consequently, if Fix $T \neq \emptyset$, then T is $\frac{1}{m}$-strongly quasi-nonexpansive and asymptotically regular.*

Proof. The operator T is $\frac{2m}{m+1}$-relaxed firmly nonexpansive as a composition of m firmly nonexpansive operators (see Corollary 2.2.43). Let now Fix $T \neq \emptyset$. Then Corollary 2.2.9 yields that T is $\frac{1}{m}$-strongly quasi-nonexpansive. The asymptotic regularity follows from Corollary 3.4.6. □

Denote $S_i := P_{C_i} P_{C_{i-1}} \ldots P_{C_1}$, $i \in I$, and $S_0 := \mathrm{Id}$. If $C := \bigcap_{i \in I} C_i \neq \emptyset$, then Fix $S_i \supseteq C$, $i \in I$, by Corollary 4.5.2. If we apply inequality (2.52) m-times, then we obtain the following estimation

$$\|Tx - z\|^2 \leq \|x - z\|^2 - \sum_{i=1}^{m} \|S_i x - S_{i-1} x\|^2 \tag{4.74}$$

for all $x \in \mathcal{H}$ and $z \in C$.

Fig. 4.26 Cyclic projection

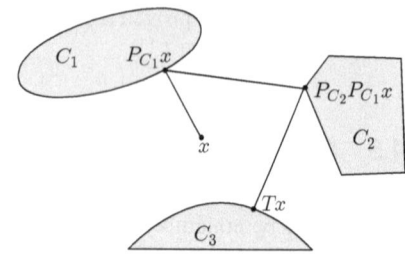

Fig. 4.27 Fixed point of a
composition of projections

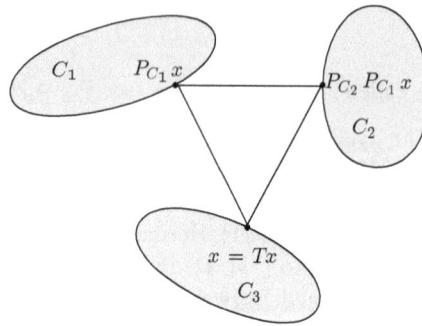

Corollary 4.5.4. *If at least one of the subsets C_i, $i \in I$, is bounded, then the cyclic projection $T := P_{C_m} P_{C_{m-1}} \dots P_{C_1} : C_m \to C_m$ has a fixed point $z \in C_m$.*

Proof. Since the metric projection is nonexpansive and $P_{C_i}(X) = C_i$, $i \in I$, the corollary follows from Theorem 2.1.13. □

A fixed point of a cyclic projection is illustrated in Fig. 4.27.

Let $C_i \subseteq \mathcal{H}$ be closed half-spaces, i.e., $C_i := H_-(a_i, \beta_i) = \{x \in \mathcal{H} : \langle a_i, x \rangle \leq \beta_i\}$, where $a_i \in \mathcal{H}$, $a_i \neq 0$ and $\beta_i \in \mathbb{R}$, $i \in I$. The following theorem was proved in [137, Lemma 1 and Proposition 13].

Theorem 4.5.5. *If $C_i \subseteq \mathbb{R}^n$ are half-spaces, $i \in I$, then the cyclic projection $T := P_{C_m} \dots P_{C_m}$ has a fixed point.*

4.5.1 Cyclic Relaxed Projection

Definition 4.5.6. Let $\lambda_i \in [0, 2]$, $i \in I$. The operator

$$T := P_{C_m, \lambda_m} P_{C_{m-1}, \lambda_{m-1}} \dots P_{C_1, \lambda_1},$$

where $P_{C_i, \lambda_i} := \mathrm{Id} + \lambda_i (P_{C_i} - \mathrm{Id})$, $i \in I$, is called an operator of *cyclic relaxed projection* (see Fig. 4.28).

Fig. 4.28 Cyclic relaxed projection

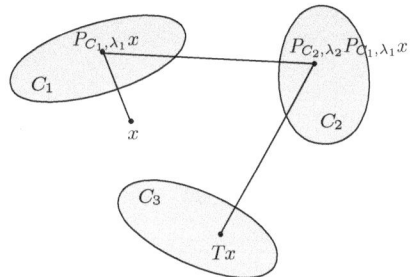

Corollary 4.5.7. *Let* $\lambda_i \in (0,2)$, $i \in I$, *and* $T := P_{C_m,\lambda_m} P_{C_{m-1},\lambda_{m-1}} \cdots P_{C_1,\lambda_1}$ *be the operator of cyclic relaxed projection. If* $C := \bigcap_{i \in I} C_i \neq \emptyset$, *then* $\operatorname{Fix} T = C$.

Proof. It follows from Theorem 2.2.21 (i) and from Remark 2.1.4 that $\operatorname{Fix} P_{C_i,\lambda_i} = C_i$, $i \in I$. The relaxed metric projection P_{C_i,λ_i} is strictly quasi-nonexpansive for $\lambda_i \in (0,2)$, $i \in I$, (see Corollary 2.2.23 (iii)). Therefore, the claim follows from Theorem 2.1.26. □

Corollary 4.5.8. *Let* $\lambda_i \in [0,2]$, $i \in I$. *The relaxed cyclic projection* $T := P_{C_m,\lambda_m} P_{C_{m-1},\lambda_{m-1}} \cdots P_{C_1,\lambda_1}$ *is* λ-*relaxed firmly nonexpansive, where*

$$\lambda = \frac{2m\lambda_{\max}}{(m-1)\lambda_{\max} + 2}$$

and $\lambda_{\max} = \max_{i \in I} \lambda_i$. *If* $\lambda_{\max} \in (0,2)$ *and* $\operatorname{Fix} T \neq \emptyset$, *then* T *is asymptotically regular.*

Proof. The operator T is λ-relaxed firmly nonexpansive as a composition of λ_i-relaxed firmly nonexpansive operators, $i \in I$ (see Theorem 2.2.42). Now, let $\lambda_{\max} \in (0,2)$ and $\operatorname{Fix} T \neq \emptyset$. Then $\lambda \in (0,2)$ and the asymptotic regularity follows from Corollary 3.4.6. □

One can prove that the cyclic projection is asymptotically regular without the assumption $\operatorname{Fix} T \neq \emptyset$ (see [19, Theorem 3.1]).

4.5.2 Cyclic-Simultaneous Projection

Definition 4.5.9. Let $C_i \subseteq \mathcal{H}$ be closed convex subsets, $i \in I$, and $w \in \Delta_m$. The operator $S : X \to \mathcal{H}$ defined by

$$S := \sum_{i=1}^{m} \omega_i S_i, \tag{4.75}$$

where $S_i := P_{C_i} \ldots P_{C_1}$, is called the *cyclic-simultaneous projection* (see Fig. 4.29).

Fig. 4.29 Cyclic-
simultaneous projection

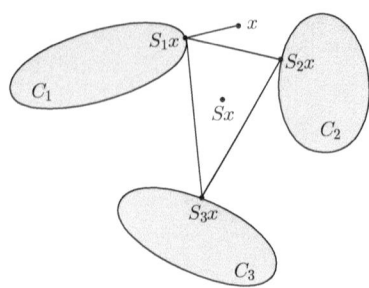

Corollary 4.5.10. *The cyclic-simultaneous projection S defined by (4.75) is β-relaxed firmly nonexpansive, where $\beta = \sum_{i=1}^{m} \omega_i \frac{2i}{i+1} \in (0,2)$. Furthermore, if Fix $S \neq \emptyset$, then S_α is strongly quasi-nonexpansive and asymptotically regular for all $\alpha \in (0, 2\beta^{-1})$.*

Proof. Let $\beta := \sum_{i=1}^{m} \omega_i \frac{2i}{i+1}$. It is clear that $\beta \in (0,2)$. Since the metric projection is firmly nonexpansive, Corollary 2.2.44 yields that S is β-RFNE. Therefore, $S_{\beta^{-1}}$ is FNE and $(S_{\beta^{-1}})_\lambda = S_{\lambda\beta^{-1}}$ is λ-RFNE for all $\lambda \in (0,2)$, or, in other words, S_α is $\alpha\beta$-RFNE for all $\alpha \in (0, 2\beta^{-1})$. Now, the second part of the corollary follows from Corollaries 2.2.9 and 3.4.6.   □

The cyclic-simultaneous projection is a special case of string-averaging projections which were studied in [91, Sect. 1], [105–107].

4.5.3   Projections with Reflection onto an Obtuse Cone

Let $C_i \subseteq \mathcal{H}$ be closed convex subsets, $i \in I$, and $K \subseteq \mathcal{H}$ be a closed convex and obtuse cone, i.e., $-K^* \subseteq K$. Suppose that $C := K \cap \bigcap_{i \in I} C_i \neq \emptyset$. Consider the following CFP: find a point $x \in C$. Denote

$$P := P_{C_m} P_{C_{m-1}} \dots P_{C_1},$$

$P_\lambda := \mathrm{Id} + \lambda(P - \mathrm{Id})$, where $\lambda \in (0,2)$, and $R_K := (P_K)_2 = 2P_K - \mathrm{Id}$, i.e., P_λ is a relaxation of the cyclic projection P and R_K is the reflection operator onto K.

Definition 4.5.11. The operator $R_K P$ is called a *projection-reflection* and the operator $R_K P_\mu$, where $\mu \in (0,2)$ is called a *relaxed projection-reflection* (see Fig. 4.30). The operator $P R_K$ is called a *reflection-projection* and the operator $P_\mu R_K$, where $\mu \in (0,2)$ is called a *reflection-relaxed projection*.

The following result can be found in [31, Lemma 2.1 (v)].

Lemma 4.5.12. *If $K \subseteq \mathcal{H}$ is a closed convex and obtuse cone, then $R_K x \in K$ for any $x \in \mathcal{H}$.*

Fig. 4.30 Projection-
reflection

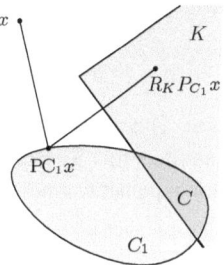

Proof. Let $x \in \mathcal{H}$ and denote $x_+ := P_K x$. By the Moreau decomposition, we have $x = x_+ + x_-$, where $x_- := P_{K^*} x$. Since K is obtuse, $-x_- \in -K^* \subseteq K$, consequently,

$$R_K x = 2P_K x - x = 2x_+ - (x_+ + x_-) = x_+ - x_- \in K + K = K$$

which completes the proof. $\square$

Bauschke and Kruk studied the properties of the reflection-projection PR_K which defines a reflection-projection method (see [31]). This operator is closely related to the projection-reflection operator $R_K P$. Similarly to the equivalence (i)$\Leftrightarrow$(ii) in Theorem 4.3.4, one can prove that

$$(x^* \in \text{Fix } R_K P_\mu \text{ and } y^* = P_\mu x^*) \iff (y^* \in \text{Fix } P_\mu R_K \text{ and } x^* = R_K y^*).$$
$$(4.76)$$

Corollary 4.5.13. *Let $K \subseteq \mathcal{H}$ be a closed convex and obtuse cone, $C_i \subseteq \mathcal{H}$ be closed convex subsets, $i \in I$, and $C := K \cap \bigcap_{i \in I} C_i \neq \emptyset$. If $\mu \in (0, \frac{m+1}{m})$, then:*

(i) $R_K P_\mu$ and $P_\mu R_K$ are nonexpansive,
(ii) $\text{Fix } R_K P_\mu = \text{Fix } P_\mu R_K = C$,
(iii) $R_K P_\mu |_K$ is β-strongly quasi-nonexpansive, where $\beta = \frac{(1-\mu)m+1}{\mu m}$.

Proof. Since the metric projection is firmly nonexpansive (see Theorem 2.2.21 (iii)), the reflection R_K is nonexpansive (see Theorem 2.2.10 (ii)) and the cyclic projection P is α-relaxed firmly nonexpansive with $\alpha = \frac{2m}{m+1}$ (see Corollary 4.5.3). Therefore, its relaxation P_ν is firmly nonexpansive for $\nu = \alpha^{-1} = \frac{m+1}{2m}$ (see Corollary 2.2.19). Let $\mu \in (0, \frac{m+1}{m})$. Then P_μ is λ-RFNE, where $\lambda = \frac{\mu}{\nu} = \alpha\mu \in (0, 2)$ (see Remark 2.1.3). Consequently, P_μ is nonexpansive (see Theorem 2.2.10 (ii)) and strongly quasi-nonexpansive (see Corollary 2.2.9). Therefore, both operators $R_K P_\mu$ and $P_\mu R_K$ are nonexpansive as compositions of nonexpansive operators. Furthermore, $\text{Fix } P_{C_i} = C_i$ (see Theorem 2.2.21 (i)),

$$\text{Fix } P_\mu = \text{Fix } P = \bigcap_{i \in I} \text{Fix } P_{C_i} = \bigcap_{i \in I} C_i = C$$

(see Theorem 2.1.26), Fix $R_K = K$ and

$$\text{Fix } R_K P_\mu = \text{Fix } P_\mu R_K = \text{Fix } R_K \cap \text{Fix } P_\mu = C$$

(see Theorem 2.1.28). Now we see that all assumptions of Theorem 2.1.51 are satisfied for $S = R_K$. Therefore, the operator $R_K P_\mu \mid_K$ is β-strongly quasi-nonexpansive, where

$$\beta = \frac{2 - \lambda}{\lambda} = \frac{2 - \alpha\mu}{\alpha\mu} = \frac{(1 - \mu)m + 1}{\mu m}$$

which completes the proof.   □

4.5.4 Cyclic Cutter

The following definition generalizes the notion of a cyclic projection.

Definition 4.5.14. Let $U_i : \mathcal{H} \to \mathcal{H}$ be cutters, $i \in I$. The operator

$$U := U_m U_{m-1} \ldots U_1$$

is called a *cyclic cutter*.

Note that the cyclic cutter $U := U_m U_{m-1} \ldots U_1$ doe not need to be a cutter even if $U_i, i \in I$, have a common fixed point (see Example 2.1.53).

Corollary 4.5.15. *Let* $U_i : \mathcal{H} \to \mathcal{H}$ *be cutters with a common fixed point, $i \in I$, and $U := U_m U_{m-1} \ldots U_1$ be a cyclic cutter. Then* Fix $U = \bigcap_{i \in I}$ Fix U_i *and U is $\frac{1}{m}$-strongly quasi-nonexpansive. Consequently, U is asymptotically regular.*

Proof. Note that a cutter is 1-strongly quasi-nonexpansive (see Theorem 2.1.39). Therefore, the equality Fix $U = \bigcap_{i \in I}$ Fix U_i follows from Theorem 2.1.26. The operator U is $\frac{1}{m}$-strongly quasi-nonexpansive as a composition of 1-strongly quasi-nonexpansive operators $U_i, i \in I$ (see Theorem 2.1.48 (ii)). Now, the asymptotic regularity of U follows from Theorem 3.4.3.   □

4.6 Landweber Operator

Let $\mathcal{H}_1$ and $\mathcal{H}_2$ be two Hilbert spaces, $A : \mathcal{H}_1 \to \mathcal{H}_2$ be a nonzero bounded linear operator and $Q \subseteq \mathcal{H}_2$ be a nonempty closed convex subset.

Definition 4.6.1. The operator $T : \mathcal{H}_1 \to \mathcal{H}_1$ defined by the equality

$$Tx := x + \frac{1}{\|A\|^2} A^*(P_Q(Ax) - Ax) \tag{4.77}$$

is called the *Landweber operator* (see Fig. 4.31).

Fig. 4.31 The Landweber
operator

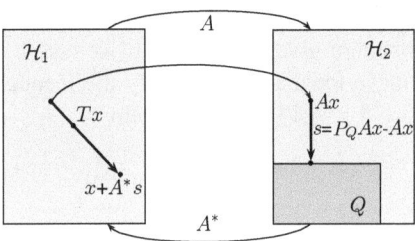

The Landweber operator is closely related to a method for the problem: find $x \in \mathcal{H}_1$
such that $Ax \in Q$. The method was proposed by Landweber for approximating
least-squares solution of a first kind integral equation [240] (see also [37, Sect. 6.1]).

Note that $\|A\| \neq 0$, because A is a nonzero operator, consequently, the operator
T is well defined. We can also take $\lambda_{\max}(A^*A)$ or $\lambda_{\max}(AA^*)$ instead of $\|A\|^2$ in
Definition 4.6.1 if A is a compact linear operator (see Theorem 1.1.27). Define a
proximity function $f : \mathcal{H}_1 \to \mathbb{R}_+$ by

$$f(x) := \frac{1}{2} \left\| P_Q(Ax) - Ax \right\|^2 \tag{4.78}$$

(cf. (1.41)). The function f is convex as a composition of a linear operator A and a
convex function $\frac{1}{2}d^2(\cdot, Q)$. Note that f has the required property: $f(x) = 0$ if and
only if $x \in C = A^{-1}(Q)$ (cf. (1.19)).

4.6.1 Main Properties

Lemma 4.6.2. *Let* $T : \mathcal{H}_1 \to \mathcal{H}_1$ *be the Landweber operator defined by (4.77) and
the corresponding proximity function* $f : \mathcal{H}_1 \to \mathbb{R}_+$ *be defined by (4.78). Then*

$$\operatorname{Fix} T = \underset{x \in \mathcal{H}_1}{\operatorname{Argmin}} f(x).$$

Proof. The function f is differentiable and $Df(x) = A^*(Ax - P_Q(Ax))$. It follows
from the necessary and sufficient optimality condition (see Corollary 1.3.3) that
$z \in \operatorname{Argmin}_{x \in \mathcal{H}_1} f(x)$ if and only if $A^*(Az - P_Q(Az)) = 0$, i.e., $Tz = z$.   □

Theorem 4.6.3. *The Landweber operator is firmly nonexpansive.*

Proof. Since P_Q is firmly nonexpansive (see Theorem 2.2.21 (iii)), the implication
(i)⇒(iv) in Theorem 2.2.10 yields that the operator $\operatorname{Id} - P_Q$ is firmly nonexpansive,
i.e.,

$$\langle (u - P_Q u) - (v - P_Q v), u - v \rangle \geq \left\| (u - P_Q u) - (v - P_Q v) \right\|^2$$

for all $u, v \in \mathcal{H}_1$. Let $G := \mathrm{Id} - T$, where $T : \mathcal{H}_1 \to \mathcal{H}_1$ is the Landweber operator given by (4.77). If we take $u := Ax$ and $v := Ay$ for $x, y \in \mathcal{H}_1$ in the above inequality and apply the inequality $\|A^*z\| \leq \|A^*\| \cdot \|z\|$ and the equality $\|A^*\| = \|A\|$, then we obtain

$$\langle G(x) - G(y), x - y \rangle$$

$$= \frac{1}{\|A\|^2} \langle A^*(\mathrm{Id} - P_Q)Ax - A^*(\mathrm{Id} - P_Q)Ay, x - y \rangle$$

$$= \frac{1}{\|A\|^2} \langle (\mathrm{Id} - P_Q)Ax - (\mathrm{Id} - P_Q)Ay, Ax - Ay \rangle$$

$$\geq \frac{1}{\|A\|^2} \left\| (\mathrm{Id} - P_Q)Ax - (\mathrm{Id} - P_Q)Ay \right\|^2$$

$$= \frac{\|A^*\|^2}{\|A\|^4} \left\| (\mathrm{Id} - P_Q)Ax - (\mathrm{Id} - P_Q)Ay \right\|^2$$

$$\geq \frac{1}{\|A\|^4} \left\| A^*(\mathrm{Id} - P_Q)Ax - A^*(\mathrm{Id} - P_Q)Ay \right\|^2$$

$$= \left\| \frac{1}{\|A\|^2} A^*(\mathrm{Id} - P_Q)Ax - \frac{1}{\|A\|^2} A^*(\mathrm{Id} - P_Q)Ay \right\|^2$$

$$= \| G(x) - G(y) \|^2,$$

i.e., G is firmly nonexpansive. By the implication (iv)$\Rightarrow$(i) in Theorem 2.2.10, the Landweber operator $T = \mathrm{Id} - G$ is also firmly nonexpansive. $\square$

Corollary 4.6.4. *Let* $\lambda \in (0, 2)$ *and* $T_\lambda : \mathcal{H}_1 \to \mathcal{H}_1$ *be a relaxation of the Landweber operator* T *with nonempty* $\mathrm{Fix}\, T$. *Then* T_λ *is* $\frac{2-\lambda}{\lambda}$-*strongly quasinonexpansive and asymptotically regular.*

Proof. By Theorem 4.6.3 the Landweber operator T is firmly nonexpansive. Therefore, the $\frac{2-\lambda}{\lambda}$-strong quasi nonexpansivity of T_λ follows from Corollary 2.2.9 and the asymptotic regularity of T_λ follows from Corollary 3.4.6. $\square$

4.6.2   Landweber Operator for Linear Systems

Let $b \in \mathcal{H}_2$. If we take $Q := \{b\}$, then $\{x \in \mathcal{H}_1 : Ax \in Q\}$ is the solution set of the linear equation $Ax = b$. Of course, $P_Q(Ax) = b$, and the Landweber operator has the form

$$Tx = x + \frac{1}{\|A\|^2} A^*(b - Ax). \tag{4.79}$$

Define the proximity function $f : \mathcal{H}_1 \to \mathbb{R}_+$ for the linear equation $Ax = b$ by $f(x) := \frac{1}{2} \|Ax - b\|^2$.

Suppose that $\mathcal{H}_1 = \mathbb{R}^n$ and $\mathcal{H}_2 = \mathbb{R}^m$ with the standard inner product and that $b \in \mathbb{R}^m$. If we take $Q := \{u \in \mathbb{R}^m : u \le b\}$, then $\{x \in \mathcal{H}_1 : Ax \in Q\}$ is the solution set of a system of linear inequalities $Ax \le b$, where A is a matrix of type $m \times n$ with rows a_i, $i \in I$, $x \in \mathbb{R}^n$ and $b \in \mathbb{R}^m$ (we suppose without loss of generality that $a_i \ne 0$, $i \in I$). In this case we have $P_Q(Ax) - Ax = -(Ax - b)_+$ (see equality (4.9)), $\|A\|^2 = \lambda_{\max}(A^\top A)$ (see Theorem 1.1.27) and

$$Tx = x - \frac{1}{\lambda_{\max}(A^\top A)} A^\top (Ax - b)_+. \tag{4.80}$$

We can also write the system of equations $Ax = b$ or inequalities $Ax \le b$ in an equivalent form $D^{\frac{1}{2}} Ax = D^{\frac{1}{2}} b$ or $D^{\frac{1}{2}} Ax \le D^{\frac{1}{2}} b$, respectively, where D is a positive definite matrix (with nonnegative elements in the second case), e.g.,

$$D := \operatorname{diag}(\frac{\omega_1}{\|a_1\|^2}, \dots, \frac{\omega_m}{\|a_m\|^2}), \tag{4.81}$$

where $w = (\omega_1, \dots, \omega_m) \in \operatorname{ri} \Delta_m$. If we apply the Landweber operator to these new systems, then we obtain

$$Tx = x - \frac{1}{\lambda_{\max}(A^\top DA)} A^\top D(Ax - b) \tag{4.82}$$

and

$$Tx = x - \frac{1}{\lambda_{\max}(A^\top DA)} A^\top D(Ax - b)_+, \tag{4.83}$$

respectively. We call the operators T given by (4.82) and (4.83), the *Landweber operators related to the matrix* D. The corresponding proximity functions obtain the form

$$f(x) = \frac{1}{2} \left\| D^{\frac{1}{2}}(Ax - b) \right\|^2 = \frac{1}{2} \|Ax - b\|_D^2$$

and

$$f(x) = \frac{1}{2} \|(Ax - b)_+\|_D^2,$$

respectively. Note that in both cases the Landweber operators and the proximity functions differ from the for the original systems.

Remark 4.6.5. The Landweber operators defined by (4.82) and (4.83) related to the matrix D given by (4.81), have a nice property. One can easily show that

$$A^\top D(Ax - b) = \sum_{i=1}^m \frac{\omega_i}{\|a_i\|^2} a_i (a_i^\top x - \beta_i)$$

and

$$A^\top D(Ax - b)_+ = \sum_{i=1}^{m} \frac{\omega_i}{\|a_i\|^2} a_i (a_i{}^\top x - \beta_i)_+.$$

Therefore, the terms $A^\top D(Ax - b)$ and $A^\top D(Ax - b)_+$ in operators T defined by (4.82) and (4.83), respectively, do not change if we rescale equations or inequalities of the system. Moreover, $\lambda_{\max}(A^\top DA)$ also does not depend on the rescaling. Therefore, the Landweber operators defined by (4.82) and (4.83) do not change after rescaling. Note that they depend only on the vector of weights $w \in \operatorname{ri} \Delta_m$. Furthermore, if A has normalized rows and $w = (\frac{1}{m}, \frac{1}{m}, \ldots, \frac{1}{m})$, then $D = \operatorname{diag}(\frac{1}{m}, \frac{1}{m}, \ldots, \frac{1}{m})$ and the Landweber operators (4.82) and (4.83) reduce to (4.79) and (4.80), respectively.

Now we state some relationships between the Landweber operators defined by (4.82) or (4.83), where D is given by (4.81), and the simultaneous projection applied to the system $Ax = b$ or $Ax \leq b$, respectively, where A is a matrix of type $m \times n$ with nonzero rows a_i, $i \in I$, $x \in \mathbb{R}^n$ and $b \in \mathbb{R}^m$.

Lemma 4.6.6. *Let* $A := [a_1, \ldots, a_m]^\top$ *be an* $m \times n$ *matrix with nonzero rows* a_i, $i \in I$, *and* $D := \operatorname{diag}(\frac{\omega_1}{\|a_1\|^2}, \ldots, \frac{\omega_m}{\|a_m\|^2})$, *where* $w = (\omega_1, \ldots, \omega_m) \in \Delta_m$. *Then*

$$0 < \lambda_{\max}(A^\top DA) \leq 1.$$

Proof. Note that $\lambda_{\max}(\alpha_1 A_1 + \alpha_2 A_2) \leq \alpha_1 \lambda_{\max}(A_1) + \alpha_2 \lambda_{\max}(A_2)$, where A_1, A_2 are positive semi-definite matrices and $\alpha_1, \alpha_2 \geq 0$. Therefore,

$$\lambda_{\max}(A^\top DA) = \lambda_{\max}\left(\sum_{i=1}^{m} \frac{\omega_i}{\|a_i\|^2} a_i a_i^\top\right)$$

$$\leq \sum_{i=1}^{m} \frac{\omega_i}{\|a_i\|^2} \lambda_{\max}(a_i a_i^\top)$$

$$= \sum_{i=1}^{m} \frac{\omega_i}{\|a_i\|^2} \lambda_{\max}(a_i^\top a_i) = 1.$$

Furthermore, $A^\top DA$ is a nonzero matrix. Therefore, $\lambda_{\max}(A^\top DA) > 0$.   □

We call the parameter

$$\sigma_L := \frac{1}{\lambda_{\max}(A^\top DA)}$$

a *Landweber extrapolation parameter* or the *Landweber step size*.

Corollary 4.6.7. *The Landweber operators T defined by (4.82) and (4.83), where D is given by (4.81), are extrapolations of the simultaneous projection operators U defined by $Ux := x - A^\top D(Ax - b)$ and by $Ux := x - A^\top D(Ax - b)_+$, respectively, i.e., $Tx = x + \sigma_L(Ux - x)$, where the Landweber step size $\sigma_L \geq 1$.*

Remark 4.6.8. Suppose that A is a matrix with normalized rows. If we take $w = (\frac{1}{m}, \frac{1}{m}, \ldots, \frac{1}{m})$ in Lemma 4.6.6, then we obtain

$$\lambda_{\max}(A^\top A) = m\lambda_{\max}(A^\top DA) \leq m.$$

One can show even more: $\lambda_{\max}(A^\top A) \leq s$, where s denotes the maximal number of nonzero elements in any column of A (see [55, Proposition 4.1] or [58, Corollary 2.8]), which is of course a stronger result than Lemma 4.6.6.

4.6.3 Extrapolated Landweber Operator for a System of Linear Equations

Consider a system of linear equations $Ax = b$, where A is an $m \times n$ matrix with nonzero rows a_i, $i \in I$, $x \in \mathbb{R}^n$ and $b \in \mathbb{R}^m$. Let $f : \mathbb{R}^n \to \mathbb{R}_+$ be a proximity function defined by

$$f(x) := \frac{1}{2}\sum_{i\in I}\omega_i\left(\frac{a_i^\top x - \beta_i}{\|a_i\|}\right)^2, \tag{4.84}$$

where $w = (\omega_1, \ldots, \omega_m) \in \Delta_m$. In this section we present an extrapolation of the Landweber operator and we show that this extrapolation is a cutter. The proximity function can be written in the form

$$f(x) = \frac{1}{2}(Ax - b)^\top D(Ax - b),$$

where $D := \mathrm{diag}(\frac{\omega_1}{\|a_1\|^2}, \ldots, \frac{\omega_m}{\|a_m\|^2})$ (see (4.64)), or, equivalently, in the form

$$f(x) = \frac{1}{2}x^\top Gx + g^\top x + c, \tag{4.85}$$

where $G = A^\top DA$, $g = -A^\top Db$ and $c = \frac{1}{2}b^\top Db$.

Let T be a simultaneous projection, i.e., $Tx = x - A^\top D(Ax - b)$. Let $x \in \mathbb{R}^n$ and

$$s(x) := Tx - x = A^\top D(b - Ax). \tag{4.86}$$

It follows from differential rules that

$$-s(x) = \nabla f(x) = Gx + g. \tag{4.87}$$

Denote $M := \operatorname{Argmin}_{x \in \mathbb{R}^n} f(x)$. It is clear that $\operatorname{Fix} T = M \neq \emptyset$ (see Theorem 4.4.8). Let $T_{\sigma,\lambda} := \operatorname{Id} + \lambda \sigma(T - \operatorname{Id})$ be a generalized relaxation of T with the step size function $\sigma = \sigma_{EL} : \mathbb{R}^n \to (0, +\infty)$ defined by

$$\sigma_{EL}(x) := \frac{\|s(x)\|^2}{s(x)^\top A^\top D A s(x)} \tag{4.88}$$

for $x \notin \operatorname{Fix} T$. We will prove that $\sigma_{EL}(x) \geq \sigma_L$ for all $x \in \mathbb{R}^n$ (see Lemma 4.6.9 below). Therefore, we call the step size σ_{EL} defined by (4.88) an *extrapolated Landweber step size* and operator $T_{\sigma_{EL}}$ with this step size—an *extrapolated Landweber operator*. We have $s(x)^\top A^\top D A s(x) > 0$ for $x \notin \operatorname{Fix} T$. Indeed, if $s(x)^\top A^\top D A s(x) = 0$, then, of course, $D^{\frac{1}{2}} A s(x) = 0$ and, by (4.86), we have

$$\|s(x)\|^2 = s(x)^\top s(x) = (b - Ax)^\top D A s(x)$$
$$= (b - Ax)^\top D^{\frac{1}{2}} D^{\frac{1}{2}} A s(x) = 0,$$

i.e., $x \in \operatorname{Fix} T$, a contradiction. Therefore, $s(x)^\top A^\top D A s(x) > 0$ and the step size $\sigma_{EL}(x)$ is well defined. Equality (4.87) yields

$$s(x)^\top A^\top D A s(x) = s(x)^\top G s(x) = (Gx + g)^\top (G^2 x + Gg)$$

which leads to the following equivalent form of the step size $\sigma_{EL}(x)$:

$$\sigma_{EL}(x) = \frac{\|Gx + g\|^2}{(Gx + g)^\top (G^2 x + Gg)}. \tag{4.89}$$

Lemma 4.6.9. *Let $x \notin \operatorname{Fix} T$ and $z \in \operatorname{Fix} T$. Then the step size $\sigma_{EL}(x)$ defined by (4.88) satisfies the following inequalities*

$$1 \leq \frac{1}{\lambda_{\max}(A^\top D A)} \leq \sigma_{EL}(x) \leq \frac{\langle z - x, Tx - x \rangle}{\|Tx - x\|^2} \tag{4.90}$$

for any $z \in \operatorname{Fix} T$. Consequently, the operator $T_{\sigma_{EL}}$ is a cutter and, for all $\lambda \in (0, 2)$, the operator $T_{\sigma_{EL},\lambda}$ is $\frac{2-\lambda}{\lambda}$-strongly quasi-nonexpansive and asymptotically regular.

Proof. Let $\sigma_{EL}(x)$ be defined by (4.88). The first inequality in (4.90) follows from Lemma 4.6.6. We have

$$\sigma_{EL}(x) = \frac{\|s(x)\|^2}{s(x)^\top A^\top D A s(x)} \geq \frac{\|s(x)\|^2}{\lambda_{\max}(A^\top D A) \|s(x)\|^2} = \frac{1}{\lambda_{\max}(A^\top D A)},$$

i.e., the second inequality in (4.90) is true. Applying (4.89), we write the third inequality in (4.90) in the form

$$\frac{\|Gx + g\|^2}{(Gx + g)^\top (G^2 x + Gg)} \leq \frac{(x - z)^\top (Gx + g)}{\|Gx + g\|^2}, \tag{4.91}$$

where $z \in \mathrm{Fix}\, T = M$. Note that $s(z) = 0$ and $Gz + g = \nabla f(z) = 0$, by (4.87) and by the sufficient optimality condition. Since G is positive semi-definite, there exists an orthogonal matrix U and a diagonal matrix $\Delta := \mathrm{diag}\, d$ with $d = (\delta_1, \delta_2 \dots, \delta_m) \geq 0$ such that $G = U^\top \Delta U$ (see Theorem 1.1.34 (iii)). Note that $\max_{i \in I} \delta_i > 0$, because A is a nonzero matrix. Let $v = (v_1, v_2, \dots, v_m) \in \mathbb{R}^m$ be such that $v^\top \Delta v > 0$ and let $x := U^\top v + z$. Then

$$Gx + g = G(U^\top v + z) + g = GU^\top v = U^\top \Delta v$$

and

$$\|Gx + g\|^2 = v^\top \Delta^2 v > 0,$$

i.e., $Gx + g \neq 0$. Note that $x \notin \mathrm{Fix}\, T$ since $x - Tx = -s(x) = Gx + g \neq 0$. Similarly, we obtain

$$(Gx + g)(G^2 x + Gx) = v^\top \Delta^3 v > 0$$

and

$$(x - z)^\top (Gx + g) = v^\top \Delta v.$$

Therefore, (4.91) can be written in equivalent forms

$$\frac{v^\top \Delta^2 v}{v^\top \Delta^3 v} \leq \frac{v^\top \Delta v}{v^\top \Delta^2 v}$$

or

$$(v^\top \Delta^2 v)^2 - (v^\top \Delta v)(v^\top \Delta^3 v) \leq 0.$$

We prove that the latter inequality is true. Denote $I' := \{i \in I : \delta_i > 0\}$. We have

$$(v^\top \Delta^2 v)^2 - (v^\top \Delta v)(v^\top \Delta^3 v)$$

$$= \left(\sum_{i \in I} \delta_i^2 v_i^2\right)^2 - \left(\sum_{i \in I} \delta_i v_i^2\right)\left(\sum_{i \in I} \delta_i^3 v_i^2\right)$$

$$= \sum_{i \in I}\sum_{j \in I}(\delta_i^2 \delta_j^2 v_i^2 v_j^2 - \delta_i \delta_j^3 v_i^2 v_j^2)$$

$$= \sum_{i \in I'}\sum_{j \in I'}(\delta_i^2 \delta_j^2 v_i^2 v_j^2 - \delta_i \delta_j^3 v_i^2 v_j^2)$$

$$= \sum_{i \in I'}\sum_{j \in I'} \delta_i^2 \delta_j^2 \left(1 - \frac{\delta_j}{\delta_i}\right) v_i^2 v_j^2$$

$$= \sum_{i,j \in I', i>j} \delta_i^2 \delta_j^2 (2 - \frac{\delta_i}{\delta_j} - \frac{\delta_j}{\delta_i}) v_i^2 v_j^2$$

$$= - \sum_{i,j \in I', i>j} \delta_i \delta_j^3 (\frac{\delta_j}{\delta_i} - 1)^2 v_i^2 v_j^2 \le 0$$

i.e., the third inequality in (4.90) is true. Note that T is a cutter as a firmly nonexpansive operator having a fixed point (see Theorem 2.2.5 (i) and Corollary 4.4.4), consequently, T is oriented (see Definition 2.4.4). By Corollary 2.4.5, the operator $T_{\sigma EL}$ is a cutter. Let $\lambda \in (0,2)$. The $\frac{2-\lambda}{\lambda}$-strong quasi nonexpansivity and asymptotic regularity of $T_{\sigma EL, \lambda}$ follow from Theorems 2.1.39 and 3.4.3.   □

4.7 Projected Landweber Operator

Let $\mathcal{H}_1$ and $\mathcal{H}_2$ be two Hilbert spaces. Consider the following split feasibility problem

$$\text{find } x \in C \text{ with } Ax \in Q,$$

if such an x exists, where $C \subseteq \mathcal{H}_1$ and $Q \subseteq \mathcal{H}_2$ are closed convex subsets and $A : \mathcal{H}_1 \to \mathcal{H}_2$ is a bounded linear operator.

Definition 4.7.1. An operator $U : \mathcal{H}_1 \to \mathcal{H}_1$ defined by

$$U := P_C (\text{Id} + \frac{1}{\|A\|^2} A^* (P_Q - \text{Id}) A) \tag{4.92}$$

is called the *projected Landweber operator* or an *oblique projection* (see Fig. 4.32).

The following result is due to Byrne [55, Proposition 2.1].

Proposition 4.7.2. *Let* $R_\lambda := P_C T_\lambda$, *where* $\lambda > 0$, *be a projected relaxation of the Landweber operator* T *defined by (4.77), i.e.,*

$$R_\lambda x := P_C (x + \frac{\lambda}{\|A\|^2} A^* (P_Q - \text{Id}) Ax), \tag{4.93}$$

and $f : C \to \mathbb{R}_+$ *be the proximity function defined by (4.78). Then*

$$\text{Fix } R_\lambda = \text{Argmin}_{x \in C} f(x).$$

Proof. The function f is convex as a composition of a linear operator A and a convex function $\frac{1}{2} d^2(\cdot, Q)$. Furthermore, f is differentiable and

$$Df(x) = A^* (Ax - P_Q(Ax)).$$

Fig. 4.32 Projected
Landweber operator

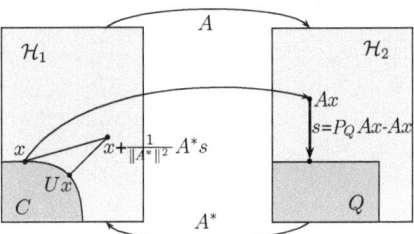

Similarly as in the proof of Corollary 1.3.5, for any $\gamma > 0$, we have

$$x \in \operatorname*{Argmin}_C f \iff -Df(x) \in N_C(x)$$

$$\iff -\gamma Df(x) \in N_C(x)$$

$$\iff x = P_C(x - \gamma Df(x))$$

$$\iff x = P_C(x + \gamma(A^*(P_Q(Ax) - Ax)))$$

$$\iff x = P_C(x + \frac{\lambda}{\|A\|^2} A^*(P_Q - \operatorname{Id})Ax) = R_\lambda x$$

$$\iff x \in \operatorname{Fix} R_\lambda,$$

which completes the proof.   □

If we take $C = \mathcal{H}_1$ and $\lambda = 1$ in Proposition 4.7.2, then we obtain Lemma 4.6.2.

Corollary 4.7.3. *Let* $\lambda \in (0, 2)$. *The projected relaxation of the Landweber operator, defined by (4.93) is* $\frac{4}{4-\lambda}$-*relaxed firmly nonexpansive. If, furthermore,* $\operatorname{Fix}(P_C T) \neq \emptyset$, *then* R_λ *is* $\frac{2-\lambda}{2}$-*strongly quasi-nonexpansive and asymptotically regular.*

Proof. By Theorem 4.6.3 the Landweber operator T is firmly nonexpansive. The $\frac{4}{4-\lambda}$-relaxed firm nonexpansivity of R_λ follows from Theorem 2.2.46 (i). Let $\operatorname{Fix}(P_C T) \neq \emptyset$. The $\frac{2-\lambda}{2}$-strong quasi nonexpansivity of R_λ follows from Theorem 2.2.46 (iii) and the asymptotic regularity of R_λ follows from Corollary 3.4.5.   □

4.8 Simultaneous Cutter

If we replace the projections P_{C_i} by cutters in the definition of a simultaneous projection (see Definition 4.4.1) and allow the weights to depend on a current point $x \in \mathcal{H}$, then we obtain a more general operator. In this section we prove that this operator is strongly quasi-nonexpansive and asymptotically regular.

Fig. 4.33 Simultaneous cutter

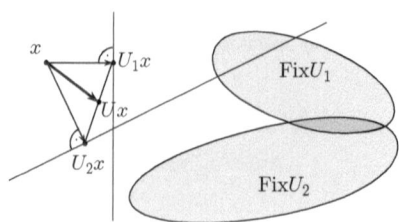

Definition 4.8.1. Let $U_i : \mathcal{H} \to \mathcal{H}$ be cutters, $i \in I$, and $w : \mathcal{H} \to \Delta_m$ be a weight function. An operator $U : \mathcal{H} \to \mathcal{H}$ defined by

$$Ux := \sum_{i \in I} \omega_i(x) U_i x \qquad (4.94)$$

is called a *simultaneous cutter* (see Fig. 4.33).

By Corollary 2.1.49, a simultaneous cutter defined by (4.94) is a cutter if $w : \mathcal{H} \to \Delta_m$ is an appropriate weight function.

Theorem 4.8.2. *Let U_i be cutters with a common fixed point, $i \in I$, and $U := \sum_{i \in I} \omega_i U_i$ be a simultaneous cutter, where the weight function $w : \mathcal{H} \to \Delta_m$ is appropriate. Then U is a cutter,*

$$\|U_\lambda x - z\|^2 \leq \|x - z\|^2 - \lambda(2 - \lambda) \sum_{i \in I} \omega_i(x) \|U_i x - x\|^2 \qquad (4.95)$$

and

$$\|U_\lambda x - z\|^2 \leq \|x - z\|^2 - \lambda(2 - \lambda) \|Ux - x\|^2 \qquad (4.96)$$

for all $\lambda \in [0,2]$, $x \in \mathcal{H}$ and $z \in \operatorname{Fix} U$. Consequently, for all $\lambda \in (0,2)$, the operator U_λ is $\frac{2-\lambda}{\lambda}$-strongly quasi-nonexpansive and asymptotically regular.

Proof. Since a cutter is strongly quasi-nonexpansive (see Theorem 2.1.39) it is strictly quasi-nonexpansive (see Remark 2.1.44 (iii)). It follows from Theorem 2.1.26 that $\operatorname{Fix} U = \bigcap_{i \in I} \operatorname{Fix} U_i$. Let $\lambda \in [0,2]$, $x \in \mathcal{H}$ and $z \in \operatorname{Fix} U$. The convexity of the function $\|\cdot\|^2$ and property (2.21) yield

$$\|U_\lambda x - z\|^2 = \left\| x + \lambda \sum_{i \in I} \omega_i(x)(U_i x - x) - z \right\|^2$$

$$= \|x - z\|^2 + \lambda^2 \left\| \sum_{i \in I} \omega_i(x)(U_i x - x) \right\|^2$$

$$- 2\lambda \sum_{i \in I} \omega_i(x)\langle z - x, U_i x - x \rangle$$

$$\leq \|x - z\|^2 + \lambda^2 \sum_{i \in I} \omega_i(x) \|U_i x - x\|^2$$

$$-2\lambda \sum_{i \in I} \omega_i(x) \|U_i x - x\|^2$$

$$= \|x - z\|^2 - \lambda(2 - \lambda) \sum_{i \in I} \omega_i(x) \|U_i x - x\|^2,$$

i.e., inequality (4.95) holds. Inequality (4.96) follows now from the convexity of the function $\|\cdot\|^2$. Let $\lambda \in (0, 2)$. The $\frac{2-\lambda}{\lambda}$-strong quasi nonexpansivity of U_λ follows from (4.96) and from the obvious equality

$$\lambda(2 - \lambda) \|Ux - x\|^2 = \frac{2 - \lambda}{\lambda} \|U_\lambda x - x\|^2.$$

The asymptotic regularity of U_λ follows now from Theorem 3.4.3.   □

4.9   Extrapolated Simultaneous Cutter

Let $U_i : \mathcal{H} \to \mathcal{H}$ be cutters with a common fixed point, $i \in I := \{1, 2, \ldots, m\}$. In this section we will show how to extend the results of Theorem 4.8.2 to a generalized relaxation of a simultaneous cutter. We denote, as in Sect. 4.8, $U := \sum_{i \in I} \omega_i U_i$, where $w : \mathcal{H} \to \Delta_m$ is a weight function. Note that the estimation of the distance $\|U_\lambda x - z\|^2$ given by (4.95) is stronger than that given by (4.96). Therefore, we can expect that the range of the relaxation parameter λ which guarantees the strong quasi nonexpansivity of U_λ with respect to $\bigcap_{i \in I}$ Fix U_i can be extended over the interval $[0, 2]$. As we will see, our expectation is true if we allow the relaxation parameter to depend on the point $x \in \mathcal{H}$. Therefore, it is more convenient to use a generalized relaxation $U_{\sigma,\lambda} : \mathcal{H} \to \mathcal{H}$ of the operator U, defined by

$$U_{\sigma,\lambda}(x) := x + \lambda\sigma(x)(Ux - x) \qquad (4.97)$$

(cf. (2.67)). Denote $C_i := $ Fix U_i and $C := \bigcap_{i \in I} C_i$. If we suppose that the weight function $w : \mathcal{H} \to \Delta_m$ is appropriate, then Theorem 2.1.26 yields Fix $U = C$.

4.9.1   Properties of the Extrapolated Simultaneous Cutter

In this section we consider a generalized relaxations of a simultaneous cutter with a step size function $\sigma : \mathcal{H} \to (0, +\infty)$ satisfying the following inequality

$$\sigma(x) \leq \sigma_w(x), \qquad (4.98)$$

Fig. 4.34 Extrapolated simultaneous cutter

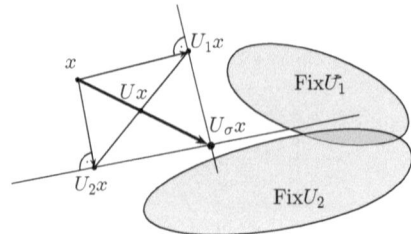

where

$$\sigma_w(x) := \frac{\sum_{i \in I} \omega_i(x) \|U_i x - x\|^2}{\left\|\sum_{i \in I} \omega_i(x) U_i x - x\right\|^2} \tag{4.99}$$

for all $x \notin \bigcap_{i \in I} \operatorname{Fix} U_i$, and the weight function $w : \mathcal{H} \to \Delta_m$ is appropriate. We suppose without loss of generality that $\sigma(x) = 1$ for all $x \in \bigcap_{i \in I} \operatorname{Fix} U_i$. The fact that w is appropriate ensures that $\operatorname{Fix} U_{\sigma,\lambda} = \operatorname{Fix} U = \bigcap_{i \in I} \operatorname{Fix} U_i \neq \emptyset$ (see Theorem 2.1.26 (i)), consequently, the step size $\sigma_w(x)$ is well defined. If $\sigma(x) \geq 1$ for all $x \in \mathcal{H}$, then the operator U_σ is an extrapolation of the simultaneous cutter $U := \sum_{i \in I} \omega_i U_i$. We call an operator U_σ with a step size function $\sigma \geq 1$ satisfying inequality (4.98) for all $x \in \mathcal{H}$ an *extrapolated simultaneous cutter* (ESC) (see Fig. 4.34). The existence of an ESC follows from the convexity of the function $\|\cdot\|^2$. It turns out that an extrapolated simultaneous cutter is a cutter.

Theorem 4.9.1. *Let $U := \sum_{i \in I} \omega_i U_i$ be a simultaneous cutter with an appropriate weight function $w : \mathcal{H} \to \Delta_m$, $U_{\sigma,\lambda}$ be a generalized relaxation of U, defined by (4.97) with the step size function $\sigma : \mathcal{H} \to (0, +\infty)$ satisfying inequality (4.98) for $x \notin \bigcap_{i \in I} \operatorname{Fix} U_i$, and with $\lambda \in [0, 2]$. Then the operator U_σ is a cutter. Consequently, for all $\lambda \in (0, 2)$, the operator $U_{\sigma,\lambda}$ is $\frac{2-\lambda}{\lambda}$-strongly quasi-nonexpansive and asymptotically regular and*

$$\|U_{\sigma,\lambda} x - z\|^2 \leq \|x - z\|^2 - \lambda(2 - \lambda)\sigma^2(x) \|U x - x\|^2 \tag{4.100}$$

for all $x \in \mathcal{H}$ and for all $z \in \bigcap_{i \in I} \operatorname{Fix} U_i$.

Proof. Since a cutter is strictly quasi-nonexpansive (see Theorem 2.1.39), it holds that $\operatorname{Fix} U_{\sigma,\lambda} = \operatorname{Fix} U = \bigcap_{i \in I} \operatorname{Fix} U_i$ for an appropriate weight function $w : \mathcal{H} \to \Delta_m$ (see Remark 2.4.2 (d) and Theorem 2.1.26). Let $x \in \mathcal{H}$ and $z \in \bigcap_{i \in I} \operatorname{Fix} U_i$. If $x \in \bigcap_{i \in I} \operatorname{Fix} U_i$, the inequality $\langle z - x, U_\sigma x - x \rangle \geq \|U_\sigma x - x\|^2$ is clear. Now let $x \notin \bigcap_{i \in I} \operatorname{Fix} U_i$. It follows from (2.21) that $\langle z - x, U_i x - x \rangle \geq \|U_i x - x\|^2, i \in I$, consequently,

$$\langle z - x, U x - x \rangle = \left\langle z - x, \sum_{i \in I} \omega_i(x)(U_i x - x) \right\rangle$$

$$= \sum_{i \in I} \omega_i(x) \langle z - x, U_i x - x \rangle$$

$$\geq \sum_{i \in I} \omega_i(x) \|U_i x - x\|^2$$

$$= \sigma_w(x) \|Ux - x\|^2.$$

Therefore, inequality (4.98) yields

$$\langle z - x, Ux - x \rangle \geq \sigma(x) \|Ux - x\|^2.$$

Multiplying both sides by $\sigma(x)$ and applying the equality $U_\sigma x - x = \sigma(x)(Ux - x)$ we obtain

$$\langle z - x, U_\sigma x - x \rangle \geq \|U_\sigma x - x\|^2.$$

Therefore, U_σ is a cutter (see Remark 2.1.31). Now Theorem 2.1.39 yields the $\frac{2-\lambda}{\lambda}$-strong quasi nonexpansivity of $U_{\sigma,\lambda}$ for all $\lambda \in (0, 2)$. The asymptotic regularity of $U_{\sigma,\lambda}$ follows from Theorem 3.4.3, and inequality (4.100) follows from the definition of $U_{\sigma,\lambda} x$.   □

The properties of extrapolations of the simultaneous projection or of simultaneous cutters were studied by Pierra [284], Dos Santos [146, Sect. 4], Cegielski [62, Sect. 4.3], Kiwiel [229, Sect. 3], Bauschke [17, Sects. 7.3 and 8.3], Combettes [118, Sects. 5.4–5.8], [120, Sect. IV], [121, Sect. 2], Crombez [127, Sect. 4], Aleyner and Reich [5, Sect. 3] and by Cegielski and Censor [70, Sect. 9.5]. In [121, Proposition 2.4] a special case of Theorem 4.9.1 was proved, where it was supposed that $w \in \mathrm{ri}\, \Delta_m$ and that $\sigma = \sigma_w$. Extrapolations of simultaneous cutters with appropriate weight functions were introduced in [70, Sect. 9.5].

4.9.2 Extrapolated Simultaneous Projection

In this section we apply the results of Sect. 4.9.1 to the generalized relaxation of the simultaneous projection $U := \sum_{i \in I} \omega_i P_{C_i}$, where $w : \mathcal{H} \to \Delta_m$ is an appropriate weight function, $C_i \subseteq \mathcal{H}$ are nonempty closed convex subsets, $i \in I$, with $C := \bigcap_{i \in I} C_i \neq \emptyset$. We use here the name 'simultaneous projection' for a more general case than in Sect. 4.4, because we allow w to depend on $x \in \mathcal{H}$. Note that U is a special case of the simultaneous cutter $U := \sum_{i \in I} \omega_i U_i$, where $U_i := P_{C_i}, i \in I$. We consider a generalized relaxation $U_{\sigma,\lambda}$ of U, where $\lambda \in [0, 2]$ and the step size function $\sigma : \mathcal{H} \to (0, +\infty)$ satisfies the inequality

$$\sigma(x) \leq \sigma_w(x), \tag{4.101}$$

where

$$\sigma_w(x) := \frac{\sum_{i \in I} \omega_i(x) \|P_{C_i} x - x\|^2}{\left\| \sum_{i \in I} \omega_i(x) P_{C_i} x - x \right\|^2} \tag{4.102}$$

Fig. 4.35 Extrapolated
simultaneous projection with
the step size $\sigma := \sigma_w$

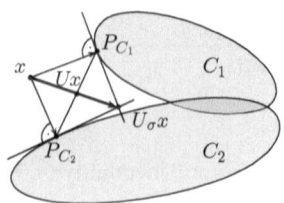

for $x \notin C$ and $\sigma_w(x) = 1$ for $x \in C$. We have $C = \text{Fix } U = \text{Fix } U_{\sigma,\lambda}$, because
w is an appropriate weight function (see Theorem 2.1.26 and Remark 2.4.2 (d)).
Therefore, σ_w is well defined. We call an operator U_σ with a step size function
$\sigma \geq 1$ satisfying inequality (4.101) an *extrapolated simultaneous projection* (see
Fig. 4.35).

Corollary 4.9.2. *Let $C \neq \emptyset$, $w : \mathcal{H} \to \Delta_m$ be an appropriate weight function and
$U : \mathcal{H} \to \mathcal{H}$ be a simultaneous projection defined by*

$$Ux := \sum_{i \in I} \omega_i(x) P_{C_i} x. \tag{4.103}$$

*Furthermore, let $U_{\sigma,\lambda}$ be the generalized relaxation of U, defined by (4.97) with
$\lambda \in [0, 2]$ and with the step size function $\sigma : \mathcal{H} \to (0, +\infty)$ satisfying inequality
(4.101). Then U_σ is a cutter. Consequently, for all $\lambda \in (0, 2)$, $U_{\sigma,\lambda}$ is $\frac{2-\lambda}{\lambda}$-strongly
quasi-nonexpansive and asymptotically regular.*

Proof. Since the metric projection P_C is a cutter with $\text{Fix } P_C = C$ (see Theorem 2.2.21 (i) and (ii)), the corollary follows directly from Theorem 4.9.1.   □

4.9.3 Extrapolated Simultaneous Projection for LFP

Consider a consistent system of linear inequalities in $\mathbb{R}^n$: $a_i^\top y \leq \beta_i$, $i \in I$, where
$a_i \in \mathbb{R}^n$, $a_i \neq 0$ and $\beta_i \in \mathbb{R}$, $i \in I$. This system can be written in the matrix form
$Ay \leq b$, where A is a matrix of type $m \times n$ with rows a_i, $i \in I$, and $b \in \mathbb{R}^m$ is a
vector with coordinates β_i, $i \in I$. If we take $C_i := H_-(a_i, \beta_i)$ and apply (4.7) and
(4.70), then we obtain the following expression for the step size $\sigma_w(x)$ defined by
(4.102):

$$\sigma_w(x) = \frac{\sum_{i \in I} \omega_i(x) \left[\frac{(a_i^\top x - \beta_i)_+}{\|a_i\|} \right]^2}{\left\| A^\top D(x)(Ax - b)_+ \right\|^2}, \tag{4.104}$$

for $x \notin C := \{y \in \mathbb{R}^n : Ay \leq b\}$, where $D(x) := \text{diag}(\frac{\omega_1(x)}{\|a_1\|^2}, \dots, \frac{\omega_m(x)}{\|a_m\|^2})$ and the
weight function $w : \mathbb{R}^n \to \Delta_m$ is appropriate. The generalized relaxation $U_{\sigma,\lambda}$ of
the simultaneous projection $U := \sum_{i \in I} \omega_i P_{C_i}$ with the step size $\sigma(x) := \sigma_w(x)$
can be presented as a λ-relaxation of the operator $T_w = U_{\sigma_w}$ defined by

$$T_w x := x - \frac{\sum_{i \in I} \omega_i(x) \left[\frac{(a_i^\top x - \beta_i)_+}{\|a_i\|} \right]^2}{\left\| A^\top D(x)(Ax - b)_+ \right\|^2} A^\top D(x)(Ax - b)_+ \qquad (4.105)$$

for $x \notin C$ and $T_w x = x$ for $x \in C$. Note that the terms $\frac{(a_i^\top x - \beta_i)_+}{\|a_i\|}$, $i \in I$, and $A^\top D(x)(Ax - b)_+$ do not change if we rescale inequalities of the system (see Remark 4.6.5), consequently, the operator T_w does not depend on the rescaling of the inequalities. Therefore, we can suppose, without loss of generality, that $\|a_i\| = 1$, $i \in I$. Then we have

$$T_w x = x - \frac{\sum_{i \in I} \omega_i(x)[(a_i^\top x - \beta_i)_+]^2}{\left\| A^\top W(x)(Ax - b)_+ \right\|^2} A^\top W(x)(Ax - b)_+, \qquad (4.106)$$

for $x \notin C$, where $W(x) := \operatorname{diag} w(x)$. As before, $C = \operatorname{Fix} U = \operatorname{Fix} T_w$. Moreover, $T_{\alpha w} = T_w$ for any function $\alpha : \mathbb{R}^n \to (0, +\infty)$. Therefore, we can take $w : \mathbb{R}^n \to \mathbb{R}_+^m \backslash \{0\}$ in (4.106) instead of $w : \mathbb{R}^n \to \Delta_m$.

Corollary 4.9.3. *Let a system of linear inequalities $a_i^\top x \leq \beta_i$, $i \in I$, be consistent and the weight function $w : \mathbb{R}^m \to \Delta_m$ be appropriate. Then the operator T_w given by (4.106) is a cutter. Consequently, for all $\lambda \in (0, 2)$, its relaxation $T_{w,\lambda}$ is $\frac{2-\lambda}{\lambda}$-strongly quasi-nonexpansive and asymptotically regular.*

Proof. Since $T_w = U_{\sigma_w}$, where U is defined by (4.103) and σ_w is defined by (4.104), the corollary follows from Corollary 4.9.2. □

4.9.4 Surrogate Projection

Similarly as in Sect. 4.9.3, we consider a consistent system of linear inequalities $Ay \leq b$. Denote $C_i := H_-(a_i, \beta_i) = \{y \in \mathbb{R}^n : a_i^\top y \leq \beta_i\}$, where $a_i \in \mathbb{R}^n$ is the ith row of the matrix A and β_i is the ith coordinate of $b \in \mathbb{R}^m$, $i \in I$. We have

$$C := \{y \in \mathbb{R}^n : Ax \leq b\} = \bigcap_{i \in I} C_i.$$

Let $v : \mathbb{R}^n \to \Delta_m$ be a weight function. Let $x \in \mathbb{R}^n$. If we multiply the particular inequalities $a_i^\top y \leq \beta_i$ of the system $Ay \leq b$ by nonnegative weights $v_i(x)$, $i \in I$, and if we add the resulting inequalities, then we obtain the inequality

$$v(x)^\top Ay \leq v(x)^\top b. \qquad (4.107)$$

Denote

$$a(x) := \sum_{i \in I} v_i(x) a_i = A^\top v(x),$$

$$\beta(x) := \sum_{i \in I} v_i(x)\beta_i = v(x)^\top b$$

and

$$C_{v(x)} := \{y \in \mathbb{R}^n : v(x)^\top Ay \leq v(x)^\top b\}.$$

It is clear that $C \subseteq C_{v(x)}$ for any weight function v. If $a(x) \neq 0$, then $C_{v(x)}$ is a half-space, $C_{v(x)} = H_-(a(x), \beta(x))$, which is called a *surrogate constraint* for the system of linear inequalities $Ax \leq b$.

Definition 4.9.4. We say that a weight function $v : \mathbb{R}^n \to \Delta_m$ is *essential* (for the system $Ax \leq b$) if for any $x \notin C$ it holds that

$$v(x)^\top (Ax - b) > 0 \qquad (4.108)$$

or, equivalently, $x \notin C_{v(x)} = H_-(a(x), \beta(x))$.

It turns out that for a consistent system $Ay \leq b$, an essential weight function v defines, for any $x \in \mathbb{R}^n$, a half-space (a surrogate constraint) containing the solution set C.

Lemma 4.9.5. *Let $C \neq \emptyset$. If $v : \mathbb{R}^n \to \Delta_m$ is an essential weight function, then for any $x \notin C$ it holds that $A^\top v(x) \neq 0$ and $C \subseteq H_-(a(x), \beta(x))$.*

Proof. Let $x \notin C$ and $v(x)^\top (Ax - b) > 0$. Suppose that $A^\top v(x) = 0$. Then $v(x)^\top b \geq 0$, because $C_{v(x)} \supseteq C \neq \emptyset$. Consequently,

$$0 < v(x)^\top (Ax - b) = (A^\top v(x))^\top x - v(x)^\top b \leq 0.$$

The contradiction proves that $a(x) := A^\top v(x) \neq 0$. Now we have $C \subseteq C_{v(x)} = H_-(a(x), \beta(x))$.   □

Let $v : \mathbb{R}^n \to \Delta_m$ be an essential weight function and $S_v x$ be the metric projection of $x \in \mathbb{R}^n$ onto the surrogate constraint $H_-(a(x), \beta(x))$. If $x \in C$, then, of course, $S_v x = x$. If $x \notin C$, then

$$S_v x = x - \frac{(a(x)^\top x - \beta(x))_+}{\|a(x)\|^2} a(x)$$

$$= x - \frac{[(A^\top v(x))^\top x - v(x)^\top b]_+}{\|A^\top v(x)\|^2} A^\top v(x)$$

$$= x - \frac{[v(x)^\top (Ax - b)]_+}{\|A^\top v(x)\|^2} A^\top v(x).$$

Inequality (4.108) yields

$$S_v x = x - \frac{v(x)^\top (Ax - b)}{\|A^\top v(x)\|^2} A^\top v(x). \qquad (4.109)$$

Fig. 4.36 Surrogate
projection

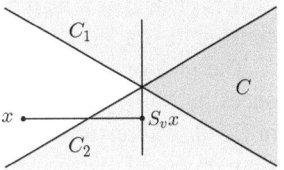

Note that $S_v x$ is well defined, because $A^\top v(x) \neq 0$ (see Lemma 4.9.5). We call
$S_v : \mathbb{R}^n \to \mathbb{R}^n$ a *surrogate projection* (see Fig. 4.36). Note that an essential weight
function v can be arbitrarily defined on C.

A surrogate projection S_v is not a metric projection, because the weights v_i
defining the half-space $H_-(a(x), \beta(x))$ depend on $x \notin C$.

Theorem 4.9.6. *Let $C \neq \emptyset$. If $v : \mathbb{R}^n \to \Delta_m$ is an essential weight function, then
Fix $S_v = C$ and S_v is a cutter. Consequently, for any $\lambda \in (0, 2)$, its relaxation $S_{v,\lambda}$
is $\frac{2-\lambda}{\lambda}$-strongly quasi-nonexpansive and asymptotically regular.*

Proof. Let v be an essential weight function. It is clear that $C \subseteq$ Fix S_v. We prove
that Fix $S_v \subseteq C$. Let $x \notin C$. Then Lemma 4.9.5 and equality (4.109) yield $S_v x \neq x$,
i.e., $x \notin$ Fix S_v. We have proved that Fix $S_v = C$. Let $x \in \mathbb{R}^n$ and $z \in C$. If $x \in C$,
then $S_v x = x$ and the inequality

$$\langle z - S_v x, x - S_v x \rangle \leq 0 \tag{4.110}$$

is obvious. Let now $x \notin C$. Since $S_v x = P_{C_v} x$ and $C \subseteq C_v$, inequality
(4.110) follows from the characterization of the metric projection (Theorem 1.2.4).
Therefore, S_v is a cutter. The second part of the theorem follows now from
Theorem 2.1.39 and from Corollary 3.4.4. $\square$

An appropriate weight function $v : \mathbb{R}^n \to \Delta_m$ does not need to satisfy condition
$A^\top v(x) \neq 0$ for $x \notin C$, consequently it does not need to be an essential weight
function.

Example 4.9.7. Let $a \in \mathbb{R}^n$, $a \neq 0$. Let $a_1 = a_2 = a$, $a_3 = -a$. Consider a
consistent system of inequalities $a_1^\top x \leq 0, a_2^\top x \leq 1, a_3^\top x \leq 1$. The weight function
$v : \mathbb{R}^n \to \Delta_3$ defined by $v(x) = (\frac{1}{4}, \frac{1}{4}, \frac{1}{2})$ for all $x \in \mathbb{R}^n$ is appropriate, because for
any $x \notin C$ we have $v_i(x)(P_{C_i} x - x) \neq 0$ for at least one i, but

$$v_1(x)a_1 + v_2(x)a_2 + v_3(x)a_3 = \frac{1}{4}a + \frac{1}{4}a - \frac{1}{2}a = 0,$$

i.e., $A^\top v(x) = 0$.

Definition 4.9.8. We say that a weight function $v : \mathbb{R}^n \to \Delta_m$ *considers only
violated constraints* if, for all $x \notin C$ and all $i \in I$, it holds

$$x \in C_i \implies v_i(x) = 0. \tag{4.111}$$

It is clear that a weight function $v : \mathbb{R}^n \to \Delta_m$ which considers only violated constraints is appropriate.

The idea of surrogate constraints for a system of linear inequalities was introduces by Merzlyakov [260] and was continued by Oko in [277], Yang and Murty in [351], where various weight functions which consider only violated constraints were proposed, by Kiwiel [229], Kiwiel and Łopuch [233] and Cegielski [67].

The following lemma gives a sufficient condition for a weight function to be essential.

Lemma 4.9.9. *Let $C \neq \emptyset$. If a weight function $v : \mathbb{R}^n \to \Delta_m$ considers only violated constraints, then v is essential and, consequently, $A^\top v(x) \neq 0$ for all $x \notin C$.*

Proof. Let a weight function v consider only violated constraints and $x \notin C$. Denote $I(x) := \{i \in I : x \notin C_i\} = \{i \in I : a_i^\top x > \beta_i\}$. Let $j \in I(x)$ be such that $v_j(x) > 0$. The existence of such j follows from the fact that $v(x) \in \Delta_m$. We have

$$v(x)^\top (Ax - b) = \sum_{i \in I} v_i(x)(a_i^\top x - \beta_i) = \sum_{i \in I(x)} v_i(x)(a_i^\top x - \beta_i)$$

$$\geq v_j(x)(a_j^\top x - \beta_j) > 0,$$

i.e., v is essential. Now the second part of the Lemma follows from Lemma 4.9.5.

$\square$

Note that $S_{\alpha v} = S_v$ for any weight function $v : \mathbb{R}^n \to \Delta_m$ and for any function $\alpha : \mathbb{R}^n \to (0, +\infty)$. Furthermore, S_v does not change if we rescale the inequalities. Therefore, we can suppose without loss of generality that $\|a_i\| = 1$, $i \in I$, and that $v : \mathbb{R}^n \to \mathbb{R}_+^m \setminus \{0\}$ instead of $v : \mathbb{R}^n \to \Delta_m$.

More general versions of the following two results were proved in [68, Theorems 5 and 8]. The first result shows that any extrapolated simultaneous projection is a surrogate projection.

Proposition 4.9.10. *Let $C \neq \emptyset$, $w : \mathbb{R}^n \to \Delta_m$ be an appropriate weight function and $v : \mathbb{R}^n \to \mathbb{R}_+^m \setminus \{0\}$ be defined by*

$$v(x) := W(x)(Ax - b)_+,$$

where $W(x) := \operatorname{diag} w(x)$, $x \in \mathbb{R}^n$. Then v considers only violated constraints. Moreover, $T_w = S_v$, where T_w and S_v are defined by (4.106) and (4.109), respectively.

Proof. If $x \in C$, then, of course, $S_v x = x = T_w x$. Now let $x \notin C$. We have $v_i(x) = \omega_i(x)(a_i^\top x - \beta_i)_+$, $i \in I$. By the fact that w is appropriate, we have $v_j(x) > 0$ for at least one $j \in I$, i.e., $v(x) \neq 0$. If $x \in C_i$, then $(a_i^\top x - \beta_i)_+ = 0$ and $v_i(x) = 0$, $i \in I$, i.e., v considers only violated constraints. By Lemma 4.9.9, the weight function v is essential and the operator S_v is well defined. Note that

$$(Ax - b)_+^\top W(x)(Ax - b) = \sum_{i \in I}(a_i^\top x - \beta_i)_+ \omega_i(x)(a_i^\top x - \beta_i)$$

$$= \sum_{i \in I}\omega_i(x)[(a_i^\top x - \beta_i)_+]^2$$

$$= (Ax - b)_+^\top W(x)(Ax - b)_+.$$

Therefore, if we take $v(x) := W(x)(Ax - b)_+$ in (4.109), then we obtain

$$T_w x = x - \frac{\sum_{i \in I}\omega_i(x)[(a_i^\top x - \beta_i)_+]^2}{\left\| A^\top W(x)(Ax - b)_+ \right\|^2} A^\top W(x)(Ax - b)_+ = S_v x$$

which completes the proof.   □

The following result gives a relationship between the operators S_v and T_w, which is, in some sense, converse to this presented in Proposition 4.9.10.

Proposition 4.9.11. *Let $C \neq \emptyset$, a weight function $v : \mathbb{R}^n \to \mathbb{R}^m_+ \backslash \{0\}$ consider only violated constraints and S_v be the surrogate projection defined by (4.109). Then there is an appropriate weight function $w : \mathbb{R}^n \to \Delta_m$ such that $S_v = T_w$, where T_w is the extrapolated simultaneous projection defined by (4.106).*

Proof. Denote $I(x) := \{i \in I : a_i^\top x - \beta_i > 0\}$ and define

$$\omega_i(x) := \begin{cases} \alpha(x)^{-1}\frac{v_i(x)}{a_i^\top x - \beta_i} & \text{for } i \in I(x) \\ 0 & \text{otherwise,} \end{cases} \tag{4.112}$$

$x \in \mathbb{R}^n$, where

$$\alpha(x) := \sum_{i \in I(x)}\frac{v_i(x)}{a_i^\top x - \beta_i}.$$

If $x \in C$, then, of course, $S_v x = x = T_w x$. Now let $x \notin C$. By Lemma 4.9.9, we have $v(x)^\top(Ax - b) > 0$ and $A^\top v(x) \neq 0$. It is clear that $w(x) \in \Delta_m$ and that w is an appropriate weight function. Note that

$$\omega_i(x)(a_i^\top x - \beta_i)_+ = \omega_i(x)(a_i^\top x - \beta_i)$$

for all $i \in I$. We have

$$A^\top W(x)(Ax - b)_+ = \sum_{i \in I}\omega_i(x)(a_i^\top x - \beta_i)_+ a_i$$

$$= \sum_{i \in I}\omega_i(x)(a_i^\top x - \beta_i)a_i$$

$$= \alpha(x)^{-1}\sum_{i \in I}v_i(x)a_i = \alpha(x)^{-1}A^\top v(x)$$

and

$$\sum_{i \in I} \omega_i(x)[(a_i^\top x - \beta_i)_+]^2 = \sum_{i \in I} \omega_i(x)(a_i^\top x - \beta_i)(a_i^\top x - \beta_i)_+$$

$$= \alpha(x)^{-1} \sum_{i \in I} v_i(x)(a_i^\top x - \beta_i)_+$$

$$= \alpha(x)^{-1} v(x)^\top (Ax - b)_+.$$

Therefore,

$$T_w x = x - \frac{\sum_{i \in I} \omega_i(x)[(a_i^\top x - \beta_i)_+]^2}{\|A^\top W(x)(Ax - b)_+\|^2} A^\top W(x)(Ax - b)_+$$

$$= x - \frac{[v(x)^\top (Ax - b)]_+}{\|A^\top v(x)\|^2} A^\top v(x) = S_v x$$

which completes the proof.   □

Let $T_w : \mathbb{R}^n \to \mathbb{R}^n$ be defined by (4.106) for an appropriate weight function w and $S_v : \mathbb{R}^n \to \mathbb{R}^n$ be defined by (4.109) for an essential weight function v. Denote $\mathcal{T} := \{T_w : w \text{ is appropriate}\}$, $\mathcal{S} := \{S_v : v \text{ considers only violated constraints}\}$ and $\mathcal{S}' := \{S_v : v \text{ is essential}\}$.

Corollary 4.9.12. *Let $C \neq \emptyset$. It holds*

$$\mathcal{T} = \mathcal{S} \subseteq \mathcal{S}'. \tag{4.113}$$

Proof. The equality in (4.113) follows from Propositions 4.9.10 and 4.9.11. The inclusion in (4.113) follows from Lemma 4.9.9.   □

One can easily prove that for a system $Ax \leq b$ containing at least two inequalities defining two different half-spaces the inclusion in (4.113) is strict. We leave the details to the reader.

Corollary 4.9.13. *Let $C \neq \emptyset$. If a weight function $v : \mathbb{R}^n \to \mathbb{R}^m \setminus \{0\}$ considers only violated constraints, then a surrogate projection $S_v : \mathbb{R}^n \to \mathbb{R}^n$ is a cutter. Consequently, for any $\lambda \in (0, 2)$, its relaxation $S_{v,\lambda}$ is $\frac{2-\lambda}{\lambda}$-strongly quasi-nonexpansive and asymptotically regular.*

Proof. The corollary follows from Lemma 4.9.9 and from Theorem 4.9.6. Since $\mathcal{T} = \mathcal{S}$, the corollary also follows directly from Corollary 4.9.3.   □

4.9.5 Surrogate Projection with Residual Selection

In this section we consider a special case of the surrogate projection S_v defined by (4.109). For a matrix A with rows a_i, $i \in I$, and for a subset $L \subseteq I$, denote by A_L

a submatrix with rows a_i, $i \in L$. For simplicity, suppose that $L = \{1, 2, \ldots, r\}$, where $r \leq m$, i.e.,

$$A = \begin{bmatrix} A_L \\ A_{I \setminus L} \end{bmatrix}.$$

We apply the same notation as above for a subvector b_L of b, for a subvector $v_L(x)$ of the vector $v(x)$ and for a subvector $r_L(x)$ of the residual vector $r(x) := Ax - b$, $x \in \mathbb{R}^n$, i.e.,

$$b = \begin{bmatrix} b_L \\ b_{I \setminus L} \end{bmatrix}, v(x) = \begin{bmatrix} v_L(x) \\ v_{I \setminus L}(x) \end{bmatrix} \text{ and } r(x) = \begin{bmatrix} r_L(x) \\ r_{I \setminus L}(x) \end{bmatrix}.$$

Let $x \notin C = \{x \in \mathbb{R}^n : Ax \leq b\}$. Let $L := L(x) \subseteq I$ be such that:

(i) A_L has full row rank, i.e., a_i, $i \in L$, are linearly independent,
(ii) $(A_L A_L^\top)^{-1}(A_L x - b_L) \geq 0$.

Note that for any $x \notin C$ the weight vector

$$v_L(x) := (A_L A_L^\top)^{-1}(A_L x - b_L)$$

is essential for the system $A_L x \leq b_L$, consequently, the weight vector $v(x)$ with $v_{I \setminus L}(x) = 0 \in \mathbb{R}^{m-r}$ is essential for the system $Ax \leq b$. Therefore, the surrogate projection $S_v x$ given by (4.109) is well defined.

We say that a subset $L(x)$ satisfying the conditions (i) and (ii) above is a *residual selection* (RS) of I. Note that such a subset exists, e.g., $L(x) := \{j(x)\}$, where $j(x) \in I(x) := \{i \in I : a_i^\top x > \beta_i\}$. Residual selections were studied in [72, Sect. 3.2] and [73, Sect. 3.1], where several constructions of maximal subsets L satisfying the conditions (i) and (ii) above were presented. If we replace the condition (ii) above by the following stronger condition:

(iii) $(A_L A_L^\top)^{-1} \geq 0$ and $A_L x - b_L \geq 0$,

then the residual selection $L(x) \subseteq I$ is called an *obtuse cone selection* (OCS). The notion OCS can be explained by the fact that for a full row matrix A_L the cone generated by the rows of A_L is obtuse (in Lin$\{a_i : i \in L\}$) if and only if $(A_L A_L^\top)^{-1} \geq 0$ (see [232, Lemma 3.1] or [65, Lemma 1.6]). A special case of an obtuse cone selection is a *regular obtuse cone selection* (ROCS), for which the condition (iii) above is replaced by

(iv) $a_i^\top a_j \leq 0$ for $i \neq j$, $i, j \in L$, and $A_L x - b_L \geq 0$

(see Fig. 4.37).

The first inequality in (iii) says that the inverse of the Gram matrix $G := A_L A_L^\top$ of rows of A_L is nonnegative. Nonsingular matrices with nonnegative inverses, called *inverse-positive matrices*, play an important role in many areas of mathematics and were studied in [35]. The properties of Gram matrices with nonnegative inverse were also studied in [232, Sect. 3] and [65, Sect. 2], where their

Fig. 4.37 Regular obtuse
cone selection

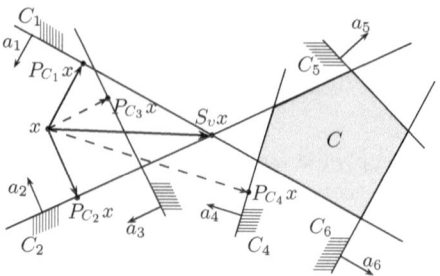

relations to obtuse cones are presented. We see that L is an obtuse cone selection if
and only if G is inverse-positive and $A_L x - b_L \geq 0$.

The first inequality in (iv) says that G has nonpositive off-diagonal elements.
Nonsingular matrices with a nonnegative inverse having this property are called *M-
matrices* or *Minkowski matrices*. The properties of such matrices were studied in
[166], [326, Sect. 4], [62, Chap. 5], [232, Sect. 3] and [65, Sect. 2]. In particular, it
follows from [166, Theorem 4.3] that a nonsingular Gram matrix with nonpositive
off-diagonals has a nonnegative inverse. This result also follows from [62, Theorem
5.4.A], from [232, Lemma 3.2], and from [65, Corollary 2.9]. Several constructions
of maximal subsets $L(x)$ being obtuse cone selection or regular obtuse cone
selection are presented in [326, Sect. 4], [62, Chap. 5], [65, Sect. 2]. In [62, Chap. 6],
[64, 230, 231] one can find applications of such selections to algorithms for convex
minimization problems.

Suppose that for any $x \notin C$ the subset $L(x) \subseteq I$ is a residual selection. If we
take $v_L(x) = (A_L A_L^\top)^{-1}(A_L x - b_L)$ and $v_{I \setminus L}(x) = 0$, then equality (4.109) obtains
the form

$$S_v x = x - A_L^\top (A_L A_L^\top)^{-1}(A_L x - b_L). \tag{4.114}$$

The operator $S_v : \mathbb{R}^n \to \mathbb{R}^n$ defined by (4.114) is called a *surrogate projection with
residual selection*.

Further properties of the residual selection as well as its applications to the
surrogate projection methods are presented in Sect. 5.13.

4.9.6   Extrapolated Simultaneous Subgradient Projection

Let $c_i : \mathcal{H} \to \mathbb{R}$ be convex continuous functions, $C_i := \{x \in \mathcal{H} : c_i(x) \leq 0\}$ be
nonempty and U_i be the subgradient projections relative to c_i, i.e.,

$$U_i(x) := \begin{cases} x - \dfrac{c_i^+(x)}{\|g_i(x)\|^2} g_i(x) & \text{if } c_i(x) > 0 \\ x & \text{if } c_i(x) \leq 0, \end{cases} \tag{4.115}$$

where $g_i(x) \in \partial c_i(x)$ is a subgradient of c_i at x, $i \in I$ (cf. (4.31)). It follows from
Lemma 4.2.5 and from Corollary 4.2.6 that Fix $U_i = C_i$ and that U_i is a cutter.

We call the operator $U := \sum_{i \in I} \omega_i U_i$, where $w : \mathcal{H} \to \Delta_m$ is a weight function, a *simultaneous subgradient projection*. In this section we apply Theorem 4.9.1 to the extrapolated simultaneous subgradient projection U_σ with the step size function $\sigma : \mathcal{H} \to (0, +\infty)$ satisfying the following inequality

$$1 \leq \sigma(x) \leq \sigma_w(x), \tag{4.116}$$

where

$$\sigma_w(x) := \frac{\sum_{i=1}^m w_i(x) \left(\frac{c_i^+(x)}{\|g_i(x)\|}\right)^2}{\left\|\sum_{i=1}^m w_i(x) \frac{c_i^+(x)}{\|g_i(x)\|^2} g_i(x)\right\|^2} \tag{4.117}$$

for all $x \notin \bigcap_{i \in I} C_i \neq \emptyset$, and the weight function w is appropriate (for simplicity, we use the convention that $\frac{(c_i(x))_+}{\|g_i(x)\|} = 0$ if $(c_i(x))_+ = 0$). Note that the step size $\sigma_w(x)$ defined by (4.117) is a special case of the step size (4.99), where U_i is the subgradient projection defined by (4.115), $i \in I$.

Corollary 4.9.14. *Let $U_{\sigma,\lambda}$ be a generalized relaxation of the simultaneous subgradient projection $U := \sum_{i \in I} \omega_i U_i$ with an appropriate weight function $w : \mathcal{H} \to \Delta_m$, with a step size function $\sigma : \mathcal{H} \to (0, +\infty)$ satisfying inequality (4.116) for $x \notin \bigcap_{i \in I} C_i \neq \emptyset$ and with $\lambda \in [0, 2]$. Then the operator U_σ is a cutter. Consequently,*

$$\|U_{\sigma,\lambda}x - z\|^2 \leq \|x - z\|^2 - \lambda(2 - \lambda)\sigma^2(x)\|Ux - x\|^2$$

$$= \|x - z\|^2 - \frac{2 - \lambda}{\lambda}\|U_{\sigma,\lambda}x - x\|^2$$

for all $x \in \mathcal{H}$, $z \in \bigcap_{i \in I} \text{Fix } U_i$ and for all $\lambda \in (0, 2)$, i.e., $U_{\sigma,\lambda}$ is $\frac{2-\lambda}{\lambda}$-strongly quasi-nonexpansive and asymptotically regular.

Proof. Since a subgradient projection U_i is a cutter and $\text{Fix } U_i = C_i$, $i \in I$, the corollary follows from Theorem 4.9.1. □

The properties of a special case of the extrapolated simultaneous subgradient projection were studied by Dos Santos in [146], where $\sigma = \sigma_w$ and an appropriate weight function w is supposed to be constant, i.e., $w \in \text{ri } \Delta_m$. The study was continued for various convex optimization problems by Cegielski [62, Sect. 4.3], Kiwiel [229, Sects. 3 and 8], [230, 231], Kiwiel and Łopuch [233, Sect. 2], Combettes [118, Sect. 5.6], [120, Sect. V] and by Cegielski and Censor [70].

4.10   Extrapolated Cyclic Cutter

Let $U_i : \mathcal{H} \to \mathcal{H}$ be cutters with a common fixed point, $i \in I := \{1, 2, \ldots, m\}$. In this section we will extend the results of Corollary 4.5.15 to the generalized relaxation of the cyclic cutter. Similarly as in Sect. 4.9, we allow the relaxation parameter to depend on $x \in \mathcal{H}$. Therefore, we work with a generalized relaxation $U_{\sigma,\lambda}$ of a cyclic cutter $U := U_m U_{m-1} \ldots U_1$, with a step size function $\sigma : \mathcal{H} \to (0, +\infty)$ and a relaxation parameter $\lambda \in [0, 2]$. It follows from Corollary 4.5.15 that Fix $U = \bigcap_{i \in I}$ Fix U_i and that U is strongly quasi-nonexpansive. We will extend this result to the generalized relaxation of U, defined by

$$U_{\sigma,\lambda} x := I + \lambda \sigma(x)(Ux - x).$$

4.10.1 Useful Inequalities

Let $S_0 := \mathrm{Id}$ and $S_i := U_i U_{i-1} \ldots U_1$, $i = 1, 2, \ldots, m$. Of course, $U = S_m$. Denote

$$u^0 := x, u^i := U_i u^{i-1} \text{ and } y^i := u^i - u^{i-1}, \qquad (4.118)$$

$i = 1, 2, \ldots, m$. We have $u^i = S_i x$.

Lemma 4.10.1. *Let $U_i : \mathcal{H} \to \mathcal{H}$ be cutters with a common fixed point, $i \in I$. The following inequalities hold*

$$\langle Ux - x, z - x \rangle \geq \sum_{i=1}^{m} \langle y^i + \ldots + y^m, y^i \rangle \geq \frac{1}{2} \sum_{i=1}^{m} \|y^i\|^2 \geq \frac{1}{2m} \left\| \sum_{i=1}^{m} y^i \right\|^2, \quad (4.119)$$

where $x \in \mathcal{H}$ and $z \in \bigcap_{i=1}^{m}$ Fix $U_i \neq \emptyset$.

Proof. (cf. [71, Lemma 7]) Let $x \in \mathcal{H}$ be arbitrary. We prove the first inequality in (4.119) by induction with respect to m. For $m = 1$ this inequality follows directly from the definition of a cutter and from inequality (2.21). Suppose that the first inequality in (4.119) is true for some $m = k$. Let $w \in \mathcal{H}$. Define $V_1 := \mathrm{Id}$, $V_i := U_i U_{i-1} \ldots U_2$ for $i = 2, 3, \ldots, k+1$, $v_1 := w$, $v^i := U_i v^{i-1}$ and $z^i := v^i - v^{i-1}$, $i = 2, 3, \ldots, k+1$. Take $w := U_1 x$ and $z \in \bigcap_{i=1}^{k+1}$ Fix U_i. Then, of course, $S_i x = V_i w$, $u^i = v^i$ and $y^i = z^i$, $i = 2, 3, \ldots, k+1$. It follows from the induction assumption that

$$\langle V_{k+1} w - w, z - w \rangle \geq \sum_{i=2}^{k+1} \langle z^i + \ldots + z^{k+1}, z^i \rangle \qquad (4.120)$$

Note that $V_{k+1} w - w = \sum_{i=2}^{k+1} z^i$ and $w - x = y^1$. Since U_1 is a cutter, it follows from (4.120) that

$$\langle S_{k+1} x - x, z - x \rangle$$
$$= \langle V_{k+1} w - x, z - x \rangle$$
$$= \langle V_{k+1} w - w, z - x \rangle + \langle U_1 x - x, z - x \rangle$$
$$\geq \langle V_{k+1} w - w, z - w \rangle + \langle V_{k+1} w - w, w - x \rangle + \|y^1\|^2$$

$$\geq \sum_{i=2}^{k+1} \langle z^i + \ldots + z^{k+1}, z^i \rangle + \langle \sum_{i=2}^{k+1} z^i, y^1 \rangle + \|y^1\|^2$$

$$= \sum_{i=2}^{k+1} \langle y^i + \ldots + y^{k+1}, y^i \rangle + \sum_{i=2}^{k+1} \langle y^i, y^1 \rangle + \|y^1\|^2$$

$$= \sum_{i=1}^{k+1} \langle y^i + \ldots + y^{k+1}, y^i \rangle.$$

Therefore, the first inequality in (4.119) is true for all $m \in \mathbb{N}$. The second inequality follows from the following equality

$$\sum_{i=1}^{m} \langle y^i + \ldots + y^m, y^i \rangle - \frac{1}{2} \sum_{i=1}^{m} \|y^i\|^2 = \frac{1}{2} \left\| \sum_{i=1}^{m} y^i \right\|^2 .$$

The third inequality in (4.119) follows from the convexity of the function $\|\cdot\|^2$.   □

4.10.2   Properties of the Extrapolated Cyclic Cutter

We use the notation from the previous subsection. Define the step size function $\sigma_{\max} : \mathcal{H} \to (0, +\infty)$ by

$$\sigma_{\max}(x) := \frac{\sum_{i=1}^{m} \langle Ux - S_{i-1}x, S_i x - S_{i-1}x \rangle}{\|Ux - x\|^2} \tag{4.121}$$

for $x \notin \text{Fix}\, U$ and $\sigma(x) = 1$ for $x \in \text{Fix}\, U$. If we take $y^i := S_i x - S_{i-1}x$, $i = 1, 2, \ldots, m$ (cf. (4.118)), then Lemma 4.10.1 yields

$$\sigma_{\max}(x) \geq \frac{\frac{1}{2} \sum_{i=1}^{m} \|S_i x - S_{i-1}x\|^2}{\|Ux - x\|^2} \geq \frac{1}{2m}, \tag{4.122}$$

where $x \notin \text{Fix}\, U$.

Note that

$$\sum_{i=1}^{m} \langle y^i + \ldots + y^m, y^i \rangle = \sum_{i=1}^{m} \langle y^1 + \ldots + y^i, y^i \rangle.$$

Therefore, the step size $\sigma_{\max}(x)$ given by (4.121) can be equivalently written in the following form

$$\sigma_{\max}(x) = \frac{\sum_{i=1}^{m} \langle S_i x - x, S_i x - S_{i-1}x \rangle}{\|Ux - x\|^2}. \tag{4.123}$$

Now consider the generalized relaxation $U_{\sigma,\lambda}$ of the cyclic cutter U with a step size function $\sigma : \mathcal{H} \rightarrow (0, +\infty)$ satisfying inequality $\sigma(x) \leq \sigma_{\max}(x)$ for all $x \notin \text{Fix}\, U$, where $\sigma_{\max}(x)$ is given by (4.121) or (4.123).

Theorem 4.10.2. *The operator U_σ is a cutter. Consequently, for all $\lambda \in (0, 2)$, the operator $U_{\sigma,\lambda}$ is $\frac{2-\lambda}{\lambda}$-strongly quasi-nonexpansive and asymptotically regular.*

Proof. (cf. [71, Lemma 8]) Let $x \in \mathcal{H}$, $\lambda \in (0, 2)$ and $z \in \bigcap_{i \in I} \text{Fix}\, U_i$. It is clear that $\text{Fix}\, U_{\sigma,\lambda} = \text{Fix}\, U = \bigcap_{i \in I} \text{Fix}\, U_i$ (see Theorem 2.1.26 and Remark 2.4.2 (d)). The first inequality in (4.119) can be written in the form

$$\langle Ux - x, z - x \rangle \geq \sigma_{\max}(x) \, \|Ux - x\|^2. \tag{4.124}$$

Consequently,

$$\langle Ux - x, z - x \rangle \geq \sigma(x) \, \|Ux - x\|^2.$$

Now Corollary 2.4.5 yields that U_σ is a cutter and $U_{\sigma,\lambda}$ is $\frac{2-\lambda}{\lambda}$-strongly quasi-nonexpansive. The asymptotic regularity of $U_{\sigma,\lambda}$ follows from Theorem 3.4.3. □

4.11   Exercises

Exercise 4.11.1. Let $B(z, \rho) \subseteq \mathcal{H}$, be a ball with the centre $z \in \mathcal{H}$ and a radius $\rho > 0$. Prove that $B(z, \rho)$ is a nonempty closed convex subset and that

$$P_{B(z,\rho)}(x) = \begin{cases} x & \text{if } \|x - z\| \leq \rho \\ z + \frac{\rho}{\|x-z\|}(x - z) & \text{if } \|x - z\| > \rho. \end{cases}$$

Exercise 4.11.2. Let $A \subseteq \mathcal{H}_1$ and $B \subseteq \mathcal{H}_2$ be closed convex subsets and $(x, y) \in A \times B$. Prove that $N_{A \times B}(x, y) = N_A(x) \times N_B(y)$.

Exercise 4.11.3. Prove that $\lambda_{\max}(\alpha_1 A_1 + \alpha_2 A_2) \leq \alpha_1 \lambda_{\max}(A_1) + \alpha_2 \lambda_{\max}(A_2)$, where A_1, A_2 are positive semi-definite matrices and $\alpha_1, \alpha_2 \geq 0$.

Exercise 4.11.4. Prove Corollary 4.6.7

Exercise 4.11.5. Let $C \subseteq \mathcal{H}$ be a closed convex subset. Prove that $P_C = P_{d(\cdot,C)}$, where $P_{d(\cdot,C)}$ denotes a subgradient projection relative to the distance function $d(\cdot, C) := \inf_{y \in C} \|\cdot - y\|$.

Chapter 5
Projection Methods

Let $X \subseteq \mathcal{H}$ be a nonempty closed convex subset and $M \subseteq X$ be a subset of solutions of a convex optimization problem defined on X. Such problems were defined in Sect. 1.3. The subset M is closed and convex and, usually, nonempty. As an example, consider the convex feasibility problem: find $x \in \bigcap_{i \in I} C_i$, where $C_i \subseteq \mathcal{H}$ are nonempty closed convex subsets, $i \in I := \{1, 2, \ldots, m\}$, if such a point exists. If the problem is consistent, we have $M = \bigcap_{i \in I} C_i \neq \emptyset$. If the problem is inconsistent, then M is defined as a subset of minimizers of a corresponding proximity function $f : \mathcal{H} \to \mathbb{R}_+$.

Let $\bar{x} \in X$ be an approximation of a solution of a convex optimization problem. An iteration in a *projection method* for this problem has the form $x^+ = U\bar{x}$, where $U : X \to X$ is an appropriate algorithmic projection operator constructed for this problem. Most of projection methods employ algorithmic operators U which are relaxations of cutters with $\operatorname{Fix} U \supseteq M$ or, equivalently, SQNE operators (see Theorem 2.1.39). By Corollary 2.2.9, any relaxed firmly nonexpansive operator $U : X \to X$ with $\operatorname{Fix} U \supseteq M$ has this property. This is the reason why firmly nonexpansive operators are often used in projection methods. Averaged operators are also often employed in these methods. Note, however, that the class of averaged operators and the class of strictly relaxed firmly nonexpansive operators coincide (see Corollary 2.2.17). Therefore, without loss of generality, we will restrict our further analysis to relaxed cutters and to relaxed firmly nonexpansive operators.

Any projection method can be presented in the form of a recurrence

$$
\begin{aligned}
& x^0 \in X \qquad \text{– arbitrary} \\
& x^{k+1} = R_k x^k,
\end{aligned}
$$

where $R_k := P_X (\operatorname{Id} + \lambda_k (T_k - \operatorname{Id}))$, $\lambda_k \in (0, 2]$, and $T_k : X \to \mathcal{H}$ is a cutter. In most cases $\operatorname{Fix} T_k \supseteq M$ or $\operatorname{Fix}(P_X T_k) \supseteq M$, $k \geq 0$. Examples of such operators were presented in Chaps. 2 and 4. If the operator $T_k = T$ and $\lambda_k = \lambda$ for all $k \geq 0$, then we say that the method is *autonomous*. Otherwise, we say that the method is *nonautonomous*. The iteration can also be presented in the form

A. Cegielski, *Iterative Methods for Fixed Point Problems in Hilbert Spaces*, Lecture Notes in Mathematics 2057, DOI 10.1007/978-3-642-30901-4_5, © Springer-Verlag Berlin Heidelberg 2012

$$x^{k+1} = P_X(x^k + \lambda_k(T_k x^k - x^k)). \tag{5.1}$$

If X is a closed affine subspace and $T_k : X \rightarrow X$, then we can omit the operator P_X in (5.1) and we obtain

$$x^{k+1} = x^k + \lambda_k(T_k x^k - x^k). \tag{5.2}$$

In autonomous methods the sequence of operators $\{T_k\}_{k=0}^\infty$ as well as the sequence of relaxation parameters (λ_k) are constant and these methods have the form

$$x^{k+1} = P_X(x^k + \lambda(T x^k - x^k)). \tag{5.3}$$

Generally, the convergence (at least weak) of sequences generated by projection methods follows from Opial's theorem (see Sect. 3.5) or from its generalizations (see Sects. 3.6 and 3.7). In this chapter we show how to apply these theorems in order to prove the convergence.

5.1   Alternating Projection Methods

The *alternating projection method* (APM), also known as the *von Neumann method* is an iterative method for finding a common point of two nonempty closed convex subsets $A, B \subseteq \mathcal{H}$, if such a point exists. One iteration of the method has the form

$$x^{k+1} = P_A P_B x^k \tag{5.4}$$

or $x^{k+1} = T x^k$, where $T := P_A P_B$ is an alternating projection. The properties of this operator were described in Sect. 4.3. In this section we consider a more general method in the form

$$x^{k+1} = P_A(x^k + \lambda_k(P_A P_B x^k - x^k)), \tag{5.5}$$

where $\lambda_k \in [0, 2]$ which we call a *relaxed alternating projection method*. We can also write $x^{k+1} = P_A T_{\lambda_k} x^k$, where T_{λ_k} is the λ_k-relaxation of the alternating projection $T := P_A P_B$. Note that if $\lambda_k = 1$, then (5.5) reduces to (5.4).

5.1.1   General Case

Corollary 5.1.1. *Let $A, B \subseteq \mathcal{H}$ be nonempty closed convex subsets with a common point and* $\liminf \lambda_k(2 - \lambda_k) > 0$. *Then, for an arbitrary $x^0 \in A$, the sequence $\{x^k\}_{k=0}^\infty$ generated by the relaxed alternating projection method (5.5) converges weakly to a point $x^* \in A \cap B$.*

Proof. The operator $T := P_A P_B$ is nonexpansive as a composition of nonexpansive operators P_A and P_B, and Fix $T = A \cap B$ (see Corollary 4.3.5 (v)). Furthermore, $T : A \to A$ is a cutter (see Lemma 4.3.2) and a demi-closed operator (see Lemma 3.2.5). Corollary 3.7.3 with $X = A$ yields now the weak convergence of x^k to a point $x^* \in A \cap B$. □

If we suppose that Fix $P_A P_B \neq \emptyset$ instead of $A \cap B \neq \emptyset$ in Corollary 5.1.1, then the weak convergence holds for a tighter range of relaxation parameters.

Corollary 5.1.2. *Let $A, B \subseteq \mathcal{H}$ be nonempty closed convex subsets with Fix $P_A P_B \neq \emptyset$ and $\lambda_k \in [\varepsilon, \frac{3}{2} - \varepsilon]$ for some $\varepsilon \in (0, \frac{3}{4})$. Then, for an arbitrary $x^0 \in A$, the sequence $\{x^k\}_{k=0}^{\infty}$ generated by the relaxed alternating projection method (5.5) converges weakly to a fixed point of the operator $T := P_A P_B$.*

Proof. By Corollary 4.3.3, the alternating projection $T := P_A P_B$ is $\frac{4}{3}$-relaxed firmly nonexpansive and the operator $S := T_{\frac{3}{4}} = \mathrm{Id} + \frac{3}{4}(T - \mathrm{Id})$ is firmly nonexpansive. Note that Fix $S = \mathrm{Fix}\, T$, $S_{\mu_k} = T_{\lambda_k}$, where $\mu_k = \frac{4}{3}\lambda_k$ (see Remarks 2.1.3 and 2.1.4) and that $\liminf_k \mu_k (2 - \mu_k) > 0$. Iteration (5.5) can be written in the form $x^{k+1} = P_A S_{\mu_k} x^k$. The weak convergence of x^k to a point $x^* \in \mathrm{Fix}\, P_A P_B$ follows now from Corollary 3.7.3. □

Since iteration (5.4) is a special case of (5.5), we obtain the following result immediately.

Corollary 5.1.3. *Let $A, B \subseteq \mathcal{H}$ be nonempty closed convex subsets. If Fix $P_A P_B \neq \emptyset$, then for an arbitrary $x^0 \in \mathcal{H}$ the sequence $\{x^k\}_{k=0}^{\infty}$ generated by the alternating projection method (5.4) converges weakly to a fixed point of the operator $T := P_A P_B$.*

Corollary 5.1.3 is a special case of [42, Theorem 1], where the weak convergence was proved for the cyclic projection method. The strong convergence of sequences generated by the (relaxed) alternating projection method does not hold in general. An example of a closed convex cone $A \subseteq \mathcal{H}$, a closed half-space $B \subseteq \mathcal{H}$, with $A \cap B = \{0\}$ and a point $x \in \mathcal{H}$ for which $(P_A P_B)^k x$ converges weakly but does not converge strongly was presented by Hundal in [213, Theorem 1]. A simplification of Hundal's example can be found in [258]. In the literature one can find several conditions for the strong convergence in Corollary 5.1.3, e.g., one of the subsets A, B is compact (see [112, Theorem 4 (a)]) or A, B are closed affine subspaces and $d(A, B)$ is attained (see [21, Theorem 4.1]). In [20, Sects. 3–6], [14, Corollary 2.4], [258, Theorem 4.1] and [235, Theorem 4.5] one can find other sufficient conditions for the strong convergence.

Recall that a composition of RFNE operators is RFNE (see Theorem 2.2.37). Therefore, Corollary 5.1.3 remains true if we replace the projections P_A and P_B by their relaxations $P_{A,\lambda} := (1 - \lambda)\,\mathrm{Id} + \lambda P_A$ and $P_{B,\mu} := (1 - \mu)\,\mathrm{Id} + \mu P_B$, where $\lambda, \mu \in (0, 2)$.

If A is a closed affine subspace, the weak convergence in Corollary 5.1.2 is guaranteed for a broader range of the relaxation parameter. Note that in this case $T_\lambda(A) \subseteq A$ and the recurrence (5.5) can be written in the equivalent form

$$x^{k+1} = x^k + \lambda_k (P_A P_B x^k - x^k). \tag{5.6}$$

The following result is due to Combettes [117, Theorem 1], where $A \subseteq \mathcal{H}$ is supposed to be a closed subspace of $\mathcal{H}$.

Theorem 5.1.4. *Let* $A \subseteq \mathcal{H}$ *be a closed affine subspace,* $B \subseteq \mathcal{H}$ *a nonempty closed convex subset and* $\liminf \lambda_k (2 - \lambda_k) > 0$. *If* Fix $P_A P_B \neq \emptyset$, *then for an arbitrary* $x^0 \in A$ *the sequence* $\{x^k\}_{k=0}^{\infty}$ *generated by the relaxed alternating projection method (5.6) converges weakly to a fixed point of the operator* $T := P_A P_B$.

Proof. The operator $T := P_A P_B$ is firmly nonexpansive (see Theorem 4.3.7). Therefore, the theorem follows directly from Corollary 3.7.3.   □

5.1.2   Alternating Projection Method for Closed Linear Subspaces

Suppose that A and B are closed subspaces of a Hilbert space $\mathcal{H}$. Then the strong convergence holds in Corollary 5.1.3. The following theorem is due to John von Neumann [271, Theorem 13.7].

Theorem 5.1.5 (von Neumann, 1933). *Let* $A, B \subseteq \mathcal{H}$ *be closed subspaces. Then, for any* $x^0 \in \mathcal{H}$, *the sequence* $\{x^k\}_{k=0}^{\infty}$ *generated by the alternating projection method (5.4) converges in norm to* $y := P_{A \cap B} x^0$.

Proof. Let $x^0 \in \mathcal{H}$ and $x^k = T^k x^0, k \geq 0$, where $T := P_A P_B$. We divide the proof into three parts.

(i) Since the metric projection is strictly quasi-nonexpansive (see Theorem 2.2.21 (iii) and Corollary 2.2.9) and $A \cap B \neq \emptyset$, we have Fix $T = A \cap B$ (see Theorem 2.1.26). The weak convergence of x^k to a point $x^* \in A \cap B$ follows from Corollary 5.1.3.

(ii) We show that $x^* = P_{A \cap B} x^0$. Denote $y^{2l} := x^l$ and $y^{2l+1} := P_B x^l, l \geq 0$. It is clear that $y^i - y^{2k} = \sum_{l=i}^{2k-1} y^l - y^{l+1}$, for all $i, k \geq 0, i \leq 2k$. Let $z \in A \cap B$ be arbitrary. By Theorem 2.2.30 (i), we have

$$\langle y^{2l} - y^{2l+1}, z \rangle = \langle x^l - P_B x^l, z \rangle = 0$$

and
$$\langle y^{2l+1} - y^{2l+2}, z \rangle = \langle P_B x^l - P_A P_B x^l, z \rangle = 0,$$

$l \geq 0$. Consequently,

$$\langle y^i - x^*, z \rangle = \lim_k \langle y^i - y^{2k}, z \rangle = \lim_k \langle \sum_{l=i}^{2k-1} y^l - y^{l+1}, z \rangle$$

$$= \lim_k \sum_{l=i}^{2k-1} \langle y^l - y^{l+1}, z \rangle = 0,$$

$i \geq 0$. In particular,

$$\langle P_B x^0 - x^*, x^* \rangle = \langle y^1 - x^*, x^* \rangle = 0$$

and

$$\langle x^0 - x^*, z \rangle = \langle y^0 - x^*, z \rangle = 0,$$

i.e., $x^* = P_{A \cap B} x^0$.

(iii) We show that $\lim_k \left\| x^k - x^* \right\| = 0$. Since the metric projection onto a closed subspace is self-adjoint (see Theorem 2.2.30 (iii)) and idempotent (see Theorem 2.2.21 (i)), we have

$$(T^*)^k T^k = \underbrace{P_B P_A \dots P_B P_A}_{k\text{-times}} \underbrace{P_A P_B \dots P_A P_B}_{k\text{-times}} = P_B \underbrace{(P_A P_B) \dots (P_A P_B)}_{(k-1)\text{-times}} = P_B T^{2k-1}.$$

Since $x^* \in \text{Fix } T$, we have

$$\langle x^k - x^*, x^k \rangle = \langle T x^{k-1} - T x^*, x^k \rangle = \langle x^{k-1} - x^*, T^* x^k \rangle.$$

If we iterate the above equalities k-times and apply the facts that $x^k \rightharpoonup x^*$ and $P_B x^* = x^*$, we obtain

$$\langle x^k - x^*, x^k \rangle = \langle x^0 - x^*, (T^*)^k x^k \rangle = \langle x^0 - x^*, (T^*)^k T^k x^0 \rangle$$
$$= \langle x^0 - x^*, P_B T^{2k-1} x^0 \rangle = \langle P_B x^0 - P_B x^*, T^{2k-1} x^0 \rangle$$
$$= \langle P_B x^0 - x^*, x^{2k-1} \rangle \to \langle P_B x^0 - x^*, x^* \rangle = 0.$$

Consequently, the weak convergence $x^k \rightharpoonup x^*$ yields

$$\lim_k \left\| x^k - x^* \right\|^2 = \lim_k (\langle x^k - x^*, x^k \rangle - \langle x^k - x^*, x^* \rangle) = 0$$

i.e., $x^k \to x^* = P_{A \cap B} x^0$. $\qquad\qquad\qquad\qquad\qquad\qquad\qquad\qquad\qquad\qquad \square$

Several proofs of the von Neumann theorem can be found, e.g., in [152], [140, Theorem 9.3] [235, Theorem 1.1] and [236].

Theorem 5.1.5 remains true if A and B are closed affine subspaces with a common point. The details are left to the reader.

One can also estimate the rate of convergence of the alternating projection method $x^{k+1} = P_A P_B x^k$. It turns out that when A and B are closed subspaces of $\mathcal{H}$ the rate depends on the angle between the subspaces. The following definition is due to Friedricks [170, Sect. 1.1] (see also [139, Sect. 6] and [140, Definition 9.4]).

Definition 5.1.6. Let $A, B \subseteq \mathcal{H}$ be closed subspaces and

$$c(A, B) := \sup\{|\langle x, y \rangle| : x \in A \cap (A \cap B)^\perp, y \in B \cap (A \cap B)^\perp, \|x\| \leq 1, \|y\| \leq 1\}$$

The value $\alpha(A, B) := \arccos c(A, B) \in [0, \frac{\pi}{2}]$ is called an *angle between the subspaces A and B*.

It follows from the Cauchy–Schwarz inequality that $c(A, B) \in [0, 1]$, consequently, the the angle between the subspaces A and B is well defined. The proof of the following theorem can be found in [11, page 379] or in [140, Sect. 9.8].

Theorem 5.1.7. *Let* $A, B \subseteq \mathcal{H}$ *be two closed subspaces, $c := c(A, B)$, $x^0 \in \mathcal{H}$, and* $\{x^k\}_{k=0}^{\infty}$ *be generated by the alternating projection method (5.4). Then*

$$\left\| x^k - P_{A \cap B} x^0 \right\| \le c^{2k-1} \left\| x^0 - P_{A \cap B} x^0 \right\|,$$

consequently $\{x^k\}_{k=0}^{\infty}$ *converges geometrically, if $c < 1$.*

In papers [290, 291, 313] the angles between subspaces are also used to obtain convergence rates for iterative projection methods. Further results on the alternating projection method can be found, e.g., in [20, 21], [140, Chap. 9] and [174]. Applications of the alternating projection method in various areas of mathematics were presented in [139].

5.2 Extrapolated Alternating Projection Methods

The sequences generated by the alternating projection method often converge very slowly. Such a slow convergence is observed, e.g., if A is a hyperplane and B is a ball which is tangent to A (see, e.g., [21, Example 5.3]) or A and B are two closed subspaces with $c(A, B)$ close to 1. In these cases the angle between $x^k - P_B x^k$ and $P_A P_B x^k - P_B x^k$ is close to 0 which causes very short steps $P_A P_B x^k - x^k$. Since the alternating projection method has many practical applications, any acceleration technique seems to be important. In this section we give several modifications of the alternating projection method which lead in practice to an acceleration of the convergence. All these methods have the form

$$x^{k+1} = P_A T_{\sigma_k, \lambda_k} x^k$$

where $x^0 \in A$, $T_{\sigma_k, \lambda_k} : A \to \mathcal{H}$ is a generalized relaxation of the alternating projection $T := P_A P_B$, defined by

$$T_{\sigma_k, \lambda_k}(x) := x + \lambda_k \sigma_k(x)(P_A P_B x - x)$$

for $x \in \mathcal{H}$, where $\{\sigma_k\}_{k=0}^{\infty} : \mathcal{H} \to (0, +\infty)$ is a sequence of step size functions and $\lambda_k \in [0, 2]$ is a relaxation parameter (cf. Sect. 2.4). Suppose, for simplicity, that $\sigma(x) = 1$ for $x \in A \cap B$. We can also write

$$x^{k+1} = P_A(x^k + \lambda_k \sigma_k(x^k)(P_A P_B x^k - x^k)). \tag{5.7}$$

If $\sigma_k(x^k) \geq 1$, $k \geq 0$, then the method described by recurrence (5.7) is called an *extrapolated alternating projection method*.

5.2.1 Acceleration Techniques for Consistent Problems

Suppose that $A \cap B \neq \emptyset$. Below we present several step size functions $\sigma :$ $A \to [1, +\infty)$ which guarantee the week convergence of sequences generated by recurrence (5.7) with $\lambda_k \in [\varepsilon, 2 - \varepsilon]$ for some $\varepsilon \in (0, 1)$. Let $T := P_A P_B$.

5.2.1.1 Gurin–Polyak–Raik Approach

Let

$$\sigma_{GPR}(x) := \frac{\|P_B x - x\|^2}{\langle P_B x - x, T x - x \rangle}, \tag{5.8}$$

for $x \in A \backslash B$. The step size (5.8) was proposed by Gurin et al. in [196, Sect. 3]. Recall that $\sigma_{GPR}(x) \geq 1$ for all $x \in A$ and that $\sigma_{GPR}(x) = 1$ if and only if $P_B x \in A \cap B$ (see Lemma 4.3.13). Therefore, recurrence (5.7) with $\sigma_k = \sigma_{GPR}(x^k)$ describes an extrapolated alternating projection method. The following theorem extends the results of [196, Theorem 4].

Corollary 5.2.1. *Let $x^0 \in A$ and the sequence $\{x^k\}_{k=0}^{\infty}$ be generated by (5.7), where*

$$1 \leq \sigma_k(x) \leq \sigma_{GPR}(x)$$

and $\liminf \lambda_k (2 - \lambda_k) > 0$. *Then x^k converges weakly to a point $x^* \in A \cap B$.*

Proof. Denote $T_k := T_{\sigma_k}$. It follows from Corollary 4.3.14 that the operator $T_{\sigma_{GPR}}$ is a cutter. Therefore, T_k are also cutters, $k \geq 0$, (see Lemma 2.4.7 (ii)). The operator T is nonexpansive as a composition of nonexpansive operators P_A and P_B. It is clear that Fix $T_k = $ Fix $T = A \cap B$. We have

$$\|T_k x^k - x^k\| = \sigma_k(x^k) \|T x^k - x^k\| \geq \|T x^k - x^k\|$$

consequently, $\|T x^k - x^k\| \to 0$ whenever $\|T_k x^k - x^k\| \to 0$. Corollary 3.7.1 with $S = T$ and $X = A$ yields now the weak convergence of x^k to a point $x^* \in A \cap B$. $\square$

Gurin et al. proved a special case of Corollary 5.2.1 with $\sigma_k = \sigma_{GPR}$ and $\lambda_k = 1$ for all $k \geq 0$ (see [196, Theorem 4]).

5.2.1.2 Bauschke–Combettes–Kruk Approach

Let $A \subseteq \mathcal{H}$ be a closed affine subspace and $x \in A$. Then it follows from Theorem 2.2.33 (i) that

$$\langle P_B x - x, Tx - x \rangle = \langle P_B x - Tx, Tx - x \rangle + \|Tx - x\|^2 = \|Tx - x\|^2.$$

Therefore, for $x \in A \backslash B$, we have

$$\sigma_{GPR}(x) = \frac{\|P_B x - x\|^2}{\langle P_B x - x, Tx - x \rangle} = \frac{\|P_B x - x\|^2}{\|Tx - x\|^2}$$

The weak convergence of sequences generated by the extrapolated alternating projection method (5.7) with the step size $\sigma_k(x) := \frac{\|P_B x - x\|^2}{\|Tx - x\|^2}$ for $x \notin \operatorname{Fix} T = A \cap B, k \geq 0$, was proved by Bauschke, Combettes and Kruk in [25, Corollary 4.11].

5.2.1.3 Bauschke–Deutsch–Hundal–Park Approach

Let $A, B \subseteq \mathcal{H}$ be closed subspaces and $x \in A$. Then it follows from Theorem 2.2.30 (i) that

$$\langle x, x - Tx \rangle = \langle x, x - P_B x \rangle + \langle x, P_B - Tx \rangle = \langle x, x - P_B x \rangle$$
$$= \|x - P_B x\|^2 + \langle P_B x, x - P_B x \rangle = \|x - P_B x\|^2.$$

Therefore, for $x \notin \operatorname{Fix} T = A \cap B$, we have

$$\sigma_{GPR}(x) = \frac{\|P_B x - x\|^2}{\|Tx - x\|^2} = \frac{\langle x, x - Tx \rangle}{\|Tx - x\|^2}.$$

The strong convergence of sequences generated by the extrapolated alternating projection method (5.7) with the step size $\sigma_k(x) := \frac{\langle x, x - Tx \rangle}{\|Tx - x\|^2}$ for $x \notin \operatorname{Fix} T = A \cap B, k \geq 0$, was proved by Bauschke et al. in [30, Theorem 3.23]. Actually, Bauschke et al. proved that the method is faster than the alternating projection method. Furthermore, they presented the rate of convergence of the method (see [30, Theorem 3.16]).

5.2.2 Acceleration Techniques for Inconsistent Problems

In this section we suppose that $\operatorname{Fix} T \neq \emptyset$, where $T := P_A P_B$, but we do not suppose that $A \cap B \neq \emptyset$. We present two step size functions $\sigma : A \rightarrow (0, +\infty)$

which guarantee the weak convergence of sequences generated by recurrence (5.7) with $\lambda_k \in [\varepsilon, 2 - \varepsilon]$ for an arbitrary $\varepsilon \in (0, 1)$. Denote

$$\delta := \delta(A, B) = \inf_{x \in A, y \in B} \|x - y\|$$

and

$$\bar{\delta}(x) := \|Tx - P_B x\|$$

for $x \in A$.

First we consider the general case, where $A, B \subseteq \mathcal{H}$ are nonempty closed convex subsets. Let

$$\sigma_{CS1}(x) := \frac{\|Tx - P_B x\|^2 - \tilde{\delta}(x) \|P_B x - x\| + \langle P_B x - x, Tx - x \rangle}{\|Tx - x\|^2}$$

for $x \notin \operatorname{Fix} T$, where $\tilde{\delta}(x) \in [\delta, \bar{\delta}(x)]$. Recall that $\sigma_{CS1}(x) \geq \frac{1}{2}$ and that the inequality is strict if $Tx \notin \operatorname{Fix} T$ (see Lemma 4.3.10).

Corollary 5.2.2. *Let $x^0 \in A$ and the sequence $\{x^k\}_{k=0}^{\infty}$ be generated by (5.7), where*

$$\frac{1}{2} \leq \sigma_k(x) \leq \sigma_{CS1}(x)$$

and $\liminf \lambda_k(2 - \lambda_k) > 0$. *Then x^k converges weakly to a point $x^* \in \operatorname{Fix} P_A P_B$.*

Proof. Denote $T_k := T_{\sigma_k} = \operatorname{Id} + \sigma_k(T - \operatorname{Id})$. It follows from Theorem 4.3.11 that $T_{\sigma_{CS1}}$ is a cutter. Therefore, T_k are also cutters, $k \geq 0$, (see Lemma 2.4.7 (ii)). It is clear that T is nonexpansive and that $\operatorname{Fix} T_k = \operatorname{Fix} T$. We have

$$\left\| T_k x^k - x^k \right\| = \sigma_k(x^k) \left\| T x^k - x^k \right\| \geq \frac{1}{2} \left\| T x^k - x^k \right\|,$$

consequently, $\left\| T x^k - x^k \right\| \to 0$ whenever $\left\| T_k x^k - x^k \right\| \to 0$. Corollary 3.7.1 with $S = T$ and $X = A$ yields now the weak convergence of x^k to a point $x^* \in \operatorname{Fix} P_A P_B$.   □

The convergence of sequences generated by the extrapolated alternating projection method (5.7) with the step size function $\sigma_k := \sigma_{CS1}$, $k \geq 0$, was proved in [76, Theorem 15 (i)].

Now we consider the case, where $A \subseteq \mathcal{H}$ is a closed affine subspace. Let

$$\sigma_{CS2}(x) := 1 + \frac{(\|Tx - P_B x\| - \tilde{\delta})^2}{\|Tx - x\|^2}$$

for $x \notin \operatorname{Fix} T$, where $\tilde{\delta}(x) \in [\delta, \bar{\delta}(x)]$. It is clear that $\sigma_{CS2}(x) \geq 1$.

Corollary 5.2.3. *Let $x^0 \in A$ and the sequence $\{x^k\}_{k=0}^{\infty}$ be generated by recurrence (5.7), where*

$$1 \leq \sigma_k(x) \leq \sigma_{CS2}(x)$$

and $\lim \inf \lambda_k(2 - \lambda_k) > 0$. Then x^k converges weakly to a point $x^* \in \mathrm{Fix}\, P_A P_B$.

Proof. The corollary can be proved similarly to Corollary 5.2.2, where one should apply Theorem 4.3.16 instead of Theorem 4.3.11. $\square$

The convergence of sequences generated by the extrapolated alternating projection method (5.7) with the step size function $\sigma_k := \sigma_{CS2}$, $k \geq 0$, was proved in [76, Theorem 15 (ii)].

In [75, 306] the problem of finding a fixed point of the alternating projection $P_A P_B$, derived from an inconsistent system of linear equations is considered and incomplete projection methods for finding $\mathrm{Fix}\, P_A P_B$ are proposed, where P_A is replaced by an 'incomplete projection' which approximates P_A in some sense.

5.2.3 Douglas–Rachford Algorithm

Let $A, B \subseteq \mathcal{H}$ be closed convex subsets. The *Douglas–Rachford algorithm* (DR) is an iterative method for finding a common point of A and B, if such a point exists. One iteration of the method has the form

$$x^{k+1} = \frac{1}{2}(R_A R_B x^k + x^k), \tag{5.9}$$

where $R_A := 2P_A - \mathrm{Id}$ and $R_B := 2P_B - \mathrm{Id}$ are reflection operators onto A and B, respectively. The iteration can also be written in the form $x^{k+1} = Tx^k$, where $T : \mathcal{H} \to \mathcal{H}$ is the averaged alternating reflection (AAR), i.e., $T := \frac{1}{2}(R_A R_B + \mathrm{Id})$ (see Sect. 4.3.5). The method is also called an *averaged alternating reflection algorithm*. The method was introduced by Douglas and Rachford in order to solve a partial differential equation describing the heat conduction problem and was studied in [26–29, 157, 161, 246], where various applications of the method were presented. Consider the following iteration

$$x^{k+1} = x^k + \frac{\lambda_k}{2}(R_A R_B x^k - x^k), \tag{5.10}$$

where $\lambda_k \in [0, 2]$. If $\lambda_k = 1$, then we obtain (5.9). We can also write $x^{k+1} = T_{\lambda_k} x^k$, where T_λ denotes a λ-relaxation of the AAR operator T. The following corollary extends the result of [26, Fact 5.9].

Corollary 5.2.4. *Let $A, B \subseteq \mathcal{H}$ be closed convex. If $\mathrm{Fix}(R_A R_B) \neq \emptyset$ and $\lim \inf_k \lambda_k(2 - \lambda_k) > 0$, then for any $x^0 \in \mathcal{H}$ the sequence $\{x^k\}_{k=0}^{\infty}$ generated by (5.10) converges weakly to a fixed point of $R_A R_B$ and the sequence $\{P_B x^k\}_{k=0}^{\infty}$ is bounded. Furthermore, if $A \cap B \neq \emptyset$, then:*

(i) *For any weak cluster point y of $\{P_B x^k\}_{k=0}^{\infty}$ it holds that $y \in A \cap B$.*
(ii) *If $\dim \mathcal{H} < \infty$, then $y := \lim_k P_B x^k$ exists and $y \in A \cap B$.*

Proof. By Corollary 4.3.17 (iv), the operator $T := \frac{1}{2}(R_A R_B + \mathrm{Id})$ is firmly nonexpansive. Therefore, Corollary 3.7.3 with $X = \mathcal{H}$ yields $x^k \rightharpoonup x^* \in \mathrm{Fix}\, T$. The boundedness of $\{P_B x^k\}_{k=0}^{\infty}$ follows from the boundedness of $\{x^k\}_{k=0}^{\infty}$ and from the nonexpansivity of P_B. Let $A \cap B \neq \emptyset$ and $z \in A \cap B$. Then Corollary 4.3.17 (i) and (iv) yields $z \in \mathrm{Fix}\, T$ and

$$\left\| x^{k+1} - z \right\|^2 = \left\| T_\lambda x^k - z \right\|^2 \leq \left\| x^k - z \right\|^2 - \lambda_k(2 - \lambda_k) \left\| T x^k - x^k \right\|^2.$$

Consequently, $\left\| x^k - z \right\|$ converges and $\lim_k \left\| T x^k - x^k \right\| = 0$, by $\liminf_k \lambda_k (2 - \lambda_k) > 0$. We have

$$\lim_k [P_A(2 P_B x^k - x^k) - P_B x^k] = \lim_k \frac{1}{2}(R_A R_B x^k - x^k) = \lim_k (T x^k - x^k) = 0.$$

(i) Let y be a weak cluster point of $\{P_B x^k\}_{k=0}^{\infty}$. Because B is weakly closed (see Theorem 1.1.40) and $\{P_B x^k\}_{k=0}^{\infty} \subseteq B$, we have $y \in B$. We prove that $y \in A$. Suppose $y \notin A$ and let a subsequence $\{P_B x^{n_k}\}_{k=0}^{\infty} \subseteq \{P_B x^k\}_{k=0}^{\infty}$ converge weakly to y. By the boundedness of $\{P_B x^k\}_{k=0}^{\infty}$ and of $\{x^k\}_{k=0}^{\infty}$ and by the nonexpansivity of P_A, the sequence $\{P_A(2 P_B x^{n_k} - x^{n_k})\}_{k=0}^{\infty}$ is bounded, consequently, it contains a subsequence $\{P_A(2 P_B x^{m_k} - x^{m_k})\}_{k=0}^{\infty}$ which converges weakly to a point $z \in \mathcal{H}$. Since A is weakly closed, $z \in A$. Because $\{P_A(2 P_B x^k - x^k) - P_B x^k\}_{k=0}^{\infty}$ converges to 0, it converges weakly to 0 and $\{P_A(2 P_B x^{m_k} - x^{m_k}) - P_B x^{m_k}\}_{k=0}^{\infty}$ also converges weakly to 0. On the other hand, $\{P_A(2 P_B x^{m_k} - x^{m_k}) - P_B x^{m_k}\}_{k=0}^{\infty}$ converges weakly to $z - y \neq 0$. A contradiction shows that $y \in A$.

(ii) Let $\dim \mathcal{H} < \infty$. The first part of the corollary and the nonexpansivity of P_B yield

$$\lim_k \left\| P_B x^k - P_B x^* \right\| \leq \lim_k \left\| x^k - x^* \right\| = 0,$$

for some $x^* \in \mathrm{Fix}\, R_A R_B$, i.e., $\lim_k P_B x^k = P_B x^* = y$. By Corollary 4.3.17 (iii), $y \in A \cap B$. $\square$

5.3 Projected Gradient Method

Several optimization problems can be reduced to a variational inequality problem, e.g., finding the metric projection onto a nonempty closed convex subset of a Hilbert space, convex minimization problem, elliptic control problem, etc. In this section we present the most popular method for solving $VIP(\mathcal{F}, C)$ where $\mathcal{F} : \mathcal{H} \to \mathcal{H}$ is a Lipschitz continuous and strongly monotone operator and $C \subseteq \mathcal{H}$ is a nonempty closed convex subset. The method is called a *projected gradient method* and has the form

$$x^{k+1} = P_C(x^k - \mu \mathcal{F} x^k). \tag{5.11}$$

The convergence of sequences generated by the method follows from the theorem below, which shows an application of the Banach fixed point theorem and the properties of the metric projection.

Theorem 5.3.1. *Let $C \subseteq \mathcal{H}$ be closed convex, $\mathcal{F} : \mathcal{H} \to \mathcal{H}$ be κ-Lipschitz continuous and η-strongly monotone over C, where $\kappa, \eta > 0$, and $\mu \in (0, \frac{2\eta}{\kappa^2})$. Then the operator $T := P_C(\mathrm{Id} - \mu \mathcal{F})$ is a contraction. Consequently, for any $x^0 \in \mathcal{H}$, the sequence $\{x^k\}_{k=0}^{\infty}$ defined by (5.11) converges to the unique solution of VIP$(\mathcal{F}, C)$ with a rate of the geometric progression.*

Proof. (cf. [358, Theorem 46.C]) Define $G = \mathrm{Id} - \mu \mathcal{F}$. Let $x, y \in \mathcal{H}$. Then

$$
\begin{aligned}
\|Gx - Gy\|^2 &= \|(x - y) - \mu(\mathcal{F}x - \mathcal{F}y)\|^2 \\
&= \|x - y\|^2 + \mu^2 \|\mathcal{F}x - \mathcal{F}y\| - 2\mu \langle \mathcal{F}x - \mathcal{F}y, x - y \rangle \\
&\le (1 + \mu^2 \kappa^2 - 2\mu \eta) \|x - y\|^2 \\
&= (1 - \tau)^2 \|x - y\|^2,
\end{aligned}
$$

where $\tau := \sqrt{1 + \mu^2 \kappa^2 - 2\mu \eta}$ (note that τ is well defined, because $\eta \le \kappa$ and $\mu \in (0, \frac{2\eta}{\kappa^2})$, and that $\tau \in (0, 1)$). Consequently,

$$\|Gx - Gy\| \le (1 - \tau) \|x - y\|. \tag{5.12}$$

Since P_C is nonexpansive, (5.12) yields

$$
\begin{aligned}
\|P_C(x - \mu \mathcal{F}x) - P_C(y - \mu \mathcal{F}y)\| &= \|P_C Gx - P_C Gy\| \\
&\le (1 - \tau) \|x - y\|,
\end{aligned}
$$

i.e., $P_C(\mathrm{Id} - \mu \mathcal{F})$ is a $(1 - \tau)$-contraction. The Banach fixed point theorem and Theorem 1.3.8 yield that $P_C(\mathrm{Id} - \mu \mathcal{F})$ has a unique fixed point $\bar{x} \in C$ and $\bar{x}$ is the unique solution of VIP$(\mathcal{F}, C)$. Furthermore, the Banach fixed point theorem yields that for any $x^0 \in \mathcal{H}$, the sequence $\{x^k\}_{k=0}^{\infty}$ generated by (5.11) converges geometrically to $\bar{x}$. $\square$

In the literature one can find several method for solving VIP$(\mathcal{F}, C)$ in particular for solving VIP$(\mathcal{F}, \mathrm{Fix}\, T)$, where $T : \mathcal{H} \to \mathcal{H}$ is a quasi-nonexpansive operator, for solving convex constrained minimization problems, for finding the metric projection onto the intersection of a finite family of convex subsets, for solving elliptic control problems, etc. For details we send the reader to [18, 52, 77, 94–97, 101, 144, 171, 200, 203, 211, 216–218, 237, 244, 250, 251, 269, 273, 274, 276, 318, 320, 338, 342, 343, 345–349, 353].

5.4 Simultaneous Projection Method

Let $\{C_i\}_{i \in I} \subseteq \mathcal{H}$ be a finite family of nonempty closed convex subsets, where $I := \{1, 2, \ldots, m\}$. We consider the convex feasibility problem of finding a point $x^* \in C := \bigcap_{i \in I} C_i$, if such a point exists. The proximity function $f : \mathcal{H} \to \mathbb{R}_+$ is defined by $f(x) := \frac{1}{2} \sum_{i \in I} d^2(x, C_i)$. The *simultaneous projection method* (SPM) for solving this problem has the following form

$$
\begin{aligned}
x^0 &\in \mathcal{H} \qquad\qquad\qquad\qquad\qquad\qquad \text{– arbitrary} \\
x^{k+1} &= x^k + \lambda_k \big(\textstyle\sum_{i \in I} \omega_i P_{C_i} x^k - x^k \big),
\end{aligned}
\tag{5.13}
$$

where $\{\lambda_k\}_{k=0}^\infty \subseteq [0, 2]$ is a sequence of relaxation parameters and $w = (\omega_1, \ldots, \omega_m) \in \mathrm{ri}\, \Delta_m$ is a vector of weights. We can also write $x^{k+1} = T_{\lambda_k} x^k$, where $T := \sum_{i \in I} \omega_i P_{C_i}$ is the operator of a simultaneous projection and $T_\lambda := \mathrm{Id} + \lambda(T - \mathrm{Id})$ denotes a relaxation of T. Recall that

$$
\mathrm{Fix}\, T = \mathrm{Fix}\, T_\lambda = \operatorname*{Argmin}_{x \in \mathcal{H}} f(x),
$$

where $\lambda > 0$ and $f : \mathcal{H} \to \mathbb{R}$ is a proximity function defined by the equality

$$
f(x) := \frac{1}{2} \sum_{i \in I} \omega_i \, \| P_{C_i} x - x \|^2
\tag{5.14}
$$

(see Theorem 4.4.6) and that $\mathrm{Fix}\, T = C$ if $C \neq \emptyset$ (see Corollary 4.4.2).

5.4.1 Convergence of the SPM

The convergence of sequences $\{x^k\}_{k=0}^\infty$ generated by the simultaneous projection method (5.13) follows from the following result which is due to Combettes [117, Theorem 4].

Corollary 5.4.1. *Let $\{x^k\}_{k=0}^\infty \subseteq \mathcal{H}$ be a sequence generated by the simultaneous projection method (5.13), where $\liminf_k \lambda_k (2 - \lambda_k) > 0$. If $\mathrm{Fix}\, T \neq \emptyset$, then x^k converges weakly to a point $x^* \in \mathrm{Fix}\, T$ and the following estimation holds*

$$
d^2(x^{k+1}, \mathrm{Fix}\, T) \leq d^2(x^0, \mathrm{Fix}\, T) - \sum_{l=0}^k \lambda_l (2 - \lambda_l) \, \big\| T x^l - x^l \big\|^2,
\tag{5.15}
$$

$k \geq 0$. *If, furthermore, $C := \bigcap_{i \in I} C_i \neq \emptyset$, then*

$$d^2(x^{k+1}, C) \le d^2(x^0, C) - \sum_{l=0}^{k} \lambda_l (2 - \lambda_l) \sum_{i \in I} \omega_i \left\| P_{C_i} x^l - x^l \right\|^2, \qquad (5.16)$$

$k \ge 0$.

Proof. Let Fix $T \ne \emptyset$. The operator T is firmly nonexpansive as a convex combination of firmly nonexpansive operators P_{C_i} (see Corollary 2.2.20). Furthermore, T is nonexpansive (see Theorem 2.2.4) and a cutter (see Theorem 2.2.5 (i)). Consequently, $T -$ Id is demi-closed at 0 (see Lemma 3.2.5). If we take $X = \mathcal{H}$, $T_k := T$ and $S := T$ in Corollary 3.7.1, then we obtain that x^k converges weakly to a point $x^* \in$ Fix T and that

$$\left\| x^{l+1} - z \right\|^2 = \left\| T_\lambda x^l - z \right\|^2 \le \left\| x^l - z \right\|^2 - \lambda_l (2 - \lambda_l) \left\| T x^l - x^l \right\|^2,$$

$l \ge 0$. Applying this inequality $(k + 1)$-times for $l = 0, 1, \ldots, k$ and for $z = P_{\text{Fix } T} x^0$, we obtain

$$d^2(x^{k+1}, \text{Fix } T) \le \left\| x^{k+1} - z \right\|^2$$

$$= \left\| T_\lambda x^k - z \right\|^2 \le \left\| x^0 - z \right\|^2 - \sum_{l=0}^{k} \lambda_l (2 - \lambda_l) \left\| T x^l - x^l \right\|^2$$

$$= d^2(x^0, \text{Fix } T) - \sum_{l=0}^{k} \lambda_l (2 - \lambda_l) \left\| T x^l - x^l \right\|^2.$$

If $C \ne \emptyset$, then estimation (5.16) can be proved in the same way as above applying Theorem 4.4.5.   □

If $C_i \subseteq \mathcal{H}$ are closed subspaces, then the convergence in Corollary 5.4.1 is strong and $x^* = P_C x^0$. This result is due to Reich [295, Theorem 1.7], where the convergence was proved for uniformly convex Banach spaces.

The strong convergence of sequences generated by the simultaneous projection method does not hold in general. Bauschke, Matoušková and Reich, relying on a result of Hundal [213, Theorem 1], gave an example of two closed convex subsets $C_1, C_2 \subseteq \mathcal{H}$ with a nonempty intersection, a simultaneous projection $T := \frac{1}{2} P_{C_1} + \frac{1}{2} P_{C_2}$ and a point x for which $T^k x$ converges weakly but does not converge strongly [32, Theorem 5.1].

If $\mathcal{H} = \mathbb{R}^n$ and C_i are hyperplanes or half-spaces, i.e., $C_i := H(a_i, \beta_i) = \{x \in \mathbb{R}^n : a_i^\top x = \beta_i\}$ or $C_i := H_-(a_i, \beta_i) = \{x \in \mathbb{R}^n : a_i^\top x \le \beta_i\}$, then it follows from equalities (4.63) and (4.70) that the simultaneous projection method (5.13) can be written in the matrix form

$$x^{k+1} = x^k - \lambda_k A^\top D(Ax^k - b) \qquad (5.17)$$

or

$$x^{k+1} = x^k - \lambda_k A^\top D(Ax^k - b)_+, \qquad (5.18)$$

respectively, where $D := \mathrm{diag}(\frac{\omega_1}{\|a_1\|^2}, \frac{\omega_2}{\|a_2\|^2}, \dots, \frac{\omega_m}{\|a_m\|^2})$.

The simultaneous projection method (5.13) was introduced by Cimmino in [116], where $\mathcal{H} = \mathbb{R}^n$, $C_i := H(a_i, \beta_i), i \in I$, $C \neq \emptyset$ and $\lambda_k = 2$ for all $k \geq 0$. It follows from (5.17) that Cimmino's method has the form

$$x^{k+1} = x^k - 2A^\top D(Ax^k - b). \qquad (5.19)$$

Cimmino proved the convergence of sequences generated by (5.19) to a solution of the system $Ax = b$ for a nonsingular $n \times n$ matrix A. Special cases of method (5.13) were studied by Auslender [12], De Pierro and Iusem [134, 135, 215], Pierra [284], Combettes [117]. In particular, the (weak) convergence of the sequences generated by recurrence (5.13) to a solution of the CFP was proved by:

(a) Auslender in [12], where $\mathcal{H} = \mathbb{R}^n$, $C \neq \emptyset$, $\omega_i = \frac{1}{m}$, $i \in I$, and $\lambda_k = 1$,
(b) Pierra in [284, Theorem 1.1 (i)], where $C \neq \emptyset$, $\omega_i = \frac{1}{m}$, $i \in I$, and $\lambda_k = 1$ for all $k \geq 0$,
(c) De Pierro and Iusem in [134, Theorem 3], where $\mathcal{H} = \mathbb{R}^n$, $C_i := H_-(a_i, \beta_i)$, $i \in I$, Fix $T \neq \emptyset$ and $\lambda_k = \lambda \in (0, 2)$ for all $k \geq 0$,
(d) Iusem and De Pierro in [215, Theorem 2], where $\mathcal{H} = \mathbb{R}^n$, and $\lambda_k = 1$ for all $k \geq 0$,
(e) Combettes in [117, Theorem 4], where $\lambda_k \in [\varepsilon, 2 - \varepsilon]$ for some $\varepsilon \in (0, 1)$.

5.4.2 Projected Simultaneous Projection Methods

In this section we consider the general convex feasibility problem of finding a point $x^* \in X$ which is closest to $C := \bigcap_{i \in I} C_i$, where $C_i \subseteq \mathcal{H}$ are nonempty closed convex subsets, $i \in I := \{1, 2, \dots m\}$. If $X \cap C \neq \emptyset$, then the problem is to find $x \in X \cap C$. The proximity function $f : X \to \mathbb{R}_+$ is defined by $f(x) := \frac{1}{2} \sum_{i \in I} d^2(x, C_i)$. The *projected simultaneous projection method* has the following form

$$\begin{aligned} x^0 &\in X &&\text{– arbitrary} \\ x^{k+1} &= P_X(x^k + \lambda_k(\textstyle\sum_{i \in I} \omega_i P_{C_i} x^k - x^k)), \end{aligned} \qquad (5.20)$$

where $\{\lambda_k\}_{k=0}^\infty \subseteq [0, 2]$ is a sequence of relaxation parameters and $w = (\omega_1, \dots, \omega_m) \in \Delta_m$ is a vector of weights. If $X = \mathcal{H}$, then method (5.20) is equivalent to the simultaneous projection method (5.13). We can also write $x^{k+1} = P_X T_{\lambda_k} x^k$, where $T := \sum_{i \in I} \omega_i P_{C_i}$ is the operator of simultaneous projection.

Corollary 5.4.2. *Let $\{x^k\}_{k=0}^\infty \subseteq X$ be a sequence generated by the projected simultaneous projection method (5.20), where $\lambda_k \in [\varepsilon, 2 - \varepsilon]$ for some $\varepsilon \in (0, 1)$.*

If Fix $P_X T \neq \emptyset$, *where* $T := \sum_{i \in I} \omega_i P_{C_i}$, *then* x^k *converges weakly to a point* $x^* \in$ Fix $P_X T$.

Proof. T is firmly nonexpansive (see Corollary 4.4.4). If we set $S_k := T, k \geq 0$, in Corollary 3.7.4, then we obtain the weak convergence of x^k to a point $x^* \in$ Fix $P_X T$. $\square$

Consider now the following recurrence

$$x^{k+1} = P_X(x^k + \mu_k(P_X U x^k - x^k)),   \tag{5.21}$$

where $x^0 \in \mathcal{H}$, $U := \sum_{i \in I} \omega_i P_{C_i}$, $w = (\omega_1, \ldots, \omega_m) \in \Delta_m$ and $\mu_k > 0$, $k \geq 0$. Note the difference between (5.21) and (5.20).

Corollary 5.4.3. *Let* $\{x^k\}_{k=0}^{\infty} \subseteq X$ *be a sequence generated by method (5.21), where* $\mu_k \in [\varepsilon, \frac{3}{2} - \varepsilon]$ *for some* $\varepsilon \in (0, \frac{3}{4})$. *If* Fix $P_X U \neq \emptyset$, *then* x^k *converges weakly to a point* $x^* \in$ Fix $P_X U$.

Proof. U is firmly nonexpansive (see Corollary 4.4.4). This fact together with the firm nonexpansivity of P_X yield that $P_X U$ is $\frac{4}{3}$-relaxed firmly nonexpansive and that $(P_X U)_{\frac{3}{4}}$ is firmly nonexpansive (see Corollary 2.2.39). Recurrence (5.21) can be written in the form

$$x^{k+1} = P_X(x^k + \lambda_k(T x^k - x^k)),$$

where $\lambda_k = \frac{4}{3}\mu_k$, $T := (P_X U)_{\frac{3}{4}} = \mathrm{Id} + \frac{3}{4}(P_X U - \mathrm{Id})$. Note that $\lambda_k \in [\varepsilon', 2 - \varepsilon']$ where $\varepsilon' = \frac{4}{3}\varepsilon \in (0, 1)$. Now, the weak convergence of x^k to a point $x^* \in$ Fix $T =$ Fix $P_X U$ follows from Corollary 3.7.3. $\square$

5.5   Cyclic Projection Methods

Let $\{C_i\}_{i \in I} \subseteq \mathcal{H}$ be a finite family of nonempty closed convex subsets, where $I := \{1, 2, \ldots, m\}$. Consider the convex feasibility problem of finding a point $x^* \in C := \bigcap_{i \in I} C_i$, if such a point exists. The *cyclic projection method* for solving this problem has the following form

$$\begin{aligned} x^0 &\in \mathcal{H} \qquad\qquad\qquad - \text{arbitrary} \\ x^{k+1} &= P_{C_m} \ldots P_{C_1} x^k. \end{aligned}   \tag{5.22}$$

If we replace the operators P_{C_i} in (5.22) by their relaxations P_{C_i, λ_i}, where $\lambda_i \in (0, 2)$, $i \in I$, then we obtain the following, more general method

$$x^0 \in \mathcal{H} \qquad\qquad\qquad \text{– arbitrary} \\ x^{k+1} = P_{C_m,\lambda_m} \cdots P_{C_1,\lambda_1} x^k, \tag{5.23}$$

which we call the *cyclic relaxed projection method*. We can also write $x^{k+1} = Tx^k$, where $T := P_{C_m,\lambda_m} \cdots P_{C_1,\lambda_1}$ is the cyclic relaxed projection. Denote $S_0 := \mathrm{Id}$ and $S_i := P_{C_i,\lambda_i} \cdots P_{C_1,\lambda_1}$ for $i = 1, 2, \ldots, m$. We have $S_m = T$.

5.5.1 Convergence

Corollary 5.5.1. *Let $\{x^k\}_{k=0}^{\infty} \subseteq \mathcal{H}$ be a sequence generated by the cyclic relaxed projection method (5.23), where $\lambda_i \in (0, 2)$, $i \in I$. If $\mathrm{Fix}\, T \neq \emptyset$, then x^k converges weakly to a point $x^* \in \mathrm{Fix}\, T$ and the following estimation holds*

$$d^2(x^{k+1}, \mathrm{Fix}\, T) \leq d^2(x^0, \mathrm{Fix}\, T) - \sum_{l=0}^{k} \alpha \left\| Tx^l - x^l \right\|^2, \tag{5.24}$$

where $\alpha = \frac{2-\lambda_{\max}}{m\lambda_{\max}}$ and $\lambda_{\max} := \max\{\lambda_i : i \in I\}$. If $C := \bigcap_{i \in I} C_i \neq \emptyset$ and $\lambda_i = 1$ for all $i \in I$, then

$$d^2(x^{k+1}, C) \leq d^2(x^0, C) - \sum_{l=0}^{k} \sum_{i \in I} \left\| S_i x^l - S_{i-1} x^l \right\|^2. \tag{5.25}$$

Proof. Let $\mathrm{Fix}\, T \neq \emptyset$. It follows from Corollary 4.5.8 that T is λ-RFNE, where $\lambda = \frac{2m\lambda_{\max}}{(m-1)\lambda_{\max}+2}$ and that T is asymptotically regular. Since T is nonexpansive, it follows from Theorem 3.5.3 that x^k converges weakly to $x^* \in \mathrm{Fix}\, T$. Corollary 2.2.9 yields the α-strong quasi nonexpansivity of T, where $\alpha := \frac{2-\lambda}{\lambda} = \frac{2-\lambda_{\max}}{m\lambda_{\max}}$. Therefore, T is α-SQNE, i.e.,

$$\left\| Tx^l - z \right\|^2 \leq \left\| x^l - z \right\|^2 - \alpha \left\| Tx^l - x^l \right\|^2$$

for all $z \in \mathrm{Fix}\, T$, $l = 0, 1, \ldots$ If we take $z = P_{\mathrm{Fix}\, T} x^0$ and apply the above inequality for $l = 0, 1, \ldots, k$, we obtain

$$d^2(x^{k+1}, \mathrm{Fix}\, T) \leq \left\| x^{k+1} - z \right\|^2$$

$$= \left\| Tx^k - z \right\|^2 \leq \left\| x^0 - z \right\|^2 - \sum_{i=0}^{k} \alpha \left\| Tx^i - x^i \right\|^2$$

$$= d^2(x^0, \mathrm{Fix}\, T) - \sum_{i=0}^{k} \alpha \left\| Tx^i - x^i \right\|^2.$$

If $C \neq \emptyset$, then we obtain estimation (5.25) in the same way as above if we apply inequality (4.74).   □

The weak convergence of sequence generated by the cyclic projection method was proved by Bregman [42, Theorem 1]. Sufficient conditions for strong convergence as well as for geometric convergence were given by Gurin et al. [196]. Other conditions for the strong convergence can be found in [14]. De Pierro and Iusem considered the method for a inconsistent convex feasibility problem in $\mathbb{R}^n$ (see [137]) and studied the convergence of sequences generated by the method. The rate of convergence for the method in a Hilbert space was studied in [142, 143, 290, 291]. The paper [23] contains a review of the results on the cyclic projection method in a Hilbert space.

The cyclic projection method for a consistent system of n linear equations in $\mathbb{R}^n$ was introduced by Stefan Kaczmarz [223]. Therefore, the cyclic projection method is also known as the *Kaczmarz method*. Kaczmarz supposed that the equations are independent or, equivalently, that the solution is uniquely defined and proved the convergence to the solution of sequences generated by the method [223]. Altman proved the equivalence of the method with the Gauss–Seidel method for the system $A^\top A x = A^\top b$ [7, Corollary in Sect. 2]. Halperin considered the cyclic projection method for a finite system of subspaces $V_i \subseteq \mathcal{H}$, $i \in I$, and proved that for any $x \in \mathcal{H}$ it holds that $\lim_{k \to \infty} \left\| (P_{V_m} P_{V_{m-1}} \ldots P_{V_1})^k x - P_V x \right\| = 0$ (see [198, Theorem 2]). Note that this generalizes the von Neumann Theorem 5.1.5.

Gordon, Bender and Herman proposed a method for solving a system of linear equations [189] which is equivalent to the Kaczmarz method, and called it an *algebraic reconstruction technique* (ART). Tanabe proved the convergence for an arbitrary system of linear equations in $\mathbb{R}^n$ [321, Corollary 9]. The Kaczmarz method in Hilbert space was studied by McCormick in [259]. Since 1970 the Kaczmarz method has been applied to computerized tomography and in radiation therapy (see, e.g., [81, 82, 100, 189]).

Further properties of the Kaczmarz method as well their modifications can be found in [9, 13, 34, 41, 81, 86, 129–132, 162, 197, 201, 239, 247, 268, 285–287].

5.5.2 Projection-Reflection Method

Let $C_i \subseteq \mathcal{H}$ be closed convex subsets, $i \in I$, and $K \subseteq \mathcal{H}$ be a closed convex and obtuse cone, i.e., $-K^* \subseteq K$. Suppose that $C := K \cap \bigcap_{i \in I} C_i \neq \emptyset$. Consider the following CFP: find $x \in C$. Denote

$$P = P_{C_m} P_{C_{m-1}} \ldots P_{C_1},$$

$P_\mu := \mathrm{Id} + \mu(P - \mathrm{Id})$, where $\mu \in (0, 2)$, and $R_K := (P_K)_2 = 2P_K - \mathrm{Id}$, i.e., P_μ is a relaxation of the cyclic projection P and R_K is the reflection operator onto K. Consider the iterative method defined by the following recurrence

Fig. 5.1 Projection-reflection method and reflection-projection method

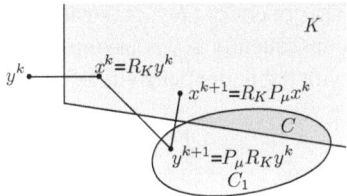

$$x^{k+1} = R_K P_\mu x^k, \qquad (5.26)$$

where $x^0 \in \mathcal{H}$. We can also write $x^{k+1} = R_\mu x^k$, where $R_\mu := R_K P_\mu$ is a μ-relaxed projection-reflection operator (see Definition 4.5.11). We call the method a *projection-reflection method* (see Fig. 5.1). Since K is obtuse, $x^k \in K$ for all $k \geq 1$ (see Lemma 4.5.12). Therefore, without loss of generality, we can suppose that $x^0 \in K$ and that $R_\mu := R_K P_\mu$ in (5.26) is restricted to K.

Corollary 5.5.2. *Let $K \subseteq \mathcal{H}$ be a closed convex and obtuse cone, $C_i \subseteq \mathcal{H}$ be closed convex subsets, $i \in I$, $C := K \cap \bigcap_{i \in I} C_i \neq \emptyset$, $\mu \in (0, \frac{m+1}{m})$ and $x^0 \in K$. Then the sequence generated by the projection-reflection method (5.26) converges weakly to a point $x^* \in C$.*

Proof. The operator $R_\mu \mid_K$ is nonexpansive and strongly quasi-nonexpansive (see Corollary 4.5.13 (i) and (iii)). Consequently $R_\mu \mid_K$ is asymptotically regular (see Theorem 3.4.3). By Corollary 4.5.13 (ii), we have Fix $R_\mu = C$. Let $\{x^k\}_{k=0}^\infty$ be generated by the projection-reflection method (5.26). We see that all assumptions of Opial's Theorem 3.5.1 are satisfied. Therefore, x^k converges weakly to $x^* \in C$. $\square$

Consider now sequences $\{y^k\}_{k=0}^\infty$ generated by the recurrence

$$y^{k+1} = P_\mu R_K y^k \qquad (5.27)$$

with $y^0 \in \mathcal{H}$. We call the method a *reflection-projection method*. The method is closely related to the projection-reflection method (5.26). If we take $x^0 := R_K y^0$, we obtain that the sequence $\{x^k\}_{k=0}^\infty$ defined by $x^k := R_K y^k$ is generated by (5.26). Furthermore, $y^{k+1} = P_\mu x^k$ (see Fig. 5.1).

Bauschke and Kruk considered the reflection-projection method (5.27) in $\mathcal{H} = \mathbb{R}^n$ with $\mu = 1$ [31] and proved that the sequences generated by the method converge to a point $x^* \in C$ (see [31, Theorem 3.1]). The following corollary generalizes this result.

Corollary 5.5.3. *Let $K \subseteq \mathcal{H}$ be a closed convex and obtuse cone, $C_i \subseteq \mathcal{H}$ be closed convex subsets, $i \in I$, $C := K \cap \bigcap_{i \in I} C_i \neq \emptyset$, $\mu \in (0, \frac{m+1}{m})$ and $y^0 \in \mathcal{H}$. Then the sequence generated by the reflection-projection method (5.27) converges weakly to a point $y^* \in C$.*

Proof. Let $\{y^k\}_{k=0}^\infty$ be generated by (5.27), $x^0 := R_K y^0$ and $\{x^k\}_{k=0}^\infty$ be generated by the projection-reflection method (5.26). We have $x^k = R_K y^k$ and $y^{k+1} = P_\mu x^k$. By Corollary 4.5.13 (i) and (ii), $P_\mu R_K$ is nonexpansive and Fix $P_\mu R_K = C$.

The operator R_μ is strongly quasi-nonexpansive (see Corollary 4.5.13 (iii)), consequently R_μ is asymptotically regular (see Theorem 3.4.3). This fact together with the nonexpansivity of P_μ yield

$$\left\| y^{k+1} - y^k \right\| = \left\| P_\mu x^k - P_\mu x^{k-1} \right\| \le \left\| x^k - x^{k-1} \right\|$$
$$= \left\| R_\mu^k x^0 - R_\mu^{k-1} x^0 \right\| \to 0.$$

Therefore,

$$\left\| (P_\mu R_K)^{k+1} y^0 - (P_\mu R_K)^k y^0 \right\| = \left\| y^{k+1} - y^k \right\| \to 0,$$

i.e., $P_\mu R_K$ is asymptotically regular. We see that all assumptions of Opial's Theorem 3.5.1 are satisfied. Therefore, y^k converges weakly to a point $y^* \in C$. □

5.6 Successive Projection Methods

Let $\{C_i\}_{i \in I} \subseteq \mathcal{H}$ be a finite family of nonempty closed convex subsets $C_i, i \in I :=$ $\{1, 2, \ldots, m\}$, with a common point. We consider the convex feasibility problem of finding a point $x^* \in C := \bigcap_{i \in I} C_i$, if such a point exists. The *successive projection method* or *sequential projection method* for solving this problem has the following form

$$
\begin{aligned}
x^0 &\in \mathcal{H} && \text{– arbitrary} \\
x^{k+1} &= x^k + \lambda_k (P_{C_{i_k}} x^k - x^k),
\end{aligned}
\tag{5.28}
$$

where $\{\lambda_k\}_{k=0}^\infty \subseteq [0, 2]$ is a sequence of relaxation parameters and $\{i_k\}_{k=0}^\infty \subseteq I$ is a *control sequence* or, shortly, *control*. Note that i_k can depend on the current approximation $x^k \in \mathcal{H}$. In this case we identify i_k with $i_k(x^k)$. It is convenient to define $i_k(x)$ for any $x \in \mathcal{H}$, i.e., to define a sequence of functions $\{i_k\}_{k=0}^\infty :$ $\mathcal{H} \to I$. We will, however, relate this sequence of functions to recurrence (5.28). If the sequence of functions $\{i_k\}_{k=0}^\infty$ is constant, i.e., $i_k = i$ for all $k \ge 0$, then the function $i : \mathcal{H} \to I$ is called a *control function* or, shortly, *control*. In this case we write $i_k = i(x^k)$. A comprehensive overview of the sequential projection methods can be found in [78] and in [108, Chap. 5].

5.6.1 Convergence

The convergence of sequences generated by (5.28) depends on the properties of the control sequence $\{i_k\}_{k=0}^\infty$. The following definition is due to Gurin et al. (see [196, Sect. 1]).

Definition 5.6.1. A control sequence $\{i_k\}_{k=0}^{\infty} : \mathcal{H} \to I$ is called an *approximately remotest set control* if for any sequence $\{x^k\}_{k=0}^{\infty}$ generated by (5.28) the following implication holds

$$\lim_k \left\| P_{C_{i_k}} x^k - x^k \right\| = 0 \implies \lim_k \max_{i \in I} \left\| P_{C_i} x^k - x^k \right\| = 0. \tag{5.29}$$

Definition 5.6.2. A control sequence $\{i_k\}_{k=0}^{\infty} : \mathcal{H} \to I$ is called an *approximately semi-remotest set control* if for any sequence $\{x^k\}_{k=0}^{\infty}$ generated by (5.28) the following implication holds

$$\lim_k \left\| P_{C_{i_k}} x^k - x^k \right\| = 0 \implies \liminf_k \left\| P_{C_i} x^k - x^k \right\| = 0 \text{ for all } i \in I. \tag{5.30}$$

Remark 5.6.3. Since I is a finite subset, the equality on the right hand side of (5.29) can be written in the following equivalent form: $\lim_k \left\| P_{C_i} x^k - x^k \right\| = 0$ for all $i \in I$. Therefore, implication (5.29) is equivalent to the following one

$$\lim_k \left\| P_{C_{i_k}} x^k - x^k \right\| = 0 \implies \lim_k \left\| P_{C_i} x^k - x^k \right\| = 0 \text{ for all } i \in I.$$

Corollary 5.6.4. *Let $C \neq \emptyset$ and $\{x^k\}_{k=0}^{\infty} \subseteq \mathcal{H}$ be a sequence generated by the successive projection method (5.28), where $\liminf_k \lambda_k (2 - \lambda_k) > 0$.*

(i) *If $\{i_k\}_{k=0}^{\infty}$ is an approximately remotest set control, then x^k converges weakly to a point $x^* \in C$.*

(ii) *If $\mathcal{H}$ is finite dimensional and $\{i_k\}_{k=0}^{\infty}$ is an approximately semi-remotest set control, then x^k converges to a point $x^* \in C$.*

Proof. Setting $X = \mathcal{H}$, $T_k := P_{C_{i_k}}$ and $S := P_{C_i}$ in Corollary 3.7.1, we obtain $x^k \rightharpoonup x^* \in \text{Fix } P_{C_i} = C_i, i \in I$, in (i) and $x^k \to x^* \in \text{Fix } P_{C_i} = C_i, i \in I$, in (ii). □

5.6.2 Control Sequences

Below we present some control functions or control sequences which satisfy implication (5.29) or at least (5.30). It follows from Corollary 5.6.4 that the successive projection methods with such controls and with relaxation parameters $\lambda_k \in [\varepsilon, 2 - \varepsilon]$ for some $\varepsilon \in (0, 1)$ generate sequences converging, at least weakly, to a point $x^* \in C$ if $C \neq \emptyset$.

Definition 5.6.5. A control function $i : \mathcal{H} \to I$ is called a *remotest set control* if

$$i(x) = \operatorname*{argmax}_{j \in I} \left\| P_{C_j} x - x \right\|.$$

Definition 5.6.6. A control sequence $\{i_k\}_{k=0}^{\infty} : \mathcal{H} \to I$ is called an *almost remotest set control* if there exists a constant $\alpha \in (0, 1]$ such that

$$\left\| P_{C_{i_k}} x^k - x^k \right\| \geq \alpha \max_{j \in I} \left\| P_{C_j} x^k - x^k \right\|.$$

Note that a remotest set control is an almost remotest set control. Furthermore, an almost remotest set control is approximately remotest set control.

Definition 5.6.7. Let $C_j := \{x \in \mathcal{H} : \langle a_j, x \rangle \leq \beta_j\}$, $j \in I$, be half-spaces, where $a_j \in \mathcal{H}$, $a_j \neq 0$, and $\beta_j \in \mathbb{R}$, $j \in I$. A control $i : \mathcal{H} \to I$ is called a *maximal residual control* if

$$i(x) = \underset{j \in I}{\mathrm{argmax}}(\langle a_j, x \rangle - \beta_j).$$

Note that if $\|a_j\| = 1$, $j \in I$, then a maximal residual control is a remotest set control.

Lemma 5.6.8. *Let* $C_j = \{x \in \mathcal{H} : \langle a_j, x \rangle \leq \beta_j\}$, $j \in I$, *be half-spaces, where* $a_j \in \mathcal{H}$, $a_j \neq 0$, *and* $\beta_j \in \mathbb{R}$, $j \in I$. *The maximal residual control is an almost remotest set control. Consequently, it is an approximately remotest set control.*

Proof. Let $\alpha := \min_{j \in I} \|a_j\|$, $\beta := \max_{j \in I} \|a_j\|$ and $i : \mathcal{H} \to I$ be a maximal residual control. We have

$$\langle a_{i(x)}, x \rangle - \beta_{i(x)} = \max_{j \in I}(\langle a_j, x \rangle - \beta_j) = \max_{j \in I}(\|a_j\| \frac{\langle a_j, x \rangle - \beta_j}{\|a_j\|})$$

$$\geq \alpha \max_{j \in I} \frac{\langle a_j, x \rangle - \beta_j}{\|a_j\|} = \alpha \max_{j \in I} \|P_{C_j} x - x\|$$

for any $x \notin \bigcap_{j \in I} C_j$. Therefore,

$$\left\| P_{C_{i(x^k)}} x^k - x^k \right\| = \frac{\langle a_{i(x^k)}, x^k \rangle - \beta_{i(x^k)}}{\|a_{i(x^k)}\|} \geq \frac{\alpha}{\beta} \max_{j \in I} \|P_{C_j} x - x\|$$

and the control function $i : \mathcal{H} \to I$ is an approximately remotest set control. $\square$

Definition 5.6.9. We say that a control sequence $\{i_k\}_{k=0}^{\infty} \subseteq I$ is *cyclic* if for all $k \geq 0$ it holds

$$\{i_k, i_{k+1}, \dots, i_{k+m-1}\} = I.$$

Without loss of generality we can suppose that the cyclic control sequence has the form $i_k = k \pmod{m} + 1$, $k \geq 0$.

Note that a successive projection method with the cyclic control is closely related to the cyclic projection method in the following sense. Let $\{x^k\}_{k=0}^{\infty}$ be generated by the cyclic projection method and $\{y^k\}_{k=0}^{\infty}$ be generated by the successive projection method with the cyclic control. If $x^0 = y^0$, then $x^k = y^{km}$ for all $k \in \mathbb{N}$.

The following definition generalizes the notion of the cyclic control (cf. [243], [244, Definition 2.3] and [78, Definition 3.4]).

Definition 5.6.10. We say that a control sequence $\{i_k\}_{k=0}^{\infty}$ is *almost cyclic* if there exists a constant $s \geq m$ (called an *almost cyclicality constant*) such that $I \subseteq \{i_k, i_{k+1}, \ldots, i_{k+s-1}\}$ for all $k \geq 0$.

An almost cyclic control is also called *quasi-periodic* (see, e.g., [8, 153]). A more general type of control, sometimes called *quasi-cyclic* is studied in [6, 331]. The following definition generalizes the notion of an almost cyclic control (cf. [2, 8, 153]).

Definition 5.6.11. We say that a control sequence $\{i_k\}_{k=0}^{\infty}$ is *repetitive* if for any $j \in I$ and for any $k \geq 0$ there exists $l \geq k$ such that $i_l = j$.

Various variants of successive projection methods with repetitive control were studied, e.g., in [2, 70, 111, 164]. In [164, Sect. 3] the repetitive control was called *admissible*.

Note that a cyclic control sequence $\{i_k\}_{k=0}^{\infty}$ is m-almost cyclic and an almost cyclic control sequence with the almost cyclicality constant $s = m$ is cyclic. Furthermore, any almost cyclic control sequence is repetitive. In the sequel we will apply a cyclic control, almost cyclic control and repetitive control also in more general recurrences than (5.28), where the projections operators will be replaced by cutters.

In the next sections we present results which are stronger than the following one (see Theorems 5.8.11, 5.8.14, 5.9.3 and 5.9.4).

Lemma 5.6.12. *Let* $\{x^k\}_{k=0}^{\infty} \subseteq \mathcal{H}$ *be generated by the successive projection method (5.28).*

(i) *If* $\{i_k\}_{k=0}^{\infty}$ *is almost cyclic, then it is an approximately remotest set control.*

(ii) *If* $\mathcal{H}$ *is finite dimensional,* $C \neq \emptyset$, $\liminf_k \lambda_k (2 - \lambda_k) > 0$ *and* $\{i_k\}_{k=0}^{\infty}$ *is repetitive, then it is an approximately semi-remotest set control.*

Proof. (i) Let the sequence $\{i_k\}_{k=0}^{\infty}$ be almost cyclic with $s \geq m$ being an almost cyclicality constant. Suppose that

$$\lim_k \left\| P_{C_{i_k}} x^k - x^k \right\| = 0. \tag{5.31}$$

Let $i \in I$ and $r_k \in \{0, 1, \ldots, s - 1\}$ be such that $i = i_{k+r_k}$, $k \geq 0$. By the triangle inequality, we have

$$\left\| x^{k+r_k} - x^k \right\| \le \sum_{l=0}^{r_k-1} \left\| x^{k+l+1} - x^{k+l} \right\|$$

$$= \sum_{l=0}^{r_k-1} \lambda_{k+l} \left\| P_{C_{i_{k+l}}} x^{k+l} - x^{k+l} \right\|$$

$$\le \sum_{l=0}^{s-2} \lambda_{k+l} \left\| P_{C_{i_{k+l}}} x^{k+l} - x^{k+l} \right\|,$$

consequently,

$$\lim_k \left\| x^{k+r_k} - x^k \right\| = 0. \tag{5.32}$$

The definition of the metric projection and the triangle inequality yield,

$$\left\| P_{C_i} x^k - x^k \right\| \le \left\| P_{C_i} x^{k+r_k} - x^k \right\|$$

$$\le \left\| P_{C_i} x^{k+r_k} - x^{k+r_k} \right\| + \left\| x^{k+r_k} - x^k \right\|$$

$$= \left\| P_{C_{i_{k+r_k}}} x^{k+r_k} - x^{k+r_k} \right\| + \left\| x^{k+r_k} - x^k \right\|.$$

Therefore, (5.31) and (5.32) yield $\lim_k \left\| P_{C_i} x^k - x^k \right\| = 0$, i.e., $\{i_k\}_{k=0}^{\infty}$ is approximately regular.

(ii) Let $\mathcal{H}$ be finite dimensional, $C \ne \emptyset$, $\liminf_k \lambda_k(2-\lambda_k) > 0$ and $\{i_k\}_{k=0}^{\infty}$ be repetitive. Since (5.28) is a special case of (3.9) with $X = \mathcal{H}$ and $T_k := P_{C_{i_k}}$, by Corollary 3.7.1, we obtain

$$\left\| x^{k+1} - z \right\|^2 \le \left\| x^k - z \right\|^2 - \lambda_k(2-\lambda_k) \left\| P_{C_{i_k}} x^k - x^k \right\|^2$$

for any $z \in C$. Therefore, $\liminf_k \lambda_k(2-\lambda_k) > 0$ yields

$$\lim_k \left\| P_{C_{i_k}} x^k - x^k \right\| = 0.$$

Let $i \in I$. Since $\{i_k\}_{k=0}^{\infty}$ is repetitive, there is a subsequence $\{n_k\}_{k=0}^{\infty} \subseteq \{k\}_{k=0}^{\infty}$ such that $i_{n_k} = i, k \ge 0$. Now we have

$$\liminf_k \left\| P_{C_i} x^k - x^k \right\| = \liminf_k \left\| P_{C_{i_{n_k}}} x^k - x^k \right\| = \lim_k \left\| P_{C_{i_k}} x^k - x^k \right\| = 0,$$

i.e., $\{i_k\}_{k=0}^{\infty}$ is an approximately semi-remotest set control sequence.   □

Since the cyclic control is almost cyclic, it is an approximately remotest set control. Therefore, Corollary 5.6.4 yields that the sequences generated by the cyclic projection method with $\liminf_k \lambda_k(2-\lambda_k) > 0$ converge (at least weakly) to a point $x^* \in C$ if $C \ne \emptyset$.

5.6.3 Examples

Example 5.6.13. Consider the successive projection method (5.28) with the cyclic control $\{i_k\}_{k=0}^{\infty}$ and with $\lambda_k \in [\varepsilon, 2-\varepsilon]$ for some $\varepsilon \in (0, 1)$. The method (also called a cyclic projection method) was studied by:

(i) Kaczmarz in [223], where $\mathcal{H} = \mathbb{R}^n$, $C_i := \{x \in \mathbb{R}^n : \langle a_i, x \rangle = \beta_i\}$, $i = 1, 2, \ldots, n$, $a_i \in \mathbb{R}^n$ are linearly independent and $\lambda_k = 1$ for all $k \geq 0$. Kaczmarz proved that for any starting point $x^0 \in \mathbb{R}^n$ the sequence $\{x^k\}_{k=0}^{\infty}$ generated by the method converges to a unique solution of the system $\langle a_i, x \rangle = \beta_i$, $i = 1, 2, \ldots, n$;

(ii) Tanabe in [321], where $\mathcal{H} = \mathbb{R}^n$, $C_i := \{x \in \mathbb{R}^n : \langle a_i, x \rangle = \beta_i\}$, $i \in I$, (with possibly an empty intersection) and $\lambda_k = 1$ for all $k \geq 0$. Tanabe proved the convergence of any sequence generated by the method (see [321, Corollary 9]);

(iii) Bregman in [42], where $C_i \subseteq \mathcal{H}$ are closed convex subsets with $C \neq \emptyset$ and $\lambda_k = 1$ for all $k \geq 0$. Bregman proved that any sequence generated by the method converges weakly to a point $x^* \in C$ (see [42, Theorem 1]);

(iv) Censor et al. in [86], where $\mathcal{H} = \mathbb{R}^n$, $C_i = \{x \in \mathbb{R}^n : \langle a_i, x \rangle = \beta_i\}$, $i \in I$, $C = \bigcap_{i \in I} C_i = \emptyset$ and $\lambda_k = \lambda \in (0, 2)$. Let $x^0 \in \mathbb{R}^n$ be arbitrary, x^k be generated by the successive projection method with cyclic control and $x^*(\lambda) = \lim_k x^{km}$. Censor et al. proved that if $\lambda \downarrow 0$, then $x^*(\lambda)$ converges to a least squares solution (see [86, Theorem 1]). Small relaxation parameters for the Kaczmarz method were also used in [201];

(v) Gurin et al. in [196, Theorem 1], where $C_i \subseteq \mathcal{H}$, $i \in I$, are closed convex subsets with $C \neq \emptyset$ and $\lambda_k \in [\varepsilon, 2 - \varepsilon]$ for some $\varepsilon \in (0, 1)$ and for all $k \geq 0$. Gurin et al. proved the strong convergence of sequences generated by the method if the subsets C_i, $i \in I$, have a special structure (e.g., they are half-spaces or are uniformly convex or C has a nonempty interior).

Example 5.6.14. Let $C \neq \emptyset$. Consider the successive projection method (5.28) with the approximately remotest set control and with $\lambda_k \in [\varepsilon, 2 - \varepsilon]$ for some $\varepsilon \in (0, 1)$. Corollary 5.6.4 yields $x^k \rightharpoonup x^* \in C$.

(i) The convergence properties of the successive projection method with the remotest set control were studied by Agmon in [1] and by Motzkin and Schoenberg in [266], where $\mathcal{H} = \mathbb{R}^n$, $C_i = \{x \in \mathbb{R}^n : \langle a_i, x \rangle \leq \beta_i\}$, $\lambda_k = \lambda \in (0, 2)$. Agmon proved that the sequences generated by the method converge linearly [1, Theorem 3]. Motzkin and Schoenberg also considered the case $\lambda = 2$ if C has a nonempty interior and proved the finite convergence in this case.

(ii) Bregman in [42] considered the convex feasibility problem with infinitely many subsets $C_i \subseteq \mathcal{H}$, $i \in I$, having a nonempty intersection, and studied successive projection methods for this problem. In [42, Theorem 2] the convergence $x^k \rightharpoonup x^* \in C$ was proved for the remotest set control (the existence of such control was supposed). Bregman also considered a special case of the problem with infinitely many half-spaces $C_i \subseteq \mathcal{H}$, $C_i := \{x \in \mathcal{H} :$

$\langle a_i, x \rangle \leq \beta_i\}, i \in I$, where $\{a_i : i \in I\}$ is bounded. In [42, Theorem 3] the convergence $x^k \rightharpoonup x^* \in C$ was proved for the maximal residual control (the existence of such control was supposed).

(iii) Goffin in [186] considered the linear feasibility problem in $\mathbb{R}^n$ and studied sufficient conditions for finite convergence of the sequential projection method with the remotest set control [186, Theorems 7.1 and 7.5].

(iv) Gurin et al. in [196] considered the convex feasibility problem with infinitely many subsets $C_i \subseteq \mathcal{H}, i \in I$, having a nonempty intersection. In [196, Theorems 1 and 2] the convergence $x^k \rightarrow x^* \in C$ was proved for an approximately remotest set control, where C_i satisfy some conditions, e.g., $C_i, i \in I$, are half-spaces or uniformly convex subsets, or the intersection has an open interior.

(v) Dye and Reich in [152] studied successive projection methods for $C_i \subseteq \mathcal{H}$, $i \in I$, being one-dimensional subspaces.

5.7   Landweber Method and Projected Landweber Method

Let $A : \mathcal{H}_1 \rightarrow \mathcal{H}_2$ be a nonzero linear bounded operator and $Q \subseteq \mathcal{H}_2$ be a nonempty closed convex subset. The *Landweber method* (LM) for finding $x \in \mathcal{H}_1$ satisfying $Ax \in Q$ (if such an x exists) is defined by the recurrence

$$x^{k+1} = x^k + \frac{\lambda_k}{\|A\|^2} A^*(P_Q(Ax^k) - Ax^k),$$

where $x^0 \in \mathcal{H}_1$ is arbitrary and $\lambda_k \in [0, 2]$. We can also write $x^{k+1} = T_{\lambda_k} x^k$. where T_λ is a λ-relaxation of the Landweber operator $T := \frac{1}{\|A\|^2} A^*(P_Q - \mathrm{Id})A$. Landweber proposed the method with $\lambda_k = \lambda \in (0, 2)$, $k \geq 0$, for solving an integral equation (see [240]). The *projected Landweber method* (PLM) is an iterative method for solving the following split feasibility problem: find $x \in C$ such that $Ax \in Q$, where $C \subseteq \mathcal{H}_1$, $Q \subseteq \mathcal{H}_2$ are nonempty closed convex subsets of Hilbert spaces $\mathcal{H}_1, \mathcal{H}_2$ and $A : \mathcal{H}_1 \rightarrow \mathcal{H}_2$ is a nonzero linear bounded operator, if such a point exists. The iterative step of the projected Landweber method has the form

$$x^{k+1} = P_C(x^k + \frac{\lambda_k}{\|A\|^2} A^*(P_Q(Ax^k) - Ax^k)), \qquad (5.33)$$

where $\lambda_k \in [0, 2]$. We can also write $x^{k+1} = P_C T_{\lambda_k} x^k$, where T_λ is the λ-relaxation of the Landweber operator T or $x^{k+1} = R_{\lambda_k} x^k$, where $R_\lambda := P_C T_\lambda$ is the projected λ-relaxation of the Landweber operator. It is clear that if $C = \mathcal{H}_1$, then the projected Landweber method reduces to the Landweber method. The projected Landweber method was studied by Byrne [55] for $\mathcal{H}_1$ and $\mathcal{H}_2$ being Euclidean spaces and for $\lambda_k = \lambda \in (0, 2)$, $k \geq 0$. Actually, Byrne called the method a *CQ algorithm* and proved the convergence of sequences generated by the method to a solution of the split feasibility problem, if such a solution exists (see [55, Theorem 2.1]). Generalized versions of the projected Landweber method were studied by Xu [344] for a multiple set split feasibility problem and by Zhao and Yang [361].

Theorem 5.7.1. *Let the projected Landweber operator defined by* (4.92) *have a fixed point. If* $\lim \inf_k \lambda_k (2 - \lambda_k) > 0$, *then, for an arbitrary* $x^0 \in \mathcal{H}_1$, *the sequence* $\{x^k\}_{k=0}^{\infty}$ *generated by the projected Landweber method* (5.33) *converges weakly to a point* $z \in \mathrm{Fix}\, U$.

Proof. It follows from Theorem 4.6.3 that the Landweber operator $T := \mathrm{Id} + \frac{1}{\|A\|^2}$ $A^*(P_Q - \mathrm{Id})A$ is firmly nonexpansive. The weak convergence of x^k to a fixed point of the projected Landweber operator U follows now from Corollary 3.7.4 (iii) with $S = T$, $X = C$ and $\mu_k = \lambda_k$, $k \geq 0$. $\square$

Variants of the projected Landweber method for the consistent split feasibility problem, i.e., such that $C \cap A^{-1}(Q) \neq \emptyset$, were studied in [292, 293, 350], where projection onto C and Q were replaced by projections onto closed convex sets $C_k \supseteq C$ and $Q_k \supseteq Q$.

Recall that $x^* \in C$ is a fixed point of the projected Landweber operator U if and only if it is a minimizer of the proximity function $f : C \to \mathbb{R}$ defined by $f(x) := \frac{1}{2} \|P_Q(Ax) - Ax\|^2$ (see Proposition 4.7.2). Consequently, Theorem 5.7.1 gives sufficient conditions for the weak convergence of sequences generated by the projected Landweber method to a solution of the split feasibility problem.

Suppose now that $\mathcal{H}_1 = \mathbb{R}^n$, $\mathcal{H}_2 = \mathbb{R}^m$ and that A is a nonzero matrix of type $m \times n$. If $Q = \{b\}$, then the projected Landweber method for finding a solution $x^* \in C$ of the linear equation $Ax = b$ has the form

$$x^{k+1} = P_C(x^k + \frac{\lambda_k}{\lambda_{\max}(A^\top A)} A^\top (b - Ax^k))$$

(cf. equality (4.79)). If A has nonzero rows $a_i \in \mathbb{R}^n$, $i \in I$, then we can write the system $Ax = b$ in an equivalent form $D^{\frac{1}{2}} Ax = D^{\frac{1}{2}} b$, where $D = \mathrm{diag}(\frac{\omega_1}{\|a_1\|^2}, \frac{\omega_2}{\|a_2\|^2}, \dots, \frac{\omega_m}{\|a_m\|^2})$ and $w = (\omega_1, \omega_2, \dots, \omega_m) \in \mathrm{ri}\,\Delta_m$. The projected Landweber method for the modified system has the form

$$x^{k+1} = P_C(x^k + \frac{\lambda_k}{\lambda_{\max}(A^\top DA)} A^\top D(b - Ax^k)) \qquad (5.34)$$

(cf. (4.82). If $Q = \{y \in \mathbb{R}^m : y \leq b\}$, then, similarly as above, the projected Landweber method for finding a solution $x^* \in C$ of the linear inequality $Ax \leq b$ has the form

$$x^{k+1} = P_C(x^k - \frac{\lambda_k}{\lambda_{\max}(A^\top A)} A^\top (Ax^k - b)_+)$$

(cf. (4.80)) and the projected Landweber method for the modified system $D^{\frac{1}{2}} Ax \le D^{\frac{1}{2}} b$ has the form

$$x^{k+1} = P_C(x^k - \frac{\lambda_k}{\lambda_{\max}(A^\top DA)} A^\top D(Ax^k - b)_+) \tag{5.35}$$

(cf. (4.83). If we take a positive definite matrix D with nonnegative elements and $C = \mathbb{R}^n$ in (5.35), then we obtain a method which was studied in [87, equality (12) and Theorem 2] (see also [322], [132, Sect. 4], [98], [220, Theorem 2], [160, 222] for related results for a system of linear equations). Note that method (5.34) for solving the system $D^{\frac{1}{2}} Ax = D^{\frac{1}{2}} b$ is an extrapolation of the simultaneous projection method (5.17) for solving the system $Ax = b$ (see Corollary 4.6.7). Similarly, method (5.35) for solving the system $D^{\frac{1}{2}} Ax \le D^{\frac{1}{2}} b$ is an extrapolation of the simultaneous projection method (5.18) for solving the system $Ax \le b$, where the extrapolation parameter (step size) $\sigma_L = \frac{1}{\lambda_{\max}(A^\top DA)} \ge 1$.

5.8 Simultaneous Cutter Methods

In this section we consider the common fixed point problem for a finite family of cutters $\mathcal{U} = \{U_i\}_{i \in I}$, where the operators $U_i : \mathcal{H} \to \mathcal{H}$, $i \in I := \{1, 2, \ldots, m\}$, have a common fixed point. We study the convergence of sequences generated by relaxations of simultaneous cutters of the form $V := \sum_{i \in J} \omega_i V_i$, where $\mathcal{V} := \{V_i\}_{i \in J}$ is a finite family of cutters $V_i : \mathcal{H} \to \mathcal{H}$, $i \in J$, with the property $\bigcap_{i \in J} \text{Fix } V_i \supseteq \bigcap_{i \in I} \text{Fix } U_i$ and $w : \mathcal{H} \to \Delta_{|J|}$ is an appropriate weight function. One iteration of the recurrence is defined by

$$x^+ = V_\lambda x = x + \lambda(\sum_{i \in J} \omega_i(x) V_i x - x),$$

where $x \in \mathcal{H}$ is the current approximation of a solution, $\lambda \in [0, 2]$ is a relaxation parameter and $w(x) = (\omega_1(x), \ldots, \omega_{|J|}(x))$. A simple example of such iteration is the simultaneous projection method applied to the convex feasibility problem. In this case $\mathcal{U} = \mathcal{V} = \{P_{C_i}\}_{i \in I}$, $w \in \Delta_m$ is a vector of constant weights and one iteration of the method has the form

$$x^+ = U_\lambda x = x + \lambda(\sum_{i \in I} \omega_i P_{C_i} x - x).$$

In this section we consider, however, more general methods, where in different iterations different families $\mathcal{V} := \{V_i\}_{i \in J}$ and different weight functions $w : \mathcal{H} \to \Delta_{|J|}$ can be applied. We will study the convergence of sequences generated by the following recurrence

$$x^0 \in \mathcal{H} \qquad\qquad - \text{arbitrary}$$
$$x^{k+1} = x^k + \lambda_k (\textstyle\sum_{i \in J_k} \omega_i^k (x^k) V_i^k x^k - x^k), \tag{5.36}$$

where $\mathcal{V}^k := \{V_i^k\}_{i \in J_k}$ is a sequence of finite families of cutters $V_i^k : \mathcal{H} \to \mathcal{H}, i \in J_k$, with the property $\bigcap_{i \in J_k} \text{Fix } V_i^k \supseteq \bigcap_{i \in I} \text{Fix } U_i \neq \emptyset$ for all $k \geq 0$, $\{\lambda_k\}_{k=0}^\infty \subseteq [0, 2]$ is a sequence of relaxation parameters, $\{w^k\}_{k=0}^\infty : \mathcal{H} \to \Delta_{m_k}$ is a sequence of appropriate weight functions, $m_k := |J_k|$ and $w^k(x) = (\omega_1^k(x), ..., \omega_{m_k}^k(x))$. We call the method defined by (5.36) a *simultaneous cutter method* (SiCM). An iteration in (5.36) can also be written in the form

$$x^{k+1} = x^k + \lambda_k (V^k x^k - x^k),$$

where $V^k x := \sum_{i \in J_k} \omega_i^k(x) V_i^k x$ for $x \in \mathcal{H}$. It is clear that V^k is a cutter (see Corollary 2.1.49). Iterations of type (5.36) were studied by several authors (see [70] and the references therein).

5.8.1 Assumptions on Weight Functions

In this section we define some classes of weight functions $w : \mathcal{H} \to \Delta_{|J|}$ related to a family of cutters $\mathcal{V} := \{V_i\}_{i \in J}$ and some classes of sequences of weight functions $\{w^k\}_{k=0}^\infty : \mathcal{H} \to \Delta_{|J_k|}$ related to a sequence of families of cutters $\mathcal{V}^k := \{V_i^k\}_{i \in J_k}$, $k \geq 0$. It turns out that if we employ sequences $\{w^k\}_{k=0}^\infty$ which belong to these classes, then we can apply convergence results presented in Sects. 3.6 and 3.7 in order to prove the convergence of sequences generated by the simultaneous cutter method. The assumption that the weight functions w^k are appropriate is not sufficient for the convergence.

Remark 5.8.1. A sequence of weight functions $\{w^k\}_{k=0}^\infty$ applied in recurrence (5.36) can be:

(a) Such that the weights $\omega_i^k(x^k)$ essentially depend on the iteration's counter k and on the current point x^k, $i \in I, k \geq 0$,
(b) A sequence of constant weight functions; in this case $\omega_i^k(x^k) = \omega_i^k$, $i \in I$, $k \geq 0$,
(c) A constant sequence; in this case $\omega_i^k(x^k) = \omega_i(x^k)$, $i \in I, k \geq 0$,
(d) A constant sequence of constant weight functions; in this case $\omega_i^k(x^k) = \omega_i$, $i \in I, k \geq 0$.

In the practical realization of a simultaneous algorithm there is no difference between cases (a), (b) and (c) and we can use the notation ω_i^k instead of $\omega_i^k(x^k)$ in (a) and instead of $\omega_i(x^k)$ in (c). Nevertheless, it is useful to distinguish the weights which depend on the current point x and the weights which depend on iteration's counter. Furthermore, some properties of the simultaneous cutter methods depend

on the properties of corresponding simultaneous cutter operators, where appropriate
weight functions play an important role.

Definition 5.8.2. Let $V := \{V_i\}_{i \in J}$ be a finite family of cutters $V_i : \mathcal{H} \to \mathcal{H}$,
$i \in J$, and $\beta > 0$ be a constant. We say that a weight function $w : \mathcal{H} \to \Delta_{|J|}$ is
β-regular with respect to the family $\{U_i\}_{i \in I}$, or, shortly, *regular* if for any $x \in \mathcal{H}$
there exists $j \in J$ such that

$$w_j(x) \left\| V_j x - x \right\|^2 \geq \beta \max_{i \in I} \left\| U_i x - x \right\|^2 . \tag{5.37}$$

Suppose that $\bigcap_{i \in J} \operatorname{Fix} V_i \supseteq \bigcap_{i \in I} \operatorname{Fix} U_i$. Note that a weight function which is
regular with respect to the family $\mathcal{U} := \{U_i\}_{i \in I}$ is appropriate with respect to the
family $V := \{V_i\}_{i \in J}$ (cf. Definition 2.1.25). It is also clear that if $V \supseteq \mathcal{U}$, then a
constant weight function w with all positive weights w_i, $i \in J$, is regular.

Example 5.8.3. Let $V := \mathcal{U}$, $I(x) := \{i \in I : x \notin \operatorname{Fix} U_i\}$ and $m(x) := |I(x)|$
be the number of elements of $I(x)$, where $x \in \mathcal{H}$. The following weight functions
$w : \mathcal{H} \to \Delta_m$, where $w(x) = (w_1(x), w_2(x), \dots, w_m(x))$, $x \in \mathcal{H}$, are regular:

(i) Positive constant weights, i.e.,

$$w(x) := w \in \operatorname{ri} \Delta_m \tag{5.38}$$

for all $x \in \mathcal{H}$. To verify that w is regular it suffices to choose $j \in \operatorname{Argmax}_{i \in I}$
$\|U_i x - x\|$ and take $\beta = \min_{i \in I} w_i$ in Definition 5.8.2.
   In particular, $w_i(x) = \frac{1}{m}$, $i \in I$, $x \in \mathcal{H}$ is regular.
(ii) Constant weights for violated constraints, i.e.,

$$w_i(x) := \begin{cases} \frac{v_i}{\sum_{j \in I(x)} v_j} & \text{for } i \in I(x) \\ 0 & \text{for } i \notin I(x), \end{cases} \tag{5.39}$$

$i \in I$, $x \in \mathcal{H}$, where $v = (v_1, \dots, v_m) \in \operatorname{ri} \Delta_m$. To verify that w is regular
it suffices to choose $j \in \operatorname{Argmax}_{i \in I} \|U_i x - x\|$ and take $\beta := \min_{i \in I} w_i$ in
Definition 5.8.2. In particular, a weight function w with

$$w_i(x) = \begin{cases} \frac{1}{m(x)} & \text{for } i \in I(x) \\ 0 & \text{for } i \notin I(x), \end{cases} \tag{5.40}$$

$i \in I$, $x \in \mathcal{H}$, is regular. The weights of the form (5.39) were applied in
[87, Sect. 2] and [215] for some variants of the simultaneous projection
method.
(iii) Weight functions $w : \mathcal{H} \to \Delta_m$, where the weights $w_i(x)$ are proportional to
$\|U_i x - x\|$, i.e.,

$$\omega_i(x) := \begin{cases} \dfrac{\|U_i x - x\|}{\sum_{j \in I} \|U_j x - x\|} & \text{for } x \notin \bigcap_{j \in I} \text{Fix } U_j \\ 0 & \text{for } x \in \bigcap_{j \in I} \text{Fix } U_j, \end{cases} \tag{5.41}$$

$i \in I$, $x \in \mathcal{H}$. To verify that w is regular it suffices to choose $j \in$ $\text{Argmax}_{i \in I} \|U_i x - x\|$ and take $\beta := \frac{1}{m}$ in Definition 5.8.2.

(iv) Weight functions $w : \mathcal{H} \to \Delta_m$ satisfying the condition

$$w_i(x) \geq \delta \text{ for all } i \in I(x) \tag{5.42}$$

$x \in \mathcal{H}$, for some constant $\delta \in (0, \frac{1}{m}]$. To verify that w is regular it suffices to choose $j(x) \in \text{Argmax}_{i \in I} \|U_i x - x\|$ and take $\beta := \delta$ in Definition 5.8.2. The weights satisfying (5.42) were applied in [120, Sect. III] and in [119, Sect. 1]. Note that the weight functions defined by (5.38) and by (5.39) satisfy (5.42).

(v) Weight functions $w : \mathcal{H} \to \Delta_m$ which consider only almost violated constraints, i.e., $\omega_i(x) = 0$ for all $i \notin J_\gamma(x)$, where

$$J_\gamma(x) := \{ j \in I : \|U_j x - x\| \geq \gamma \max_{i \in I} \|U_i x - x\| \} \tag{5.43}$$

for some $\gamma \in (0, 1]$, $i \in I$, $x \in \mathcal{H}$. To verify that w is regular it suffices to choose $j := j(x) \in J_\gamma(x)$ with $\omega_j(x) \geq \frac{1}{m}$ and take $\beta := \frac{\gamma^2}{m}$ in Definition 5.8.2. The existence of such j follows from the fact that $\omega_i(x) \geq 0$ for all $i \in J_\gamma(x)$ and $\sum_{i \in J_\gamma(x)} \omega_i(x) = 1$.

In particular, the following weight functions satisfy (5.43):

(a)

$$\omega_i(x) := \begin{cases} 1 & \text{if } i = \text{argmax}_{j \in I} \|U_j x - x\| \\ 0 & \text{in other cases,} \end{cases} \tag{5.44}$$

$i \in I$, $x \in \mathcal{H}$, where $U_i = P_{C_i}$ for a closed convex subset $C_i \subseteq \mathcal{H}$, $i \in I$. In this case w defines a remotest set control (cf. Definition 5.6.5).

(b)

$$\omega_i(x) := \begin{cases} 1 & \text{if } i = j(x) \\ 0 & \text{in other cases,} \end{cases} \tag{5.45}$$

where $j(x) \in J_\gamma(x)$ for some $\gamma \in (0, 1]$, $i \in I$, $x \in \mathcal{H}$. If $U_i = P_{C_i}$ for a closed convex subset $C_i \subseteq \mathcal{H}$, $i \in I$, then any sequence of weight functions $\{w_k\}_{k=0}^\infty$ of this form (with γ independent of k) is an almost remotest set control (cf. Definition 5.6.6).

The next definitions extend the notion of a regular weight function to a sequence of weight functions.

Definition 5.8.4. Let $\mathcal{V}^k := \{V_i^k\}_{i \in J_k}$, be a sequence of families of cutters $V_i^k :$ $\mathcal{H} \to \mathcal{H}$, $i \in J_k$, $k \geq 0$. We say that a sequence of weight functions $w^k : \mathcal{H} \to \Delta_{|J_k|}$ is *regular* (with respect to the family $\mathcal{U} := \{U_i\}_{i \in I}$), if there exists a sequence

$\{\beta_k\}_{k=0}^{\infty} \subseteq (0, +\infty)$ with $\liminf_k \beta_k > 0$ such that w^k applied to the family $\mathcal{V}^k$ is β_k-regular with respect to the family $\mathcal{U}$, $k \geq 0$.

It is clear that for a regular weight function w applied to the family $\mathcal{U}$ the constant sequence $w^k := w$ is regular.

Definition 5.8.5. Let $\mathcal{V}^k := \{V_i^k\}_{i \in J_k}$ be a sequence of cutters $V_i^k : \mathcal{H} \to \mathcal{H}$, $i \in J_k$, $k \geq 0$, and the sequence $\{x^k\}_{k=0}^{\infty}$ be generated by recurrence (5.36). We say that a sequence of appropriate weight functions $w^k : \mathcal{H} \to \Delta_{|J_k|}$ applied to the sequence of families $\{\mathcal{V}^k\}_{k=0}^{\infty}$ is:

(i) *Approximately regular* (with respect to the family $\mathcal{U} := \{U_i\}_{i \in I}$) if there exists a sequence $\{i_k\}_{k=0}^{\infty}$ with $i_k \in J_k$ for all $k \geq 0$, such that the following implication holds

$$\lim_k \omega_{i_k}^k(x^k) \left\| V_{i_k}^k x^k - x^k \right\|^2 = 0 \implies \lim_k \left\| U_i x^k - x^k \right\| = 0, \text{ for all } i \in I,$$

(5.46)

(ii) *Approximately semi-regular* (with respect to the family $\mathcal{U} := \{U_i\}_{i \in I}$) if there exists a sequence $\{i_k\}_{k=0}^{\infty}$ with $i_k \in J_k$ for all $k \geq 0$, such that the following implication holds

$$\lim_k \omega_{i_k}^k(x^k) \left\| V_{i_k}^k x^k - x^k \right\|^2 = 0 \implies \liminf_k \left\| U_i x^k - x^k \right\| = 0, \text{ for all } i \in I.$$

(5.47)

A sequence of weight functions $\{w^k\}_{k=0}^{\infty}$ applied to recurrence (5.36) is also called a *control sequence* or a *control*. If $\{w^k\}_{k=0}^{\infty}$ satisfies (i) or (ii), then we also say that the *control* in recurrence (5.36) is *approximately regular* or *approximately semi-regular*.

Remark 5.8.6. Let x^k be generated by recurrence (5.36).

(i) By Corollary 3.7.1 (i), the sequence $\{x^k\}_{k=0}^{\infty}$ is Fejér monotone with respect to $\bigcap_{i \in I} \text{Fix } U_i$, because $V^k := \sum_{i \in J_k} \omega_i^k V_i^k$ is a cutter as a convex combination of cutters (see Corollary 2.1.49) and $\bigcap_{i \in I} \text{Fix } U_i \subseteq \bigcap_{i \in J_k} \text{Fix } V_i^k$. Therefore, x^k is bounded. Corollary 2.1.37 yields

$$\left\| V_i^k x^k - x^k \right\| \leq \left\| z - x^k \right\|$$

for any $z \in \bigcap_{i \in I} \text{Fix } U_i$, consequently, $\left\| V_i^k x^k - x^k \right\|$ is bounded. Since $\omega_i^k(x^k)$ is also bounded, we have the following equivalence

$$\lim_k \omega_{i_k}^k(x^k) \left\| V_{i_k}^k x^k - x^k \right\| = 0 \iff \lim_k \omega_{i_k}^k(x^k) \left\| V_{i_k}^k x^k - x^k \right\|^2 = 0 \quad (5.48)$$

for any $i_k \in J_k$, $k \geq 0$. Therefore, implication (5.46) is equivalent to the following one

$$\lim_k \omega_{i_k}^k (x^k) \left\| V_{i_k}^k x^k - x^k \right\| = 0 \implies \lim_k \left\| U_i x^k - x^k \right\| = 0 \text{ for all } i \in I,$$
(5.49)

and implication (5.47) is equivalent to the following one

$$\lim_k \omega_{i_k}^k (x^k) \left\| V_{i_k}^k x^k - x^k \right\| = 0 \implies \liminf_k \left\| U_i x^k - x^k \right\| = 0 \text{ for all } i \in I.$$
(5.50)

(ii) The following condition yields the approximate regularity of $\{w^k\}_{k=0}^{\infty}$: There is a sequence $\{i_k\}_{k=0}^{\infty}$ with $i_k \in J_k$ for all $k \geq 0$, such that

$$\liminf_k \omega_{i_k}^k (x^k) > 0$$

and

$$\lim_k \left\| V_{i_k}^k x^k - x^k \right\| = 0 \implies \lim_k \left\| U_i x^k - x^k \right\| = 0 \text{ for all } i \in I. \quad (5.51)$$

In particular, implication (5.51) holds if

$$\left\| V_{i_k}^k x^k - x^k \right\| \geq \beta \max_{i \in I} \left\| U_i x^k - x^k \right\| \quad (5.52)$$

for all $k \geq 0$ and for a constant $\beta > 0$. If $J_k = I$, $k \geq 0$, then the following condition also implies the approximate regularity of $\{w^k\}_{k=0}^{\infty}$: $\liminf_k \omega_i^k (x^k) > 0$ and for all $i \in I$ with $\omega_i^k (x^k) > 0$ it holds

$$\left\| V_i^k x^k - x^k \right\| \geq \beta \left\| U_i x^k - x^k \right\| \text{ for all } k \geq 0. \quad (5.53)$$

If $U_i = P_{C_i}$ for closed convex subsets $C_i \subseteq \mathcal{H}$, $i \in I$, then we say that an algorithm satisfying the latter condition is *linearly focusing* (cf. [22, Definition 4.8]).

Lemma 5.8.7. *Let* $\mathcal{V}^k := \{V_i^k\}_{i \in J_k}$ *be a sequence of families of cutters* $V_i^k : \mathcal{H} \to \mathcal{H}$, $i \in J_k$, $k \geq 0$, *and a sequence* $\{x^k\}_{k=0}^{\infty}$ *be generated by recurrence (5.36). If the sequence of weight functions* $w^k : \mathcal{H} \to \Delta_{|J_k|}$:

 (i) *Is regular, then* w^k *is approximately regular;*
(ii) *Contains a regular subsequence* $\{w^{n_k}\}_{k=0}^{\infty}$, *then* w^k *is approximately semiregular.*

Proof. (i) Let a sequence $\{w^k\}_{k=0}^{\infty}$ applied to the sequence of families $\{\mathcal{V}^k\}_{k=0}^{\infty}$ be β_k-regular, where $\liminf_k \beta_k > 0$. Then, for an arbitrary k, there exists $i_k \in J_k$ such that

$$\omega_{i_k}^k (x^k) \left\| V_{i_k}^k x^k - x^k \right\|^2 \geq \beta_k \max_{i \in I} \left\| U_i x^k - x^k \right\|^2.$$

If $\omega_{i_k}^k (x^k) \left\| V_{i_k}^k x^k - x^k \right\|^2 \to 0$, then, of course,

$$\lim_k \beta_k \max_{i \in I} \left\| U_i x^k - x^k \right\|^2 = 0$$

and, due to the positivity of $\liminf_k \beta_k$, we have

$$\lim_k \max_{i \in I} \left\| U_i x^k - x^k \right\|^2 = 0.$$

Consequently, $\lim_k \max_{i \in I} \left\| U_i x^k - x^k \right\| = 0$, i.e., $\lim_k \left\| U_i x^k - x^k \right\| = 0$ for all $i \in I$.

(ii) Let a subsequence $\{w^{n_k}\}_{k=0}^{\infty} \subseteq \{w^k\}_{k=0}^{\infty}$ applied to the sequence of families $\mathcal{V}^{n_k}$, $k \geq 0$, be regular, i.e., for an arbitrary k, there exists $i_{n_k} \in J_{n_k}$ such that

$$\omega_{i_{n_k}}^{n_k} (x^{n_k}) \left\| V_{i_{n_k}}^{n_k} x^{n_k} - x^{n_k} \right\|^2 \geq \beta_{n_k} \max_{i \in I} \left\| U_i x^{n_k} - x^{n_k} \right\|^2$$

and $\liminf_k \beta_{n_k} > 0$. If

$$\lim_k \omega_{i_k}^k (x^k) \left\| V_{i_k}^k x^k - x^k \right\|^2 = 0$$

then, of course,

$$\lim_k \omega_{i_{n_k}}^{n_k} (x^{n_k}) \left\| V_{i_{n_k}}^{n_k} x^{n_k} - x^{n_k} \right\|^2 = 0,$$

consequently,

$$\lim_k \beta_{n_k} \max_{i \in I} \left\| U_i x^{n_k} - x^{n_k} \right\|^2 = 0.$$

By the positivity of $\liminf_k \beta_{n_k}$, we have

$$\lim_k \max_{i \in I} \left\| U_i x^{n_k} - x^{n_k} \right\|^2 = 0,$$

i.e.,

$$\liminf_k \max_{i \in I} \left\| U_i x^k - x^k \right\|^2 = 0$$

or, in other words, $\liminf_k \left\| U_i x^k - x^k \right\| = 0$ for all $i \in I$. $\square$

Now we present theorems which give sufficient conditions for important control sequences to be approximately (semi-)regular. Consider a special case of recurrence (5.36), where $J_k = I$, Fix $V_i^k \supseteq$ Fix U_i for all $k \geq 0$ and for all $i \in I$. We can write recurrence (5.36) in the form

$$x^{k+1} = x^k + \lambda_k \sum_{i \in I} \omega_i^k (V_i^k x^k - x^k), \tag{5.54}$$

where $\lambda_k \in [0, 2]$, $k \geq 0$. Denote, for simplicity, $\omega_i^k = \omega_i^k (x^k)$. Let $I_k \subseteq I$ be nonempty and $\omega_i^k = 0$ for $i \notin I_k$, $k \geq 0$. Then (5.54) is equivalent to

$$x^{k+1} = x^k + \lambda_k \sum_{i \in I_k} \omega_i^k (V_i^k x^k - x^k), \tag{5.55}$$

where we can suppose without loss of generality that $w^k : \mathcal{H} \to \Delta_{|I_k|}$. Recurrence (5.55) describes two stage algorithms, where $\{I_k\}_{k=0}^\infty$ is a control sequence for the outer stage and $\{w^k\}_{k=0}^\infty$ is a control sequence for inner stages. The application of such algorithms is convenient if $m = |I|$ is a big number. In this case we can split I onto smaller subsets. The control I_k is, actually, a subset of I on which the algorithm works in kth iteration. Consider the following condition for sequences generated by recurrence (5.55): For any $k \geq 0$, there is $l_k \in I_k$ such that

$$\lim_k \omega_{l_k}^k \left\| V_{l_k}^k x^k - x^k \right\|^2 = 0 \implies \lim_k \max_{i \in I_k} \left\| P_{\mathrm{Fix}\, U_i} x^k - x^k \right\| = 0. \tag{5.56}$$

Note that the latter condition is weaker than the approximate regularity of $\{w^k\}_{k=0}^\infty$ with respect to the family $\{P_{\mathrm{Fix}\, U_i}\}_{i \in I}$, because $I_k \subseteq I$, consequently $\max_{i \in I_k} \left\| P_{\mathrm{Fix}\, U_i} x^k - x^k \right\| \leq \max_{i \in I} \left\| P_{\mathrm{Fix}\, U_i} x^k - x^k \right\|$.

Definition 5.8.8. (cf. [22, Definition 3.18]) We say that the sequence $\{I_k\}_{k=0}^\infty \subseteq I$ is p-intermittent, where $p \in \mathbb{N}$, or *intermittent* if $I = I_k \cup I_{k+1} \cup \ldots \cup I_{k+p-1}$ for any $k \geq 0$.

A simple example of an intermittent sequence is an almost cyclic control (cf. Definition 5.6.10).

Theorem 5.8.9. *Let $\{x^k\}_{k=0}^\infty$ be generated by (5.54), $\{I_k\}_{k=0}^\infty \subseteq I$ be intermittent. Suppose that for any $k \geq 0$ there is $l_k \in I_k$, $k \geq 0$, satisfying condition (5.56). If $w_i^k = 0$ for all $i \notin I_k$, $k \geq 0$, then $\{w^k\}_{k=0}^\infty$ is approximately regular with respect to the family of cutters $\mathcal{U} := \{U_i\}_{i \in I}$.*

Proof. Let $w_i^k = 0$ for all $i \notin I_k$, $k \geq 0$. Define $i_k := \operatorname{argmax}_{i \in I_k} \left\| V_i^k x^k - x^k \right\|$, $k \geq 0$. Suppose that $\lim_k \omega_{l_k}^k \left\| V_{l_k}^k x^k - x^k \right\|^2 = 0$. Denote $C_i := \mathrm{Fix}\, U_i$, $i \in I$. By property (5.56), we have

$$\lim_k \max_{i \in I_k} \left\| P_{C_i} x^k - x^k \right\| = 0. \tag{5.57}$$

Corollary 2.1.37, the assumption $\mathrm{Fix}\, V_i^k \supseteq \mathrm{Fix}\, U_i = C_i$ for all $i \in I$ and $k \geq 0$, and the definition of the metric projection yield

$$\left\| V_i^k y - y \right\| \leq \left\| P_{\mathrm{Fix}\, V_i^k} y - y \right\| \leq \left\| P_{C_i} y - y \right\|,$$

$y \in \mathcal{H}$, consequently,

$$\left\| V_{i_k}^k x^k - x^k \right\| = \max_{i \in I_k} \left\| V_i^k x^k - x^k \right\| \leq \max_{i \in I_k} \left\| P_{C_i} x^k - x^k \right\|. \tag{5.58}$$

Inequality (5.58) and equality (5.57) imply

$$\lim_k \left\| V_{i_k}^k x^k - x^k \right\| = 0. \tag{5.59}$$

Let $i \in I$. Since I_k is intermittent, there is $p \geq 1$ such that $i \in I_{k+r_k}$ for some $r_k \in \{0, 1, \ldots, p-1\}$. By the triangle inequality, the convexity of the norm, the definition of $i_{k+l}, l = 0, 1, \ldots, p-1$, the equality $\sum_{j \in I_{k+l}} \omega_j^{k+l} = 1$ and by (5.59), we have

$$\left\| x^{k+r_k} - x^k \right\| \leq \sum_{l=0}^{r_k-1} \left\| x^{k+l+1} - x^{k+l} \right\|$$

$$= \sum_{l=0}^{r_k-1} \lambda_{k+l} \left\| \sum_{j \in I_{k+l}} \omega_j^{k+l} (V_j^{k+l} x^{k+l} - x^{k+l}) \right\|$$

$$\leq \sum_{l=0}^{r_k-1} \lambda_{k+l} \sum_{j \in I_{k+l}} \omega_j^{k+l} \left\| V_j^{k+l} x^{k+l} - x^{k+l} \right\|$$

$$\leq \sum_{l=0}^{p-1} \lambda_{k+l} \sum_{j \in I_{k+l}} \omega_j^{k+l} \left\| V_{i_{k+l}}^{k+l} x^{k+l} - x^{k+l} \right\|$$

$$= \sum_{l=0}^{p-1} \lambda_{k+l} \left\| V_{i_{k+l}}^{k+l} x^{k+l} - x^{k+l} \right\| \to 0$$

as $k \to \infty$, i.e.,

$$\lim_k \left\| x^{k+r_k} - x^k \right\| = 0. \tag{5.60}$$

Further, by the definition of the metric projection and by the triangle inequality,

$$\left\| P_{C_i} x^k - x^k \right\| \leq \left\| P_{C_i} x^{k+r_k} - x^k \right\| \leq \left\| P_{C_i} x^{k+r_k} - x^{k+r_k} \right\| + \left\| x^{k+r_k} - x^k \right\|. \tag{5.61}$$

Since $i \in I_{k+r_k}$, equality (5.57) yields $\lim_k \left\| P_{C_i} x^{k+r_k} - x^{k+r_k} \right\| = 0$. Now inequalities (5.61) and equality (5.60) imply $\lim_k \left\| P_{C_i} x^k - x^k \right\| = 0$. Since U_i are cutters, Corollary 2.1.37 yields $\lim_k \left\| U_i x^k - x^k \right\| = 0$, $i \in I$, i.e., condition (5.46) is satisfied, which means that $\{w^k\}_{k=0}^\infty$ is approximately regular. □

Let now

$$I_k := \{ i \in I : \omega_i^k > 0 \}. \tag{5.62}$$

Definition 5.8.10. We say that a control sequence $\{w^k\}_{k=0}^\infty$ is *p-intermittent, where* $p \geq 1$, or *intermittent* if the sequence $\{I_k\}_{k=0}^\infty$, where I_k is defined by (5.62), is *p*-intermittent, i.e., $I = I_k \cup I_{k+1} \cup \ldots \cup I_{k+p-1}$.

If $|I_k| = 1$ or, equivalently, $w^k = e_{i_k}$, $k \geq 0$, for a sequence $\{i_k\}_{k=0}^{\infty} \subseteq I$, then the fact that $\{w^k\}_{k=0}^{\infty}$ is intermittent means that $\{i_k\}_{k=0}^{\infty}$ is an almost cyclic control sequence.

Suppose now that $\{x^k\}_{k=0}^{\infty}$ is generated by recurrence (5.54) having the following property: For any subsequence $\{x^{n_k}\}_{k=0}^{\infty} \subseteq \{x^k\}_{k=0}^{\infty}$ it holds

$$\lim_k \left\| V_i^{n_k} x^{n_k} - x^{n_k} \right\| = 0 \implies \lim_k \left\| P_{\text{Fix } U_i} x^{n_k} - x^{n_k} \right\| = 0, \tag{5.63}$$

for all $i \in I$ (actually, it suffices to consider only $i \in I_k$ in (5.63)).

Theorem 5.8.11. *Let $\{x^k\}_{k=0}^{\infty}$ be generated by recurrence (5.54) having property (5.63). If $\{w^k\}_{k=0}^{\infty}$ is intermittent and $\omega_i^k \geq \delta$ for all $i \in I_k$, $k \geq 0$ and for some constant $\delta > 0$, then $\{w^k\}_{k=0}^{\infty}$ is approximately regular with respect to the family of cutters $\mathcal{U} := \{U_i\}_{i \in I}$.*

Proof. Let $\{w^k\}_{k=0}^{\infty}$ be p-intermittent, where $p \geq 1$, and $\delta > 0$ be such that $\omega_i^k \geq \delta$ for all $i \in I_k$, $k \geq 0$. Define

$$i_k = \underset{i \in I_k}{\operatorname{argmax}} \, \omega_i^k \left\| V_i^k x^k - x^k \right\|,$$

$k \geq 0$. Suppose that $\lim_k \omega_{i_k}^k \left\| V_{i_k}^k x^k - x^k \right\|^2 = 0$. By equivalence (5.48), we have

$$\lim_k \omega_{i_k}^k \left\| V_{i_k}^k x^k - x^k \right\| = 0. \tag{5.64}$$

Let $i \in I$ and $r_k \in \{0, 1, \ldots, p-1\}$ be such that $i \in I_{k+r_k}$. By the triangle inequality, the convexity of the norm, the definition of i_{k+l}, $l = 0, 1, \ldots, p-1$, and by (5.64), we have

$$\left\| x^{k+r_k} - x^k \right\| \leq \sum_{l=0}^{r_k-1} \left\| x^{k+l+1} - x^{k+l} \right\|$$

$$= \sum_{l=0}^{r_k-1} \lambda_{k+l} \left\| \sum_{j \in I_{k+l}} \omega_j^{k+l} V_j^{k+l} x^{k+l} - x^{k+l} \right\|$$

$$\leq \sum_{l=0}^{r_k-1} \lambda_{k+l} \sum_{j \in I_{k+l}} \omega_j^{k+l} \left\| V_j^{k+l} x^{k+l} - x^{k+l} \right\|$$

$$\leq m \sum_{l=0}^{p-1} \lambda_{k+l} \omega_{i_{k+l}}^{k+l} \left\| V_{i_{k+l}}^{k+l} x^{k+l} - x^{k+l} \right\| \to 0$$

as $k \to \infty$, consequently,

$$\lim_k \left\| x^{k+r_k} - x^k \right\| = 0. \tag{5.65}$$

Further, by the definition of the metric projection and by the triangle inequality, we have

$$\left\| P_{C_i} x^k - x^k \right\| \leq \left\| P_{C_i} x^{k+r_k} - x^k \right\| \tag{5.66}$$

$$\leq \left\| P_{C_i} x^{k+r_k} - x^{k+r_k} \right\| + \left\| x^{k+r_k} - x^k \right\|,$$

where $C_i = \mathrm{Fix}\, U_i$, $i \in I$. Since $i \in I_{k+r_k}$, we have $\omega_i^{k+r_k} \geq \delta$. By the definition of i_{k+r_k}, we have

$$\lim_k \left\| V_i^{k+r_k} x^{k+r_k} - x^{k+r_k} \right\| \leq \delta^{-1} \lim_k \omega_i^{k+r_k} \left\| V_i^{k+r_k} x^{k+r_k} - x^{k+r_k} \right\|$$

$$\leq \delta^{-1} \lim_k \omega_{i_{k+r_k}}^{k+r_k} \left\| V_{i_{k+r_k}}^{k+r_k} x^{k+r_k} - x^{k+r_k} \right\| = 0,$$

i.e., $\lim_k \left\| V_i^{k+r_k} x^{k+r_k} - x^{k+r_k} \right\| = 0$. By condition (5.63),

$$\lim_k \left\| P_{C_i} x^{k+r_k} - x^{k+r_k} \right\| = 0.$$

Inequalities (5.66) and equality (5.65) imply now $\lim_k \left\| P_{C_i} x^k - x^k \right\| = 0$. Since U_i are cutters, we have $\lim_k \left\| U_i x^k - x^k \right\| = 0$ (see Corollary 2.1.37), $i \in I$, i.e., condition (5.46) is satisfied, which means that $\{w^k\}_{k=0}^\infty$ is approximately regular. $\square$

Remark 5.8.12. Since a sequence generated by (5.54) is bounded as a Fejér monotone sequence, it has a weak cluster point y^*. Therefore, property (5.63) and the demi-closedness of $P_{\mathrm{Fix}\, U_i} - \mathrm{Id}$ at 0 yields that $y^* \in \mathrm{Fix}\, U_i$. This leads to the assumption that algorithm (5.54) is focusing (see [22, Definition 3.7], where this property was applied to firmly nonexpansive operators). Furthermore, property (5.63) applied to operators V_i^k being metric projections leads to the assumption that algorithm (5.54) is strongly focusing (see [22, Definition 4.8]).

Consider now the following recurrence

$$x^{k+1} = x^k + \lambda_k \sum_{i \in I_k} \omega_i^k (U_i x^k - x^k), \tag{5.67}$$

where $I_k := \{i \in I : \omega_i^k > 0\}$, $k \geq 0$, which is a special case of (5.54) with $\mathcal{V}^k = \mathcal{U}$, $k \geq 0$. Define $K_i := \{k \geq 0 : i \in I_k\}$, $i \in I$.

Definition 5.8.13. We say that $\{I_k\}_{k=0}^\infty$ is *repetitive* if for any $i \in I$ the subset K_i is infinite. We say that a control sequence $\{w^k\}_{k=0}^\infty$ is *repetitive* if the sequence $\{I_k\}_{k=0}^\infty$ is repetitive.

An extended classification of control sequences for infinite I can be found in [119, Sect. 3].

If $\mid I_k \mid = 1$ or, equivalently, $w^k = e_{i_k}$, $k \geq 0$, for a sequence $\{i_k\}_{k=0}^{\infty} \subseteq I$, then $\{w^k\}_{k=0}^{\infty}$ is repetitive if and only if $\{i_k\}_{k=0}^{\infty}$ is a repetitive control in the sense of Definition 5.6.11.

Note that Theorems 5.8.9 and 5.8.11 hold for any sequence of relaxation parameters $\lambda_k \in [0, 2]$ and without the assumption that $\bigcap_{i \in I} \operatorname{Fix} U_i \neq \emptyset$. It turns out that a repetitive control is approximately semi-regular if $\bigcap_{i \in I} \operatorname{Fix} U_i \neq \emptyset$ and if we restrict the choice of λ_k to an interval $[\varepsilon, 2 - \varepsilon]$ for some $\varepsilon \in (0, 1)$.

Theorem 5.8.14. *Let $U_i : \mathcal{H} \to \mathcal{H}$, $i \in I$, be cutters with a common fixed point and $\{x^k\}_{k=0}^{\infty}$ be generated by recurrence (5.67), where $\liminf_k \lambda_k (2 - \lambda_k) > 0$ and $\omega_i^k \geq \delta$ for all $i \in I_k$, $k \geq 0$ and for some $\delta > 0$. If the control sequence $\{w^k\}_{k=0}^{\infty}$ is repetitive, then $\{w^k\}_{k=0}^{\infty}$ is approximately semi-regular.*

Proof. By Theorem 4.8.2, we have

$$\left\| x^{k+1} - z \right\|^2 \leq \left\| x^k - z \right\|^2 - \lambda_k (2 - \lambda_k) \sum_{i \in I_k} \omega_i^k \left\| U_i x^k - x^k \right\|^2 .$$

If we iterate the above inequality k-times, we obtain

$$\left\| x^{k+1} - z \right\|^2 \leq \left\| x^0 - z \right\|^2 - \sum_{l=0}^{k} \lambda_l (2 - \lambda_l) \sum_{i \in I_l} \omega_i^l \left\| U_i x^l - x^l \right\|^2 .$$

Consequently,

$$\sum_{k=0}^{\infty} \lambda_k (2 - \lambda_k) \sum_{i \in I_k} \omega_i^k \left\| U_i x^k - x^k \right\|^2 < \infty.$$

Since the sum of an absolutely convergent series does not depend on the order of summands, we have

$$\sum_{i=1}^{m} \sum_{k \in K_i} \lambda_k (2 - \lambda_k) \omega_i^k \left\| U_i x^k - x^k \right\|^2$$

$$= \sum_{k=0}^{\infty} \lambda_k (2 - \lambda_k) \sum_{i \in I_k} \omega_i^k \left\| U_i x^k - x^k \right\|^2 < \infty.$$

Therefore,

$$\sum_{k \in K_i} \lambda_k (2 - \lambda_k) \omega_i^k \left\| U_i x^k - x^k \right\|^2 < \infty,$$

$i \in I$, and

$$\lim_{k \to \infty, k \in K_i} \lambda_k (2 - \lambda_k) \omega_i^k \left\| U_i x^k - x^k \right\|^2 = 0,$$

$i \in I$. Since $\liminf_k \lambda_k(2 - \lambda_k) > 0$, it holds $\liminf_{k \to \infty, k \in K_i} \lambda_k(2 - \lambda_k) > 0$, consequently,

$$\lim_{k \to \infty, k \in K_i} \omega_i^k \left\| U_i x^k - x^k \right\|^2 = 0,$$

for all $i \in I$. The assumption $\omega_i^k \geq \delta > 0$ for all $i \in I_k$, $k \geq 0$, yields now $\lim_{k \to \infty, k \in K_i} \left\| U_i x^k - x^k \right\| = 0$, $i \in I$, i.e., $\liminf_k \left\| U_i x^k - x^k \right\| = 0$ for all $i \in I$ which means that $\{w^k\}_{k=0}^{\infty}$ is approximately semi-regular.   □

5.8.2   Convergence Theorem

Theorem 5.8.15. *Suppose that:*

(a) $U_i : \mathcal{H} \to \mathcal{H}$, $i \in I$, *are cutters with a common fixed point,*
(b) $U_i - \mathrm{Id}$ *are demi-closed at 0, $i \in I$,*
(c) $\mathcal{V}^k = \{V_i^k\}_{i \in J_k}$ *are families of cutters $V_i^k : \mathcal{H} \to \mathcal{H}$, $i \in J_k$, with the property*
    $\bigcap_{i \in J_k} \mathrm{Fix}\, V_i^k \supseteq \bigcap_{i \in I} \mathrm{Fix}\, U_i$, $k \geq 0$,
(d) $\{w^k\}_{k=0}^{\infty} : \mathcal{H} \to \Delta_{|J_k|}$ *is a sequence of appropriate weight functions,*
(e) $\liminf_k \lambda_k(2 - \lambda_k) > 0$,
(f) $\{x^k\}_{k=0}^{\infty}$ *is generated by recurrence (5.36).*

If the sequence of weight functions $\{w^k\}_{k=0}^{\infty}$ applied to the sequence of families $\mathcal{V}^k$:

(i) *Is approximately regular with respect to the family $\mathcal{U} := \{U_i\}_{i \in I}$, then x^k converges weakly to a common fixed point of U_i, $i \in I$;*
(ii) *Is approximately semi-regular with respect to the family $\mathcal{U} := \{U_i\}_{i \in I}$ and $\mathcal{H}$ is finite dimensional, then x^k converges to a common fixed point of U_i, $i \in I$.*

Proof. (cf. [70, Theorem 9.27]) Let $V^k : \mathcal{H} \to \mathcal{H}$ be defined by

$$V^k x := \sum_{i \in J_k} \omega_i^k(x) V_i^k x$$

and T_k be the λ_k-relaxation of the operator V^k, i.e.,

$$T_k x := V_{\lambda_k}^k x = x + \lambda_k(V^k x - x). \tag{5.68}$$

The operators V^k are cutters (see Corollary 2.1.49), consequently, T_k are strongly quasi-nonexpansive (see Theorem 2.1.39). By Theorem 2.1.26, we have

$$\mathrm{Fix}\, T_k = \mathrm{Fix}\, V^k = \bigcap_{i \in J_k} \mathrm{Fix}\, V_i^k \supseteq \bigcap_{i \in I} \mathrm{Fix}\, U_i,$$

consequently,

$$\bigcap_{k=0}^{\infty} \operatorname{Fix} T_k \supseteq \bigcap_{i \in I} \operatorname{Fix} U_i.$$

Let $\varepsilon > 0$ be such that $\liminf_k \lambda_k > \varepsilon$ and $\liminf_k (2 - \lambda_k) > \varepsilon$ and let $z \in \bigcap_{i \in I} \operatorname{Fix} U_i$. For a sufficiently large k we have $2 - \lambda_k \geq \frac{\varepsilon}{2}$. Now, it follows from Theorem 4.8.2 that, for sufficiently large k,

$$
\begin{aligned}
\left\| x^{k+1} - z \right\|^2 &= \left\| T_k x^k - z \right\|^2 \\
&\leq \left\| x^k - z \right\|^2 - \lambda_k (2 - \lambda_k) \sum_{i \in J_k} \omega_i^k (x^k) \left\| V_i^k x^k - x^k \right\|^2 \\
&\leq \left\| x^k - z \right\|^2 - \lambda_k (2 - \lambda_k) \left\| V^k x^k - x^k \right\|^2 \\
&= \left\| x^k - z \right\|^2 - \frac{2 - \lambda_k}{\lambda_k} \left\| T_k x^k - x^k \right\|^2 \\
&\leq \left\| x^k - z \right\|^2 - \frac{\varepsilon}{4} \left\| T_k x^k - x^k \right\|^2 .
\end{aligned}
$$

Therefore, $\left\| x^k - z \right\|$ decreases, consequently, $\left\| V^k x^k - x^k \right\| \to 0$, and, for an arbitrary $i_k \in J_k$,

$$\omega_{i_k}^k (x^k) \left\| V_{i_k}^k x^k - x^k \right\|^2 \leq \sum_{i \in J_k} \omega_i^k (x^k) \left\| V_i^k x^k - x^k \right\|^2 \to 0 \qquad (5.69)$$

as $k \to \infty$.

(i) Suppose that $\{w^k\}_{k=0}^{\infty}$ is approximately regular with respect to the family $\mathcal{U} := \{U_i\}_{i \in I}$. Let $i_k \in J_k$, $k \geq 0$, be such that implication (5.46) holds. Then (5.69) yields $\lim_k \left\| U_i x^k - x^k \right\| = 0$ for all $i \in I$. Therefore, condition (3.11) is satisfied for $T_k := V^k$ and for $S := U_i$, $i \in I$. We have proved that all assumptions of Corollary 3.7.1 (ii) are satisfied for $X = \mathcal{H}$, $T_k := V^k$ and for $S := U_i$, $i \in I$. Therefore, x^k converges weakly to a common fixed point of U_i, $i \in I$.

(ii) Suppose that $\mathcal{H}$ is finite dimensional and $\{w^k\}_{k=0}^{\infty}$ is approximately semi-regular with respect to the family $\mathcal{U} := \{U_i\}_{i \in I}$. Let $i_k \in J_k$, $k \geq 0$, be such that implication (5.47) holds. Then (5.69) yields $\liminf_k \left\| U_i x^k - x^k \right\| = 0$ for all $i \in I$. Therefore, condition (3.12) is satisfied for $T_k := V^k$ and for $S := U_i$, $i \in I$. We have proved that all assumptions of Theorem 3.7.1 (iii) are satisfied for $X = \mathcal{H}$, $T_k := V^k$ and for $S := U_i$, $i \in I$. Therefore, x^k converges to a common fixed point of U_i, $i \in I$. $\qquad \square$

Remark 5.8.16.   (i) We can also consider recurrence (5.36), where λ_k depends on $x \in \mathcal{H}$, i.e.,

$$x^0 \in \mathcal{H} \qquad\qquad\qquad\qquad\qquad - \text{arbitrary}$$
$$x^{k+1} = x^k + \lambda_k(x^k)(\textstyle\sum_{i \in J_k} \omega_i^k(x^k)V_i^k x^k - x^k), \tag{5.70}$$

where $\lambda_k : \mathcal{H} \to (0, 2]$ is a sequence of relaxation functions. It follows from the proof of Theorem 5.8.15 that the theorem remains true if we suppose that $\liminf_k \lambda_k(x) \geq \varepsilon$ and $\limsup_k \lambda_k(x) \leq 2 - \varepsilon$ for all $x \in \mathcal{H}$ and for some $\varepsilon \in (0, 1)$.

(ii) We can apply different relaxation parameters for different operators V_i^k, $i \in I$, in recurrence (5.36) (we suppose, for simplicity, $J_k = I$ for all $k \geq 0$) which leads to the following recurrence

$$x^0 \in \mathcal{H} \qquad\qquad\qquad\qquad\qquad - \text{arbitrary}$$
$$x^{k+1} = x^k + \textstyle\sum_{i \in I} \mu_i^k(v_i^k(x^k)V_i^k x^k - x^k), \tag{5.71}$$

where $\{\mu^k\}_{k=0}^{\infty} \subseteq (0, 2]$ is a sequence of relaxation vectors, i.e., $\mu^k = (\mu_1^k, \dots, \mu_m^k)$, and $\{v^k\}_{k=0}^{\infty}$ is a sequence of weight functions $v^k : \mathcal{H} \to \Delta_m$. Note, however, that recurrence (5.71) can be reduced to (5.70) (or to (5.36) if we apply constant weight functions v^k) if we substitute the sequence of weight functions $\{v^k\}_{k=0}^{\infty}$ and the sequence of relaxation vectors $\{\mu^k\}_{k=0}^{\infty}$ to the following sequences $\{w^k\}_{k=0}^{\infty} : \mathcal{H} \to \Delta_m$ and $\{\lambda_k\}_{k=0}^{\infty} : \mathcal{H} \to (0, 2]$, where

$$\lambda_k(x) = \sum_{i \in I} \mu_i^k v_i^k(x) \tag{5.72}$$

and

$$\omega_i^k(x) = \mu_i^k v_i^k(x)/\lambda_k(x), \tag{5.73}$$

$i \in I$ and $x \in \mathcal{H}$ (cf. Remark 4.4.3). Note that this transformation maintains the assumptions on weight functions and on relaxation parameters presented before. In particular, if $\liminf_k \mu_i^k \geq \varepsilon$ and $\limsup_k \mu_i^k \leq 2 - \varepsilon$ for all $i \in I$ and for some $\varepsilon \in (0, 1)$, then $\liminf_k \lambda_k(x) \geq \varepsilon$ and $\limsup_k \lambda_k(x) \leq 2 - \varepsilon$ for all $x \in \mathcal{H}$. Furthermore, if v^k is an approximately regular sequence of weight functions and $\liminf_k \mu_k(2 - \mu_k) > 0$, then w^k is also approximately regular. Therefore, we apply recurrence (5.36) instead of (5.71) in Theorem 5.8.15 in order to simplify the notation.

(iii) If we replace the assumption on the demi-closedness of $U_i - \text{Id}$ at 0, $i \in I$ and the assumption on the approximate regularity of $\{w^k\}_{k=0}^{\infty}$ in Theorem 5.8.15 by the following one:

$\lim_k \omega_{i_k}^k(x^k) \left\| V_{i_k}^k x^k - x^k \right\|^2 = 0 \Longrightarrow$ all weak cluster points of $\{x^k\}_{k=0}^{\infty}$ lie in $\bigcap_{i \in I} \text{Fix } U_i$,

then the theorem remains true (cf. Remark 3.7.5).

5.8.3 Examples

Example 5.8.17. (Iusem and De Pierro [215]) Consider recurrence (5.36), where $J_k = I$, $\lambda_k = 1$, $V_i^k = P_{C_i}$, for all $k \geq 0$, $C_i \subseteq \mathcal{H}$ are closed convex subsets, $i \in I$, the sequence $\{w^k\}_{k=0}^{\infty}$ of weight functions $w^k : \mathcal{H} \to \Delta_m$ is constant, i.e., $w^k = w$, and w has the form

$$
\omega_i(x) := \begin{cases} \frac{v_i}{\sum_{j \in I(x)} v_j} & \text{for } i \in I(x) \\ 0 & \text{for } i \notin I(x), \end{cases}
$$

where $v = (v_1, v_2, \ldots, v_m) \in \operatorname{ri} \Delta_m$ and $I(x) := \{i \in I : x \notin C_i\}$. Suppose that $C := \bigcap_{i \in I} C_i \neq \emptyset$. Since w^k is regular (see Example 5.8.3 (ii)), it follows from Theorem 5.8.15 that $x^k \rightharpoonup x^* \in C$. The convergence was proved by Iusem and de Pierro [215, Corollary 4] for $\mathcal{H} = \mathbb{R}^n$. Iusem and de Pierro have also considered the case $C = \emptyset$ and have proved the local convergence of x^k to a fixed point of the operator $T := \sum_{i \in I} \omega_i P_{C_i}$ with $\operatorname{Fix} T \neq \emptyset$ for a starting point $x^0 \in \bigcup_{z \in \operatorname{Fix} T} B(z, \frac{1}{2} \min_{i \in I} \|P_{C_i} z - z\|)$, see [215, Theorem 3]). Note that this convergence can be proved simply as follows. If $x^0 \in B(z, r)$ for some $z \in \operatorname{Fix} T$, where $r := \frac{1}{2} \min_{i \in I} \|P_{C_i} z - z\|$, then $\omega_i(x^0) = v_i$, $i \in I$. The quasi nonexpansivity of T (see Corollary 4.4.4) yields that $x^1 \in B(z, r)$. One can prove by induction with respect to k that $\omega_i(x^k) = v_i$ for all $i \in I$ and for all k. The convergence follows now from Corollary 5.4.1.

Example 5.8.18. (Aharoni and Censor [3]) Consider recurrence (5.36), where $\mathcal{H} = \mathbb{R}^n$, $J_k = I$, $V_i^k = P_{C_i}$ for closed convex subsets $C_i \subseteq \mathbb{R}^n$, $i \in I$, with $C := \bigcap_{i \in I} C_i \neq \emptyset$, $\lambda_k \in [\varepsilon, 2 - \varepsilon]$ for some $\varepsilon \in (0, 1)$, $k \geq 0$, $w^k \in \Delta_m$ with $\sum_{k=0}^{\infty} \omega_i^k = +\infty$, $i \in I$. By Theorem 4.4.5, we have

$$
\left\| x^{k+1} - z \right\|^2 \leq \left\| x^k - z \right\|^2 - \lambda_k (2 - \lambda_k) \sum_{i=1}^{m} \omega_i^k \left\| P_{C_i} x^k - x^k \right\|^2
$$

If we iterate this inequality k-times, we obtain

$$
\left\| x^{k+1} - z \right\|^2 \leq \left\| x^0 - z \right\|^2 - \sum_{l=0}^{k} \lambda_l (2 - \lambda_l) \sum_{i=1}^{m} \omega_i^l \left\| P_{C_i} x^l - x^l \right\|^2.
$$

Consequently,

$$
\sum_{i=1}^{m} \sum_{k=0}^{\infty} \lambda_k (2 - \lambda_k) \omega_i^k \left\| P_{C_i} x^k - x^k \right\|^2
$$

$$
= \sum_{k=0}^{\infty} \lambda_k (2 - \lambda_k) \sum_{i=1}^{m} \omega_i^k \left\| P_{C_i} x^k - x^k \right\|^2 < +\infty
$$

and

$$\sum_{k=0}^{\infty} \lambda_k (2 - \lambda_k) \omega_i^k \left\| P_{C_i} x^k - x^k \right\|^2 < +\infty$$

for any $i \in I$. The assumption $\liminf_k \lambda_k (2 - \lambda_k) > 0$ yields now

$$\sum_{k=0}^{\infty} \omega_i^k \left\| P_{C_i} x^k - x^k \right\|^2 < +\infty,$$

$i \in I$. Since $\sum_{k=0}^{\infty} \omega_i^k = +\infty$, we have $\liminf_k \left\| P_{C_i} x^k - x^k \right\| = 0$, $i \in I$, i.e., w^k is approximately semi-regular. Theorem 5.8.15 (ii) yields now the convergence $x^k \to x^* \in C$. (See [3, Theorem 1] for a different proof of convergence).

Example 5.8.19. (Butnariu and Censor [53]). Consider recurrence (5.36), where $\mathcal{H} = \mathbb{R}^n$, $J_k = I$, $V_i^k = P_{C_i}$, $C_i \subseteq \mathcal{H}$ are closed convex subsets, $k \geq 0$, $i \in I$, $\liminf_k \lambda_k > 0$, $\limsup_k \lambda_k < 2$, $w^k \in \Delta_m$ has a subsequence converging to a point $w^* \in \operatorname{ri} \Delta_m$. Suppose that $C := \bigcap_{i \in I} C_i \neq \emptyset$. Let $\varepsilon > 0$ be such that $\omega_i^* > \varepsilon$ for all $i \in I$. Then there exists a subsequence $\{w^{n_k}\}_{k=0}^{\infty} \subseteq \{w^k\}_{k=0}^{\infty}$ such that $\omega_i^{n_k} > \frac{\varepsilon}{2}$ for all $i \in I$ and $k \geq 0$, consequently, $\{w^{n_k}\}_{k=0}^{\infty}$ is $\frac{\varepsilon}{2}$-regular (see Example 5.8.3 (i)). Therefore, $\{w^k\}_{k=0}^{\infty}$ is approximately semi-regular (see Lemma 5.8.7 (ii)). Theorem 5.8.15 (ii) yields now $x^k \to x^* \in C$. If we suppose that all cluster points w of $\{w^k\}_{k=0}^{\infty}$ have the property $\omega_i \geq \delta$ for all $i \in I$ and for some $\delta \in (0, \frac{1}{m}]$, then $\{w^k\}_{k=0}^{\infty}$ is regular (see Example 5.8.3 (iv)) and the weak convergence holds in general Hilbert spaces $\mathcal{H}$.

Example 5.8.20. (Flåm and Zowe [168]) Consider recurrence (5.36), where $J_k = I$ for all k, $\liminf_k \lambda_k > 0$ and $\limsup_k \lambda_k < 2$ and V_i^k are cutters with $\operatorname{Fix} V_i^k \supseteq C_i$ for closed convex subsets $C_i \subseteq \mathcal{H}$ with $C := \bigcap_{i \in I} C_i \neq \emptyset$, satisfying the inequality

$$\left\| V_i^k x^k - x^k \right\| \geq \alpha \left\| P_{C_i} x^k - x^k \right\|, \tag{5.74}$$

$i \in I, k \geq 0$, for some constant $\alpha > 0$. Furthermore, suppose that the sequence of weights $\{w^k\}_{k=0}^{\infty}$ satisfies the following conditions:

(i) $\limsup_k \omega_i^k > 0$, $i \in I$,
(ii) $\omega_i^k (P_{C_i} x^k - x^k) \neq 0 \Rightarrow \omega_i^k > \delta > 0$.

If we take $U_i := P_{C_i}$, $i \in I$, then it follows from (5.74) and from (i) to (ii) that the sequence of weights $\{w^k\}_{k=0}^{\infty}$ is regular. We see that all assumptions of Theorem 5.8.15 (i) are satisfied. Therefore, $x^k \to x^* \in C$. The convergence was proved by Flåm and Zowe [168, Theorem 1] for $\mathcal{H} = \mathbb{R}^n$. Actually, Flåm and Zowe considered recurrence (5.71), but it can be reduced to (5.36), as it was explained in Remark 5.8.16 (ii).

Example 5.8.21. Consider recurrence (5.55) with $V_i^k := P_{C_i}$, $i \in I, k \geq 0$, for closed convex subsets $C_i \subseteq \mathcal{H}$ having a common fixed point, where $\lambda_k \in [\varepsilon, 2 - \varepsilon]$ for some $\varepsilon \in (0, 1)$, $I_k := \{i \in I : \omega_i^k > 0\}$ and $\omega_i^k \geq \delta$ for all $i \in I_k$ such that $x^k \notin C_i$, $k \geq 0$ (recurrences of this form with countable I were considered by Combettes and Puh [125, Sect. 3]). Then condition (5.56) is automatically satisfied (it suffices to take $l_k := \text{argmax}_{i \in I_k} \| P_{C_i} x^k - x^k \|$). If $\{I_k\}_{k=0}^\infty$ is intermittent, then $\{w^k\}_{k=0}^\infty$ is approximately regular (see Theorem 5.8.9) and it follows from Theorem 5.8.15 (i) that $x^k \rightharpoonup x^* \in C = \bigcap_{i \in I} C_i$.

Example 5.8.22. (Simultaneous subgradient projection method) Consider recurrence (5.36), where $J_k = I$ for all $k \geq 0$, $\liminf_k \lambda_k > 0$ and $\limsup_k \lambda_k < 2$ and $V_i^k = V_i := P_{c_i}$ is a subgradient projection relative to a continuous convex function $c_i : \mathcal{H} \to \mathbb{R}$ which is globally Lipschitz continuous on bounded subsets, $i \in I, k \geq 0$. Recall that V_i are cutters (see Corollary 4.2.6) and $V_i - \text{Id}$ are demiclosed at 0 (see Theorem 4.2.7). Denote $C_i := S(c_i, 0)$, $i \in I$, and suppose that $C := \bigcap_{i \in I} C_i \neq \emptyset$. Let $\{w^k\}_{k=0}^\infty$ be a sequence of appropriate weight functions.

(a) If $\{w^k\}_{k=0}^\infty$ is approximately regular, then it follows from Theorem 5.8.15 (i) that x^k converges weakly to a point $x^* \in C$.
(b) If $\{w^k\}_{k=0}^\infty$ is approximately semi-regular and $\mathcal{H}$ is finite dimensional, then it follows from Theorem 5.8.15 (ii) that x^k converges to a point $x^* \in C$.

Example 5.8.23. (Censor and Elfving [87]) Consider a consistent system of linear inequalities $Ax \leq b$ and a method defined by the recurrence

$$x^{k+1} = x^k + \lambda_k (T_k x^k - x^k),$$

where $T_k : \mathbb{R}^n \to \mathbb{R}^n$ is given by

$$T_k x := x - \frac{1}{\lambda_{\max}(A^\top M_k A)} A^\top M_k (Ax - b)_+,$$

with $M_k := D_k M D_k$ for a positive definite matrix M with nonnegative elements and a diagonal matrix $D_k := \text{diag}(d_1^k, d_2^k, \ldots, d_m^k)$ given by

$$d_i^k := \begin{cases} 1 \text{ if } i \in I^k \\ 0 \text{ otherwise,} \end{cases}$$

and $I_k := \{i \in I : a_i^\top x^k - \beta_i > 0\}$. Denote by A_k the submatrix of A with rows a_i, $i \in I_k$, by b_k the subvector of b with coordinates β_i, $i \in I_k$, and by R_k the principal submatrix of M with rows and columns $i \in I_k$. Further, denote $C := \{x \in \mathbb{R}^n : Ax \leq b\}$, $C_k := \{x \in \mathbb{R}^n : A_k x \leq b_k\}$. Of course, $C \subseteq C_k$. Then T_k is the Landweber operator for the system $A_k x \leq b_k$ relative to the positive definite matrix R_k. Therefore, $\text{Fix } T_k = C_k$ (see Lemma 4.6.2) and T_k is firmly nonexpansive (see Theorem 4.6.3), consequently, T_k is a cutter. Let U be an orthogonal matrix, $G = \text{diag}(g_1, g_2, \ldots, g_m)$ be such that $g_i > 0$, $i \in I$, and $M = U^\top G U$ (the existence of U and G follows from Theorem 1.1.34 (iii)). Denote $G_k := D_k G D_k$. It is clear that $G_k = \text{diag}(g_1^k, g_2^k, \ldots, g_m^k)$ where

$$g_i^k := \begin{cases} g_i \text{ if } i \in I^k \\ 0 \text{ otherwise.} \end{cases}$$

We leave it to the reader to check that

$$M_k = D_k U^\top G U D_k = U^\top G_k U \tag{5.75}$$

and

$$M_k^{\frac{1}{2}} = U^\top G_k^{\frac{1}{2}} U. \tag{5.76}$$

Denote $y(x) = (\eta_1(x), \eta_2(x), \ldots, \eta_m(x))^\top = UAx$. By Theorem 1.1.27, we have

$$\lambda_{\max}(A^\top M_k A) = \left\| M_k^{\frac{1}{2}} A \right\|^2 = \sup_{\|x\|=1} \left\| M_k^{\frac{1}{2}} Ax \right\|^2 = \sup_{\|x\|=1} \left\| U^\top G_k^{\frac{1}{2}} UAx \right\|^2$$

$$= \sup_{\|x\|=1} \left\| G_k^{\frac{1}{2}} y(x) \right\|^2 = \sup_{\|x\|=1} \sum_{i \in I} g_i^k (\eta_i(x))^2$$

$$\leq \sup_{\|x\|=1} \sum_{i \in I} g_i (\eta_i(x))^2 = \sup_{\|x\|=1} \left\| G^{\frac{1}{2}} y(x) \right\|^2$$

$$= \sup_{\|x\|=1} \left\| G^{\frac{1}{2}} UAx \right\|^2 = \sup_{\|x\|=1} \left\| U^\top G^{\frac{1}{2}} UAx \right\|^2$$

$$= \sup_{\|x\|=1} \left\| M^{\frac{1}{2}} Ax \right\|^2 = \left\| M^{\frac{1}{2}} A \right\|^2 = \lambda_{\max}(A^\top M A).$$

Let $S : \mathbb{R}^n \to \mathbb{R}^n$ be the Landweber operator for the system $Ax \leq b$ related to the matrix M, i.e., S has the form

$$Sx := x - \frac{1}{\lambda_{\max}(A^\top M A)} A^\top M (Ax^k - b)_+.$$

Note that $M_k(Ax^k - b)_+ = M(Ax^k - b)_+$. Because the system $Ax \leq b$ is equivalent to $M^{\frac{1}{2}} Ax \leq M^{\frac{1}{2}} b$, Lemma 4.6.2 yields that Fix $S = \{x \in \mathbb{R}^n : M^{\frac{1}{2}} Ax \leq M^{\frac{1}{2}} b\} = C$. Therefore,

$$\left\| T_k x^k - x^k \right\| = \frac{1}{\lambda_{\max}(A^\top M_k A)} \left\| A^\top M_k (Ax^k - b)_+ \right\|$$

$$\geq \frac{1}{\lambda_{\max}(A^\top M A)} \left\| A^\top M_k (Ax^k - b)_+ \right\|$$

$$= \frac{1}{\lambda_{\max}(A^\top M A)} \left\| A^\top M (Ax^k - b)_+ \right\|$$

$$= \left\| Sx^k - x^k \right\|,$$

i.e., condition (3.11) is satisfied and $C = $ Fix $S \subseteq$ Fix T_k.

We see that all assumptions of Corollary 3.7.1 (iii) are satisfied. Therefore, $x^k \to x^* \in$ Fix $S = \{x \in \mathbb{R}^n : Ax \leq b\}$.

5.8.4   Block Iterative Projection Methods

Consider the convex feasibility problem: find a common point of a finite family of nonempty closed convex subsets $C_i \subseteq \mathcal{H}$, $i \in I$. If m is a large number, it is convenient to split the set I into smaller blocks and to perform the operations of the projection method in two stages: first on blocks and then on the family of blocks.

Let $I = L_1 \cup L_2 \cup \ldots \cup L_r$, where L_j are nonempty subsets which are pairwise disjoint, $j \in J = \{1, 2, \ldots, r\}$. Then each $i \in I$ can be transformed in a unique way to a pair (j, l) such that $j \in J$ and $l \in L_j$. The subset C_i can now be presented as C_{jl} and the convex feasibility problem obtains the form: $x^* \in \bigcap_{j \in J} \bigcap_{l \in L_j} C_{jl}$.

Consider the following two stage recurrence

$$x^{k+1} = U_{j_k}^k x^k, \tag{5.77}$$

where $j_k \in J$ is a control for the outer loop and $U_{j_k}^k$ is a relaxed simultaneous projection which "employs" only projections $P_{C_{j_k,l}}$ for $l \in L_{j_k}$, i.e., $U_{j_k}^k$ has the form

$$U_{j_k}^k := \mathrm{Id} + \lambda_k \Big(\sum_{l \in L_{j_k}} \omega_i^k P_{C_{j_k,l}} - \mathrm{Id} \Big), \tag{5.78}$$

where $\lambda_k \in [0, 2]$ and $w^k : \mathcal{H} \to \Delta_{|L_{j_k}|}$ is a sequence of weight functions applied in the inner loop. The sequence $\{(j_k, w^k)\}_{k=0}^{\infty}$ defines a two stage control. In particular, the following operator $U_{j_k}^k$ can be used in the inner loop of a two stage recurrence (5.77):

$$U_{j_k}^k := P_{C_{j_k,l_k}} \tag{5.79}$$

where $l_k \in L_{j_k}$ is a control sequence for the inner loop. Note that $P_{C_{j_k,l_k}}$ is a special case of the operator $U_{j_k}^k$ defined by (5.78), where $\lambda_k = 1$ and $w^k = e_{l_k}$.

We can identify the weight function $w^k : \mathcal{H} \to \Delta_{|L_{j_k}|}$ with a weight function $v^k : \mathcal{H} \to \Delta_m$ defined by

$$v_i^k := \begin{cases} \omega_i^k & \text{if } i \in L_{j_k} \\ 0 & \text{otherwise.} \end{cases} \tag{5.80}$$

5.8.4.1   Double Layer Control Sequence

Consider the two stage recurrence (5.77), where the control sequence: $\{j_k\}_{k=0}^{\infty} \subseteq J$ applied in the outer stage is an almost cyclic control, i.e.,

$$J \subseteq \{j_k, j_{k+1}, \ldots, j_{k+p-1}\} \tag{5.81}$$

for some $p \geq r$ and, for all k, $U_{j_k}^k$ is defined by (5.78), where the weight function $w^k : \mathcal{H} \rightarrow \Delta_{|L_{j_k}|}$ is α-regular with respect to the family $\{P_{C_{j_k,l}}\}_{l \in L_{j_k}}$, where $\alpha \in (0, 1]$, i.e., there exists $l_k \in L_{j_k}$ such that

$$\omega_{l_k}^k \left\| P_{C_{j_k, l_k}} x^k - x^k \right\|^2 \geq \alpha \max_{l \in L_{j_k}} \left\| P_{C_{j_k, l}} x^k - x^k \right\|^2, \tag{5.82}$$

$k \geq 0$. Examples of such weight functions were presented in Sect. 5.8.1. We call the sequence $\{(j_k, w^k)\}_{k=0}^{\infty}$ defined above a *double layer control*. A special case of this control is the following: $\alpha = 1$, $\lambda_k = 1$, $w^k = e_{l_k} \in \text{ext} \, \Delta_{|L_{j_k}|}$ and $l_k \in L_{j_k}$ is a remotest set control, i.e., $l_k = l_k(x^k) \in L_{j_k}$ is such that

$$\left\| P_{C_{j_k, l_k}} x^k - x^k \right\| = \max_{l \in L_{j_k}} \left\| P_{C_{j_k, l}} x^k - x^k \right\|. \tag{5.83}$$

The double layer control $\{(j_k, w^k)\}_{k=0}^{\infty}$ can be identified with a sequence of weight functions $\{v^k\}_{k=0}^{\infty}$ defined by (5.80).

Corollary 5.8.24. *The double layer control is approximately regular with respect to the family* $\mathcal{U} := \{P_{C_i}\}_{i \in I}$.

Proof. Recurrence (5.77), where $U_{j_k}^k$ is defined by (5.78), is a special case of (5.55), where $I_k = L_{j_k}$, $\mathcal{V}^k = \mathcal{U} := \{P_{C_i}\}_{i \in I}$. Furthermore, (5.82) yields condition (5.56). Since $\{j_k\}_{k=0}^{\infty}$ is almost cyclic, $\{I_k\}_{k=0}^{\infty}$ is intermittent. The claim follows now from Theorem 5.8.9.   □

Theorem 5.8.25. *Let* $C \neq \emptyset$ *and* x^k *be generated by recurrence (5.77) with a double layer control and with a sequence of relaxation parameters* $\{\lambda_k\}_{k=0}^{\infty}$ *such that* $\liminf \lambda_k (2 - \lambda_k) > 0$. *Then for an arbitrary* $x^0 \in \mathcal{H}$ *the sequence* $\{x^k\}_{k=0}^{\infty}$ *converges weakly to a point* $x^* \in C$.

Proof. It follows from Corollary 5.8.24 that the double layer control is approximately regular. The claim follows now from Theorem 5.8.15 (i).   □

Several block iterative projection methods are presented in [3, 53, 81, 229], [118, Chap. 5], [108, Sect. 5.6], [5, 59, 159].

5.9 Sequential Cutter Methods

In this section we consider the common fixed point problem for a finite family of cutters $U_i : \mathcal{H} \rightarrow \mathcal{H}$, $i \in I := \{1, 2, \ldots, m\}$, having a common fixed point, and sequences generated by the following recurrence

$$\begin{aligned} x^0 &= x \in \mathcal{H} && - \text{arbitrary} \\ x^{k+1} &= x^k + \lambda_k (U_{i_k} x^k - x^k) \end{aligned} \tag{5.84}$$

where $\{\lambda_k\}_{k=0}^{\infty} \subseteq (0, 2]$ is a sequence of relaxation parameters and $\{i_k\}_{k=0}^{\infty} \subseteq I$ is a control sequence.

Definition 5.9.1. Let $\{x^k\}_{k=0}^{\infty}$ be generated by recurrence (5.84). A control sequence $\{i_k\}_{k=0}^{\infty} : \mathcal{H} \to I$ is called:

(i) *Approximately regular* if the following implication holds

$$\lim_k \left\| U_{i_k} x^k - x^k \right\| = 0 \Longrightarrow \lim_k \left\| U_i x^k - x^k \right\| = 0 \text{ for all } i \in I, \quad (5.85)$$

(ii) *Approximately semi-regular* if the following implication holds

$$\lim_k \left\| U_{i_k} x^k - x^k \right\| = 0 \Longrightarrow \liminf_k \left\| U_i x^k - x^k \right\| = 0 \text{ for all } i \in I.$$

Note that recurrence (5.84) is a special case of (5.36), where $w^k = e_{i_k}$, $k \geq 0$. Therefore, a control sequence $\{i_k\}_{k=0}^{\infty}$ is approximately (semi-)regular means that $\{e_{i_k}\}_{k=0}^{\infty}$ is an approximately (semi-)regular sequence of weight functions with respect to the family $\mathcal{U} := \{U_i\}_{i \in I}$.

If U_i are metric projections, then an approximately (semi-)regular control coincides with an approximately (semi-)remotest set control (cf. Definitions 5.6.1 and 5.6.2). If, furthermore, $\{i_k\}_{k=0}^{\infty}$ is an almost remotest set control (see Definition 5.6.6), then the sequence $\{e_{i_k}\}_{k=0}^{\infty}$ is a regular sequence of weight functions (see Example 5.8.3 (v)).

5.9.1 Convergence Theorem

If we take $\mathcal{V}^k = \mathcal{U} := \{U_i\}_{i \in I}$ and $w^k := e_{i_k}$, $k \geq 0$, in Theorem 5.8.15, then we obtain the following result.

Corollary 5.9.2. Let $U_i : \mathcal{H} \to \mathcal{H}$, be cutters with a common fixed point and such that $U_i - \mathrm{Id}$ are demi-closed at 0, $i \in I$, and the sequence $\{x^k\}_{k=0}^{\infty}$ be generated by recurrence (5.84), where $\liminf \lambda_k (2 - \lambda_k) > 0$. If the control sequence $\{i_k\}_{k=0}^{\infty}$ is:

(i) *Approximately regular*, then x^k converges weakly to a common fixed point of U_i, $i \in I$.

(ii) *Approximately semi-regular and $\mathcal{H}$ is finite dimensional*, then x^k converges to a common fixed point of U_i, $i \in I$.

5.9.2 Control Sequences for Sequential Cutter Methods

Corollary 5.9.2 shows the importance of approximate (semi-)regularity of the control sequence $\{i_k\}_{k=0}^{\infty}$ for the convergence of the sequences generated by the sequential cutter method. In this section we present some control sequences which are approximately (semi-)regular. The results presented in this section are special cases of Theorems 5.8.11 and 5.8.14, where $\mathcal{V}^k = \mathcal{U} = \{U_i\}_{i \in I}$ and $w^k = e_{i_k}$.

Corollary 5.9.3. *Let* $U_i : \mathcal{H} \to \mathcal{H}$ *be cutters having the following property: for any bounded sequence* $\{y^k\}_{k=0}^\infty \subseteq \mathcal{H}$

$$\lim_k \| U_i y^k - y^k \| = 0 \Longrightarrow \lim_k \| P_{\mathrm{Fix}\, U_i} y^k - y^k \| = 0, \qquad (5.86)$$

$i \in I$. *Let* $\{x^k\}_{k=0}^\infty$ *be generated by recurrence (5.84). If the control sequence* $\{i_k\}_{k=0}^\infty$ *is almost cyclic, then* $\{i_k\}_{k=0}^\infty$ *is approximately regular.*

A family of operators $\{U_i\}_{i \in I}$ has property (5.86) if, e.g.,

$$\| U_i y - y \| \geq \delta \, \| P_{\mathrm{Fix}\, U_i} y - y \| \qquad (5.87)$$

for a constant $\delta > 0$, and for arbitrary $i \in I$ and $y \in \mathcal{H}$. Of course, the latter property is satisfied for operators U_i being strict relaxations of metric projections P_{C_i} onto nonempty closed convex subsets C_i, $i \in I$. Therefore, Corollary 5.9.3 yields in particular that an almost cyclic control sequence applied to a successive projection method (5.28), where $\liminf_k \lambda_k (2 - \lambda_k) > 0$, is an approximately remotest set control sequence (see Lemma 5.6.12 (i)).

Corollary 5.9.4. *Let* $U_i : \mathcal{H} \to \mathcal{H}$, $i \in I$ *be cutters with a common fixed point and* $\{x^k\}_{k=0}^\infty$ *be generated by recurrence (5.84), where* $\liminf_k \lambda_k (2 - \lambda_k) > 0$. *If the control sequence* $\{i_k\}_{k=0}^\infty$ *is repetitive, then* $\{i_k\}_{k=0}^\infty$ *is approximately semi-regular.*

5.9.3 Examples

Example 5.9.5. (Aharoni, Berman and Censor [2]) Consider method (5.84), where $\mathcal{H}$ is finite dimensional and the cutters U_i satisfy the demi-closedness principle, $i \in I$, (in [2] it was supposed that U_i have property (5.87), $i \in I$), $\lambda_k \in [\eta, 2-\eta]$ for some $\eta \in (0, 1)$ and the control $\{i_k\}_{k=0}^\infty$ is repetitive. By Corollary 5.9.4 the control $\{i_k\}_{k=0}^\infty$ is approximately semi-regular. Note that for a cutter U_i with property (5.87), the operator $U_i - \mathrm{Id}$ is closed at 0, because, by the nonexpansivity of the metric projection, the operator $P_{\mathrm{Fix}\, U_i} - \mathrm{Id}$ is closed at 0, $i \in I$ (see Lemma 3.2.5). Therefore, Corollary 5.9.2 (ii) yields the convergence $x^k \to x^* \in \bigcap_{i \in I} \mathrm{Fix}\, U_i$. The convergence was proved by Aharoni et al. [2, Theorem 1] (see also [108, Sect. 5.5]), where the method was called the (δ, η)-*algorithm*. An interesting convergence result for method (5.84) with $\lambda_k = 1$ and $U_i : \mathbb{R}^n \to \mathbb{R}^n$ being paracontractions (continuous strictly quasi-nonexpansive operators) and with repetitive control was obtained by Elsner et al. [164, Theorem 1] (see also [163, Theorem 1] for linear case). This result is not covered by Corollaries 5.9.2 (ii) and 5.9.4. Conversely, the convergence of sequences in $\mathbb{R}^n$ generated by (5.84) with repetitive control does not follow from the [164, Theorem 1], because the cutters are not supposed to be continuous.

Example 5.9.6. Consider method (5.84), where $\liminf_k \lambda_k(2 - \lambda_k) > 0$ and U_i, $i \in I$, are firmly nonexpansive operators with a common fixed point. Then $U_i - \mathrm{Id}$ are demi-closed at 0, $i \in I$.

(i) If U_i have property (5.86), $i \in I$, and the control sequence $\{i_k\}_{k=0}^\infty$ is almost cyclic, then Corollaries 5.9.2 (i) and 5.9.3 yield $x^k \rightharpoonup x^* \in \bigcap_{i \in I} \mathrm{Fix}\, U_i$ (see also [331, Theorem 2] and [6], where a quasi-cyclic order is considered, which is slightly more general than the almost cyclic control).

(ii) If $\mathcal{H}$ is finite dimensional and the control sequence $\{i_k\}_{k=0}^\infty$ is repetitive, then Corollaries 5.9.2 (ii) and 5.9.4 yield $x^k \to x^* \in \bigcap_{i \in I} \mathrm{Fix}\, U_i$ (see also [22, Example 3.14], [331, Theorem 1] and [6, Theorem 3.1] for related results).

Example 5.9.7. (Sequential subgradient projection method) Consider method (5.84), where $\liminf_k \lambda_k(2 - \lambda_k) > 0$ and U_i, $i \in I$, are subgradient projections, i.e., $U_i = P_{c_i}$, where $c_i : \mathcal{H} \to \mathbb{R}$ are continuous convex functions, $i \in I$, which are globally Lipschitz continuous on bounded subsets (this holds if, e.g., $\mathcal{H} = \mathbb{R}^n$). Denote $C_i := S(c_i, 0)$, $i \in I$, and suppose that $C := \bigcap_{i \in I} C_i \neq \emptyset$. Since U_i are cutters (see Corollary 4.2.6) and $U_i - \mathrm{Id}$ are demi-closed at 0 (see Theorem 4.2.7), it follows from Corollary 5.9.2 (i) that x^k converges weakly to a point $x^* \in C$ if the control sequence $\{i_k\}_{k=0}^\infty$ is approximately regular (e.g., if $\{i_k\}_{k=0}^\infty$ is almost cyclic). Similarly, it follows from Corollary 5.9.2 (ii) that x^k converges to a point $x^* \in C$, if $\mathcal{H}$ is finite dimensional and the control sequence $\{i_k\}_{k=0}^\infty$ is approximately semi-regular (e.g., if $\{i_k\}_{k=0}^\infty$ is repetitive). A special case of the sequential subgradient projection method was presented in [102], where $\mathcal{H} = \mathbb{R}^n$ and the control is almost cyclic. In [136] a finitely convergent modification of the method of [102] was presented. In [80] the method of [102] was applied for solving interval linear inequalities. For a survey on subgradient projection methods, see [79].

5.10 Extrapolated Simultaneous Cutter Methods

We consider the common fixed point problem for a finite family of cutters $U_i : \mathcal{H} \to \mathcal{H}$, $i \in I := \{1, 2, \ldots, m\}$, with $\bigcap_{i \in I} \mathrm{Fix}\, U_i \neq \emptyset$. In this section we study convergence of sequences which are generated by generalized relaxations of simultaneous cutters of the form

$$Tx := V_{\sigma,\lambda} x = x + \lambda \sigma(x)(Vx - x), \tag{5.88}$$

where $\lambda \in [0, 2]$, $\sigma : \mathcal{H} \to (0, \infty)$ is a step size function, $V : \mathcal{H} \to \mathcal{H}$ is a simultaneous cutter, i.e., $Vx := \sum_{i \in J} v_i(x) V_i x$, $\mathcal{V} := \{V_i\}_{i \in J}$ is a finite family of cutters, $\bigcap_{i \in J} \mathrm{Fix}\, V_i \supseteq \bigcap_{i \in I} \mathrm{Fix}\, U_i$, and $v : \mathcal{H} \to \Delta_{|J|}$ is a weight function with

$v(x) = (v_1(x), \ldots, v_{|J|}(x)) \in \Delta_{|J|}$. In particular, we can take $\mathcal{V} = \mathcal{U} := \{U_i\}_{i \in I}$. In this case

$$T x = U_{\sigma,\lambda} x = x + \lambda \sigma(x)(Ux - x),$$

where $Ux := \sum_{i \in I} \omega_i(x) U_i x$ and $w : \mathcal{H} \to \Delta_m$. A sequence of operators T_k of the form (5.88) describes, for any $x^0 \in \mathcal{H}$, the recurrence $x^{k+1} = T_k x^k$, i.e.,

$$x^{k+1} = x^k + \lambda_k \sigma_k(x^k)(\sum_{i \in J_k} v_i^k(x^k) V_i^k x^k - x^k), \tag{5.89}$$

where $\lambda_k \in [0,2]$, $\{\sigma_k\}_{k=0}^{\infty} : \mathcal{H} \to (0, \infty)$ is a sequence of step size functions, $\{v^k\}_{k=0}^{\infty} : \mathcal{H} \to \Delta_{|J_k|}$ is a sequence of weight functions, i.e., $v^k(x) = (v_1^k(x), \ldots, v_{|J_k|}^k(x)) \in \Delta_{|J_k|}$ and $\mathcal{V}^k := \{V_i^k\}_{i \in J_k}$, $k \geq 0$, is a sequence of families of cutters with $\bigcap_{i \in J_k} \text{Fix } V_i^k \supseteq \bigcap_{i \in I} \text{Fix } U_i$ for all $k \geq 0$. The iterative method described by (5.89) is called an *extrapolated simultaneous cutter method* (ESCM). If we take, in particular, $\mathcal{V}^k := \mathcal{U}$ for all $k \geq 0$, we obtain the following recurrence

$$x^{k+1} = x^k + \lambda_k \sigma_k(x^k)(\sum_{i \in I}^{k} v_i^k(x^k) U_i x^k - x^k), \tag{5.90}$$

where $\{v^k\}_{k=0}^{\infty} : \mathcal{H} \to \Delta_m$ is a sequence of weight functions and $v^k(x) = (v_1^k(x), \ldots, v_m^k(x))$, $x \in \mathcal{H}$. We can appropriately apply Definition 5.8.5 and Remark 5.8.6 also for sequences generated by recurrence (5.89).

If U_i are projections, then the methods defined by recurrence (5.90) are known in the literature as *extrapolated simultaneous projection methods* or extrapolated methods of parallel projections. Various variants of extrapolated simultaneous cutter (in particular projection) methods were studied in [146, 284, 351], [62, Chaps. 4 and 5], [175, 229, 247], [17, Chap. 8], [118, Chap. 5], [89, 90, 98, 99, 119, 120, 176, 230–233, 304, 305], [10, Theorem 1 and Corollary 4.4], [67–70, 92, 190].

5.10.1 Assumptions on Step Sizes

Let $\mathcal{V} := \{V_i\}_{i \in J}$ be a finite family of cutters with a common fixed point and $v : \mathcal{H} \to \Delta_{|J|}$ be an appropriate weight function. Define a step size function $\sigma_v : \mathcal{H} \to \mathbb{R}$ by the equality

$$\sigma_v(x) := \begin{cases} \frac{\sum_{i \in J} v_i(x) \|V_i x - x\|^2}{\|\sum_{i \in J} v_i(x) V_i x - x\|^2} & \text{if } x \notin \bigcap_{i \in J} \text{Fix } V_i, \\ 1 & \text{if } x \in \bigcap_{i \in J} \text{Fix } V_i. \end{cases}$$

Since cutters are strictly quasi-nonexpansive (see Theorem 2.1.39), it follows from Theorem 2.1.26 that $\sum_{i \in J} v_i(x) V_i x \neq x$ for $x \notin \bigcap_{i \in J} \text{Fix } V_i$, consequently, σ_v is well defined. Furthermore, it follows from the convexity of the function $\|\cdot\|^2$ that $\sigma_v(x) \geq 1$ for all $x \in \mathcal{H}$.

Definition 5.10.1. Let $V_i : \mathcal{H} \to \mathcal{H}$, $i \in J$, be cutters with a common fixed point and $v : \mathcal{H} \to \Delta_{|J|}$ be a weight function which is appropriate with respect to the family $\mathcal{V} := \{V_i\}_{i \in J}$. We say that a step size function $\sigma : \mathcal{H} \to (0, +\infty)$ is α-admissible with respect to the family $\{V_i\}_{i \in J}$, where $\alpha \in (0, 1]$, or, shortly, admissible, if

$$\alpha \sigma_v(x) \leq \sigma(x) \leq \sigma_v(x) \tag{5.91}$$

for all $x \notin \bigcap_{i \in J} \text{Fix } V_i$.

5.10.2 Convergence Theorem

Theorem 5.10.2. *Suppose that:*

(a) $U_i : \mathcal{H} \to \mathcal{H}$, $i \in I$, are cutters with a common fixed point,
(b) $U_i - \text{Id}$, $i \in I$, are demi-closed at 0,
(c) $\mathcal{V}^k := \{V_i^k\}_{i \in J_k}$ are families of cutters $V_i^k : \mathcal{H} \to \mathcal{H}$, $i \in J_k$, with the properties

$$\bigcap_{i \in J_k} \text{Fix } V_i^k \supseteq \bigcap_{i \in I} \text{Fix } U_i \tag{5.92}$$

and

$$\max_{i \in J_k} \left\| V_i^k x - x \right\| \leq \gamma \max_{i \in I} \| U_i x - x \| \tag{5.93}$$

for all $x \in \mathcal{H}$, $k \geq 0$, and for some constant $\gamma > 0$.
(d) $\{v^k\}_{k=0}^{\infty} : \mathcal{H} \to \Delta_{|J_k|}$ *is a sequence of appropriate weight functions,*
(e) *The step size $\sigma_k : \mathcal{H} \to (0, +\infty)$ is α-admissible with respect to $\mathcal{V}^k$, $k \geq 0$, for some $\alpha \in (0, 1]$,*
(f) $\liminf_k \lambda_k (2 - \lambda_k) > 0$,
(g) $\{x^k\}_{k=0}^{\infty}$ *is generated by recurrence (5.89).*

Then:

(i) *If the sequence of weight functions $\{v^k\}_{k=0}^{\infty}$ applied to the sequence of families $\mathcal{V}^k$ is regular with respect to the family $\mathcal{U} := \{U_i\}_{i \in I}$, then x^k converges weakly to a common fixed point of U_i, $i \in I$.*
(ii) *If the sequence of weight functions $\{v^k\}_{k=0}^{\infty}$ applied to the sequence of families $\mathcal{V}^k$ contains a subsequence of weight functions which is regular with respect to the family $\mathcal{U} := \{U_i\}_{i \in I}$ and $\mathcal{H}$ is finite dimensional, then x^k converges to a common fixed point of U_i, $i \in I$.*

Proof. (cf. [70, Theorem 9.35]) Let $V^k : \mathcal{H} \to \mathcal{H}$ be defined by

$$V^k x := \sum_{i \in J_k} v_i^k(x) V_i^k x$$

and T_k be a generalized relaxation of V^k, i.e.,

$$T_k x := V^k_{\sigma_k, \lambda_k} x = x + \lambda_k \sigma_k(x)(V^k x - x). \tag{5.94}$$

The operators V^k are cutters (see Corollary 2.1.49). By Theorem 2.1.26, we have $\operatorname{Fix} T_k = \operatorname{Fix} V^k = \bigcap_{i \in J_k} \operatorname{Fix} V^k_i$ (note that a cutter is strictly quasi-nonexpansive by Theorem 2.1.39). By (5.92), $\bigcap_{k=0}^{\infty} \operatorname{Fix} T_k \supseteq \bigcap_{i \in I} \operatorname{Fix} U_i$. Let $\varepsilon \in (0, 1)$ and $k_0 \geq 0$ be such that $\lambda_k \in [\varepsilon, 2 - \varepsilon]$ for all $k \geq k_0$. By Theorem 4.9.1, the operator $V^k_{\sigma_k}$ is a cutter, consequently T_k is a λ_k-relaxed cutter. Theorem 2.1.39 implies

$$\left\| x^{k+1} - z \right\|^2 \leq \left\| x^k - z \right\|^2 - \frac{2 - \lambda_k}{\lambda_k} \left\| T_k x^k - x^k \right\|^2$$

for all $z \in \bigcap_{i=1}^{m} \operatorname{Fix} U_i$ and all $k \geq k_0$. Consequently, $\{x^k\}$ is bounded, $\left\| x^k - z \right\|$ is monotone and $\left\| T_k x^k - x^k \right\| \to 0$.

(i) Let $\beta > 0$, $k_1 \geq k_0$ and $j_k \in \{1, 2, \dots, J_k\}$ be such that

$$v^k_{j_k}(x) \left\| V^k_{j_k} x - x \right\|^2 \geq \beta \max_{i \in I} \left\| U_i x - x \right\|^2$$

for any $x \in \mathcal{H}$ and $k \geq k_1$. Since σ_k is α-admissible, the norm $\|\cdot\|$ is a convex function and $\left\| V^k_j x - x \right\| \leq \gamma \max_{i \in I} \left\| U_i x - x \right\|$ for all $j \in J_k$, we have

$$\left\| T_k x^k - x^k \right\| = \lambda_k \sigma_k(x^k) \left\| V^k x^k - x^k \right\|$$

$$\geq \lambda_k \alpha \frac{\sum_{i \in J_k} v^k_i(x^k) \left\| V^k_i x^k - x^k \right\|^2}{\left\| \sum_{i \in J_k} v^k_i(x^k) V^k_i x^k - x^k \right\|^2} \left\| V^k x^k - x^k \right\|$$

$$= \lambda_k \alpha \frac{\sum_{i \in J_k} v^k_i(x^k) \left\| V^k_i x^k - x^k \right\|^2}{\left\| \sum_{i \in J_k} v^k_i(x^k) V^k_i x^k - x^k \right\|}$$

$$\geq \lambda_k \alpha \frac{v^k_{j_k}(x^k) \left\| V^k_{j_k} x^k - x^k \right\|^2}{\sum_{i \in J_k} v^k_i(x^k) \left\| V^k_i x^k - x^k \right\|}$$

$$\geq \lambda_k \alpha \frac{\beta \max_{i \in I} \left\| U_i x^k - x^k \right\|^2}{\left(\sum_{i \in J_k} v^k_i(x^k) \right) \gamma \max_{i \in I} \left\| U_i x^k - x^k \right\|}.$$

$$= \frac{\varepsilon \alpha \beta}{\gamma} \max_{i \in I} \left\| U_i x^k - x^k \right\|$$

for all $k \geq k_0$. Consequently,

$$\left\| T_k x^k - x^k \right\| \geq \frac{\varepsilon \alpha \beta}{\gamma} \max_{i \in I} \left\| U_i x^k - x^k \right\| \tag{5.95}$$

and $\left\| U_i x^k - x^k \right\| \to 0$ for all $i \in I$. Therefore, condition (3.6) is satisfied for $U_k = T_k$ and $S = U_i$, $i \in I$. We have proved that all assumptions of Theorem 3.6.2 (i) are satisfied for $S = U_i$, $i \in I$. Therefore, x^k converges weakly to a common fixed point of U_i, $i \in I$.

(ii) Suppose that $\mathcal{H}$ is finite dimensional and $\{v^k\}_{k=0}^{\infty}$ contains a β-regular subsequence $\{v^{n_k}\}_{k=0}^{\infty}$. Let $\beta > 0$, $k_1 \geq k_0$ and $j_{n_k} \in I$ be such that

$$v_{j_{n_k}}^{n_k} (x) \left\| V_{j_{n_k}}^{n_k} x - x \right\|^2 \geq \beta \max_{i \in I} \left\| U_i x - x \right\|^2 .$$

Similarly as in (i) one proves that

$$\left\| T_{n_k} x^{n_k} - x^{n_k} \right\| \geq \frac{\varepsilon \alpha \beta}{\gamma} \max_{i \in I} \left\| U_i x^{n_k} - x^{n_k} \right\| .$$

Consequently, $\liminf_k \left\| U_i x^k - x^k \right\| = 0$ for all $i \in I$. If we take $U_k = T_k$ and $S = U_i$, $i \in I$, in Theorem 3.6.2 (ii), we obtain the convergence of x^k to a fixed point of U_i for all $i \in I$.   □

The operators T_k defined by (5.94) do not need to be continuous, because we have not assumed that the operators V_i^k as well as the weight functions v_i^k are continuous. This allows application of operators from an essentially broader family than the nonexpansive ones. We have supposed in Theorem 5.10.2 that $\{v^k\}_{k=0}^{\infty}$ is regular (in (i)) or that $\{v^k\}_{k=0}^{\infty}$ contains a regular subsequence (in (ii)). These assumptions seem not to be too restrictive. The assumption on the step sizes guarantees that T_k is a relaxed cutter. All these assumptions allow using algorithms which can treat the violated and the nonviolated constraints in a more flexible way and provide long steps.

Remark 5.10.3. In [119, Sect. 1] and [120, Sects. IV and V] Combettes presented three methods which are quite similar to ESCM: two extrapolated methods of parallel projections (EMOPP and EPPM2) and an extrapolated method of parallel subgradient projection (EMOPSP):

(i) In EMOPP (see [119, Definition 1.2]) it is supposed that $J_k = I$, $V_j^k = P_{C_i^k}$, where $P_{C_i^k}$ is the metric projection onto a closed convex superset C_i^k of C_i with

$$d(x^k, C_i^k) \geq \eta d(x^k, C_i), \tag{5.96}$$

for all $i \in I$, $k \geq 0$, and for some constant $\eta > 0$, and the weights $v^k \in \Delta_m$ satisfy condition (5.42), $k \geq 0$.

(ii) In EPPM2 (see [120, Sect. IV.C]) it is supposed that $J_k = I$, $V_j^k = P_{C_i^k}$, where $P_{C_i^k}$ is the metric projection onto a closed and convex superset C_i^k of C_i, $i \in I$,

$k \geq 0$, and $v \in \Delta_m$. The supersets C_i^k of C_i are chosen in such a way that the following condition is satisfied: for any subsequence $\{x^{n_k}\}_{k=0}^{\infty} \subseteq \{x^k\}_{k=0}^{\infty}$ and for all $i \in I$ it holds

$$(x^{n_k} \rightharpoonup x \text{ and } \lim_k \left\| P_{C_i^{n_k}} x^{n_k} - x^{n_k} \right\| = 0) \Longrightarrow x \in C_i. \qquad (5.97)$$

(iii) In EMOPSP (see [120, Sect. V]) it is supposed that $J_k \subseteq I$, $J_k \neq \emptyset$, $V_i^k = P_{f_i}$ are subgradient projections relative to a (lower semi-)continuous convex function $f_i : \mathcal{H} \to \mathbb{R}$,

$$v_i^k \geq \delta > 0 \text{ for } x_k \notin C_i, i \in J_k. \qquad (5.98)$$

The above methods EMOPP and EMOPSP are covered by ESCM, where V_j^k are cutters, $j \in J_k$, $k \geq 0$, satisfying (5.92) and (5.93) and $\{v^k\}_{k=0}^{\infty}$ is a sequence of appropriate weight functions. Note that (5.96) with v_i^k satisfying $v_i^k \geq \delta > 0$ for $x_k \notin C_i$, $i \in J_k$, implies that $\{v^k\}_{k=0}^{\infty}$ is a regular sequence of weight functions (see Example 5.8.3 (iv) and Definition 5.8.4). In EPPM2 condition (5.97) is slightly weaker (only in an infinite dimensional case) than the approximate regularity of v^k satisfying (5.98), but Theorem 5.10.2 enables application of more general sequences of weight functions than the ones satisfying (5.98). Furthermore, Theorem 5.10.2 does not require $C_i^k \supseteq C_i$, $i \in J_k$, $k \geq 0$. The theorem requires only the inclusion $\bigcap_{i \in J_k} C_i^k \supseteq C := \bigcap_{i \in I} C_i$ to be satisfied, $k \geq 0$.

5.10.3 *Extrapolated Simultaneous Subgradient Projection Method*

Let $c_i : \mathcal{H} \to \mathbb{R}$ be convex functions which are globally Lipschitz continuous on bounded subsets, $C_i := \{x \in \mathcal{H} : c_i(x) \leq 0\}$, $i \in I$ and $C := \bigcap_{i \in I} C_i \neq \emptyset$. The *extrapolated simultaneous subgradient projection method* (ESSPM) is a special case of an ESCM defined by recurrence (5.90), where U_i are subgradient projections, i.e., U_i are defined by (4.115), $i \in I$. By Corollary 4.2.6, Lemma 4.2.5 and by Theorem 4.2.7, U_i are cutters, Fix $U_i = C_i$ and $U_i - \mathrm{Id}$ are demi-closed at 0, $i \in I$. Let $v^k : \mathcal{H} \to \Delta_m$ be an appropriate weight function and a step size function $\sigma_k : \mathcal{H} \to (0, +\infty)$ be defined by

$$\sigma_k(x) := \frac{\sum_{i \in I} v_i^k(x) \left(\frac{(c_i(x))_+}{\|g_i(x)\|} \right)^2}{\left\| \sum_{i \in I} v_i^k(x) \frac{(c_i(x))_+}{\|g_i(x)\|^2} g_i(x) \right\|^2},$$

$k \geq 0$ (for simplicity, we use the convention that $\frac{(c_i(x))_+}{\|g_i(x)\|} = 0$ if $(c_i(x))_+ = 0$). The ESSPM is defined by

$$x^{k+1} = x^k - \lambda_k \sigma_k(x^k) \sum_{i \in I} v_i^k(x^k) \frac{(c_i(x^k))_+}{\|g_i(x^k)\|^2} g_i(x^k), \qquad (5.99)$$

where $x^0 \in \mathcal{H}$ is arbitrary and $\lambda_k \in (0, 2)$, $k \geq 0$. We can also write

$$x^{k+1} = U_{\sigma_k, \lambda_k}^k x^k,$$

where

$$U_{\sigma_k, \lambda_k}^k := \mathrm{Id} + \lambda_k \sigma_k (U^k - \mathrm{Id})$$

is a generalized relaxation of the simultaneous subgradient projection

$$U^k := \sum_{i \in I} v_i^k P_{c_i},$$

$k \geq 0$.

Theorem 5.10.4. *Let $x^0 \in \mathcal{H}$ be arbitrary and x^k be generated by recurrence (5.99), where $\liminf_k \lambda_k (2 - \lambda_k) > 0$.*

(i) *If $\{v^k\}_{k=0}^\infty$ is regular, then x^k converges weakly to a common fixed point of U_i, $i \in I$.*

(ii) *If $\{v^k\}_{k=0}^\infty$ contains a subsequence of weight functions which is regular and $\mathcal{H}$ is finite dimensional, then x^k converges to a common fixed point of U_i, $i \in I$.*

Proof. The step sizes σ_k are 1-admissible. Therefore, the theorem follows from Theorem 5.10.2.   □

The ESSPM was studied by Dos Santos in [146], where positive constant weights $w \in \mathrm{ri}\, \Delta_m$ were considered and the convergence in the finite dimensional case was proved [146, Sect. 5]. Furthermore, the method was studied by Combettes in [120, Sect. V] (cf. Remark 5.10.3 (iii)).

5.11   Extrapolated Cyclic Cutter Method

Let $U_i : \mathcal{H} \to \mathcal{H}$ be cutters with a common fixed point, $U = U_m U_{m-1} \ldots U_1$ be a cyclic cutter and $x^0 \in \mathcal{H}$. Consider sequences $\{x^k\}_{k=0}^\infty$ generated by the recurrence

$$x^{k+1} = x^k + \lambda_k \sigma_k(x^k)(Ux^k - x^k), \qquad (5.100)$$

where $\lambda_k \in (0, 2]$, the step size functions $\sigma_k : \mathcal{H} \to (0, +\infty)$ satisfy the inequality

$$\alpha \leq \sigma_k(x) \leq \sigma_{\max}(x) \qquad (5.101)$$

for all $x \in \mathcal{H}$ with $\alpha \in (0, \frac{1}{2m}]$,

$$\sigma_{\max}(x) := \frac{\sum_{i \in I} \langle Ux - S_{i-1}x, S_i x - S_{i-1}x \rangle}{\|Ux - x\|^2},$$

where $S_i := U_i U_{i-1} \dots U_1$ for $i = 1, 2, \dots, m$ and $S_0 := \mathrm{Id}$ (cf. (4.121)). The existence of the step size $\sigma_k(x)$ follows from inequality (4.122). Since

$$m \frac{\sum_{i \in I} \|S_i x - S_{i-1} x\|^2}{\|Ux - x\|^2} \geq 1$$

(see again (4.122)) we can use in particular step size functions σ_k satisfying the inequality

$$m\alpha \frac{\sum_{i \in I} \|S_i x - S_{i-1} x\|^2}{\|Ux - x\|^2} \leq \sigma_k(x) \leq \sigma_{\max}(x) \qquad (5.102)$$

for all $x \in \mathcal{H}$. We can also write recurrence (5.100) in the form $x^{k+1} = U_{\sigma_k, \lambda_k} x^k$, where $U_{\sigma_k, \lambda_k} := \mathrm{Id} + \lambda_k \sigma_k (U - \mathrm{Id})$ is a generalized relaxation of the cyclic cutter U with the step size function σ_k satisfying inequalities (5.101).

5.11.1 Convergence

In [71] one can find a slightly weaker version of the following theorem.

Theorem 5.11.1. *Let $\{x^k\}_{k=0}^\infty$ be generated by (5.100), where $\sigma_k(x)$ satisfies (5.101) and let $z \in \mathrm{Fix}\, U$. Then the following inequality holds*

$$\left\|x^{k+1} - z\right\|^2 \leq \left\|x^k - z\right\|^2 - \lambda_k(2 - \lambda_k)\alpha^2 \left\|Ux^k - x^k\right\|^2. \qquad (5.103)$$

Moreover:

(i) *If $U - \mathrm{Id}$ is demi-closed at 0 and $\liminf_k \lambda_k(2 - \lambda_k) > 0$, then $x^k \rightharpoonup x^* \in \mathrm{Fix}\, U$.*

(ii) *If the step size function σ_k satisfies inequalities (5.102), then*

$$\left\|x^{k+1} - z\right\|^2 \leq \left\|x^k - z\right\|^2 - \lambda_k(2 - \lambda_k)m^2\alpha^2 \frac{(\sum_{i=1}^m \|S_i x^k - S_{i-1}x^k\|^2)^2}{\|Ux^k - x^k\|^2}. \qquad (5.104)$$

If, furthermore, $U_i - \mathrm{Id}$ are demi-closed at 0, $i = 1, 2, \dots, m$, and $\liminf_k \lambda_k (2 - \lambda_k) > 0$, then $x^k \rightharpoonup x^* \in \mathrm{Fix}\, U$.

Proof. Since a cutter is strictly quasi-nonexpansive (see Theorem 2.1.39), we have $\mathrm{Fix}\, U = \bigcap_{i \in I} \mathrm{Fix}\, U_i$. Note that $U_{\sigma_k, \lambda_k} x^k - x^k = \lambda_k \sigma_k(x^k)(Ux^k - x^k)$.

By Theorem 4.10.2, the operator U_{σ_k} is a cutter and

$$
\begin{aligned}
\left\| x^{k+1} - z \right\|^2 = \left\| U_{\sigma,\lambda} x^k - z \right\|^2 &\leq \left\| x^k - z \right\|^2 - \frac{2 - \lambda_k}{\lambda_k} \left\| U_{\sigma_k,\lambda_k} x^k - x^k \right\|^2 \\
&= \left\| x^k - z \right\|^2 - \lambda_k (2 - \lambda_k) \sigma_k^2(x^k) \left\| U x^k - x^k \right\|^2 \\
&\leq \left\| x^k - z \right\|^2 - \lambda_k (2 - \lambda_k) \alpha^2 \left\| U x^k - x^k \right\|^2 .
\end{aligned}
$$

Therefore, $\left\| x^k - z \right\|$ is decreasing, x^k is bounded and

$$
\left\| U x^k - x^k \right\| \to 0. \tag{5.105}
$$

Let $x^* \in \mathcal{H}$ be a weak cluster point of the sequence $\{x^k\}_{k=0}^\infty$ and $\{x^{n_k}\}_{k=0}^\infty \subseteq \{x^k\}_{k=0}^\infty$ be a subsequence which converges weakly to x^*.

(i) Suppose that $U - \mathrm{Id}$ is demi-closed at 0. Condition (5.105) yields $x^* \in \mathrm{Fix}\, U$. The weak convergence of the whole sequence $\{x^k\}_{k=0}^\infty$ to x^* follows now from Lemma 3.3.3.

(ii) Suppose that the step size function σ_k satisfies inequality (5.102). Then inequality (5.104) can be proved similarly to (5.103). Consequently,

$$
\left\| S_i x^k - S_{i-1} x^k \right\| \to 0 \tag{5.106}
$$

for $i = 1, 2, \ldots, m$, because $\liminf_k \lambda_k (2 - \lambda_k) > 0$ and $U x^k - x^k$ is bounded. Suppose that $U_i - \mathrm{Id}$ are demi-closed at 0, $i = 1, 2, \ldots, m$. Condition (5.106) for $i = 1$ yields

$$
\left\| U_1 x^{n_k} - x^{n_k} \right\| = \left\| S_1 x^{n_k} - S_0 x^{n_k} \right\| \to 0.
$$

Due to the demi-closedness of $U_1 - \mathrm{Id}$ at 0, we have $U_1 x^* = x^*$. Since

$$
\left\| (U_1 x^{n_k} - U_1 x^*) - (x^{n_k} - x^*) \right\| = \left\| U_1 x^{n_k} - x^{n_k} \right\| \to 0
$$

and $x^{n_k} \rightharpoonup x^*$, we have $U_1 x^{n_k} \rightharpoonup U_1 x^* = x^*$. By (5.106) for $i = 2$,

$$
\left\| U_2 (U_1 x^{n_k}) - U_1 x^{n_k} \right\| = \left\| S_2 x^{n_k} - S_1 x^{n_k} \right\| \to 0.
$$

Since $U_2 - \mathrm{Id}$ is demi-closed at 0 and $U_1 x^{n_k} \rightharpoonup x^*$, $U_2 x^* = x^*$. In a similar way we obtain that $U_i x^* = x^*$ for $i = 3, \ldots, m$. Therefore,

$$
U x^* = U_m \ldots U_1 x^* = x^*.
$$

We conclude that the subsequence $\{x^{n_k}\}_{k=0}^\infty$ converges weakly to a fixed point of the operator U. The weak convergence of the whole sequence $\{x^k\}_{k=0}^\infty$ to $x^* \in \mathrm{Fix}\, U$ follows now from Lemma 3.3.3. $\square$

Even if we take $\sigma := \sigma_{\max}$ and $\lambda = 2$, the generalized relaxation $U_{\sigma,\lambda}$ needs not to be an extrapolation of U because $\sigma_{\max}$ needs not to be greater or equal to $\frac{1}{2}$. We only know that $\sigma_{\max} \geq \frac{1}{2m}$ (see (4.122)). Note, however, that U is $\frac{1}{m}$-SQNE (see Corollary 4.5.15), consequently, it is a $\frac{2m}{m+1}$-relaxed cutter (see Corollary 2.1.43) and $U_{\frac{1+m}{2m}}$ is a cutter (see Remark 2.1.3). Since $U_{\sigma_{\max}}$ is a cutter (see Theorem 4.10.2), Lemma 2.4.7 (i) yields that U_{σ} is also a cutter, where $\sigma = \max\{\frac{1+m}{2m}, \sigma_{\max}\}$. Therefore, $U_{\sigma,\lambda}$ with $\lambda \in (\frac{2m}{1+m}, 2]$ is an extrapolation of U.

Interesting schemes for accelerated cyclic projections for subspaces of $\mathcal{H}$ can be found in [177] and in [141].

5.11.2   Accelerated Kaczmarz Method for a System
of Linear Equations

Consider a consistent system of linear equations

$$\langle a_i, x \rangle = \beta_i, i \in I, \tag{5.107}$$

where $a_i \in \mathcal{H}$, $a_i \neq 0$ and $\beta_i \in \mathbb{R}$, $i \in I$. Suppose for simplicity that $\|a_i\| = 1$. Let $C_i := H(a_i, \beta_i) = \{x \in \mathcal{H} : \langle a_i, x \rangle = \beta_i\}$, $i \in I$. By assumption, $C := \bigcap_{i \in I} C_i \neq \emptyset$. Let $U_i := P_{C_i}$, $i \in I$. We have $U_i x = x - (\langle a_i, x \rangle - \beta_i)a_i$, $i \in I$ (see (4.1)). Of course, Fix $U_i = C_i$ and U_i a cutter as the metric projection (see Theorem 2.2.21 (ii)), $i \in I$. Since the metric projection is strictly quasi-nonexpansive (see Corollary 2.2.24), Theorem 2.1.26 yields Fix $U = \bigcap_{i \in I}$ Fix $U_i = \bigcap_{i \in I} C_i = C$. We apply method (5.100) with $\sigma_k := \sigma_{\max}$ and $\lambda_k = 1$ for all $k \geq 0$ to system (5.107). By inequality (5.103), any sequence generated by (5.100) is Fejér monotone with respect to C. By (4.123),

$$\sigma_{\max}(x) = \frac{\sum_{i=1}^{m} \langle S_i x - x, S_i x - S_{i-1} x \rangle}{\|Ux - x\|^2}.$$

Denote $u^0 := x$ and $u^i := U_i u^{i-1} = S_i x$, $i = 1, 2, \ldots, m$. Since $\langle U_i y, a_i \rangle = \beta_i$ for any $y \in \mathcal{H}$ and for all $i \in I$, we have

$$\langle S_i x, a_i \rangle = \langle U_i u^{i-1}, a_i \rangle = \beta_i,$$

$i \in I$, consequently,

$$\sum_{i=1}^{m} \langle S_i x - x, S_i x - S_{i-1} x \rangle = \sum_{i=1}^{m} \langle S_i x - x, U_i u^{i-1} - u^{i-1} \rangle$$

$$= \sum_{i=1}^{m} \langle S_i x - x, (\beta_i - \langle a_i, u^{i-1} \rangle) a_i \rangle$$

$$= \sum_{i=1}^{m} (\beta_i - \langle a_i, u^{i-1} \rangle)(\langle S_i x, a_i \rangle - \langle x, a_i \rangle)$$

$$= \sum_{i=1}^{m} (\beta_i - \langle a_i, x \rangle)(\beta_i - \langle a_i, u^{i-1} \rangle)$$

and we obtain the following form for the step size

$$\sigma_{\max}(x) = \frac{\sum_{i=1}^{m} (\beta_i - \langle a_i, x \rangle)(\beta_i - \langle a_i, u^{i-1} \rangle)}{\|Ux - x\|^2}, \tag{5.108}$$

where $x \notin \text{Fix } U$. It follows from Theorem 5.11.1 that for any $x^0 \in \mathcal{H}$, the sequence $\{x^k\}_{k=0}^{\infty}$ generated by the recurrence $x^{k+1} = U_{\sigma_k} x^k$ with $\sigma_k = \sigma_{\max}$ is Fejér monotone with respect to C and converges weakly to a solution of the system (5.107). Moreover, by Theorem 3.8.4, the sequence converges strongly.

Remark 5.11.2. For any $u \in \mathcal{H}$ and $z \in C_i$, we have $\langle U_i u - u, z - u \rangle = \|U_i u - u\|^2$. Therefore, it follows from the proof of Lemma 4.10.1 that in this case the first inequality in (4.119) is, actually, an equality. Consequently, we have an equality in (4.124) and the operator U_σ with the step size function given by (4.123) has the property $\langle U_\sigma x - x, z - U_\sigma x \rangle = 0$ for all $z \in C$, or, equivalently, $\langle U_\sigma x - x, z - x \rangle = \|U_\sigma x - x\|^2$. Consequently, the step size function $\sigma := \sigma_{\max}$ is optimal in the following sense:

$$\|U_\sigma x - z\| = \min_\alpha \|x + \alpha(Ux - x) - z\|.$$

One can expect that this property leads in practice to an acceleration of the convergence to a solution of the system (5.107), of sequences generated by the recurrence $x^{k+1} = U_{\sigma_{\max}} x^k$ in comparison to the classical Kaczmarz method.

A different acceleration scheme for the Kaczmarz method was presented in [247, Sect. 3.1].

5.12 Surrogate Constraints Methods

In this section we consider a consistent system of linear inequalities $Ay \leq b$, where A is a matrix of type $m \times n$ and $b \in \mathbb{R}^m$. Let $a_i \in \mathbb{R}^n$ denote the ith row of the matrix A and $\beta_i \in \mathbb{R}$ denote the ith coordinate of b, i.e., $A = [a_1, \ldots, a_m]^\top$ and $b = [\beta_1, \ldots, \beta_m]^\top$. Suppose, for simplicity, that $\|a_i\| = 1$, $i \in I$. Denote $C_i := H_-(a_i, \beta_i) = \{y \in \mathbb{R}^n : a_i^\top y \leq \beta_i\}$, $i \in I$, and $C := \bigcap_{i \in I} C_i$. The *surrogate constraints method* (SCM) is described by the recurrence

$$x^{k+1} = x^k - \lambda_k \frac{v^k(x^k)^\top (Ax^k - b)}{\|A^\top v^k(x^k)\|^2} A^\top v^k(x^k), \tag{5.109}$$

where $x^0 \in \mathbb{R}^n$, $\lambda_k \in [0, 2]$ and $\{v^k\}_{k=0}^\infty$ is a sequence of essential weight functions, i.e., weight functions $v^k : \mathbb{R}^n \to \mathbb{R}^m \backslash \{0\}$ satisfying (4.108) for any $x \notin C$. It follows from Lemma 4.9.5 that recurrence (5.109) is well defined. We can also write

$$x^{k+1} = S_{v^k, \lambda_k} x^k,$$

where $\{S_{v^k}\}_{k=0}^\infty$ is a sequence of surrogate projection operators $S_{v^k} : \mathbb{R}^n \to \mathbb{R}^n$ defined by (4.109) and $S_{v^k, \lambda_k} := \mathrm{Id} + \lambda_k (S_{v^k} - \mathrm{Id})$ is the λ_k-relaxation of the operator S_{v^k}, $k \geq 0$. Recall that $S_{v^k} x^k$ is the projection of x^k onto a surrogate constraint $H_k := H_-(a^k(x^k), \beta_k(x^k))$, where $a^k(x^k) := A^\top v^k(x^k)$ and $\beta_k(x^k) := v^k(x^k)^\top b$ (see Sect. 4.9.4) and that S_{v^k} is a cutter with $\mathrm{Fix}\, S_{v^k} = C$ (see Theorem 4.9.6).

Various versions of the surrogate projection method were studied by Merzlyakov [260], Oko [277], Yang and Murty [351], Kiwiel [229–231], Kiwiel and Łopuch [233], Cegielski [67, 68] and by Dudek [150].

5.12.1 Proper Control

Definition 5.12.1. Let a sequence $\{x^k\}_{k=0}^\infty$ be generated by the surrogate constraints method (5.109). We say that a sequence of weight functions (or the control) $\{v^k\}_{k=0}^\infty$ is *proper* if

$$\lim_k v^k(x^k)^\top (Ax^k - b) = 0 \Longrightarrow \liminf_k \|(Ax^k - b)_+\| = 0 \qquad (5.110)$$

Now we apply Definition 5.8.5 (ii) to $\mathcal{V}^k := \mathcal{V} = \mathcal{U} = \{P_{C_i}\}_{i \in I}$ with $C_i = H_-(a_i, \beta_i)$, $i \in I$. We see that $\{v^k\}_{k=0}^\infty$ is approximately semi-regular if and only if there exists a sequence $\{i_k\}_{k=0}^\infty \subseteq I$ such that the following implication holds

$$\lim_k v_{i_k}^k (x^k)^\top [(a_{i_k}^\top x^k - \beta_{i_k})_+]^2 = 0 \Longrightarrow \liminf_k \max_{i \in I} (a_i^\top x^k - \beta_i)_+ = 0 \quad (5.111)$$

It follows from Remark 5.8.6 (i) and from the equivalence of all norms in $\mathbb{R}^n$ that condition (5.111) is equivalent to

$$\lim_k v_{i_k}^k (x^k)^\top (a_{i_k}^\top x^k - \beta_{i_k})_+ = 0 \Longrightarrow \liminf_k \|(Ax^k - b)_+\| = 0. \qquad (5.112)$$

If v^k considers only violated constraints (see Definition 4.9.8), then

$$v^k(x^k)^\top (Ax^k - b) = v^k(x^k)^\top (Ax^k - b)_+ = \sum_{i \in I} v_i^k (x^k)^\top (a_i^\top x^k - \beta_i)_+.$$

Therefore, the assumption that a sequence of weight functions $\{v^k\}_{k=0}^\infty$ which consider only violated constraints is approximately semi-regular is stronger than

(5.110). This fact follows from a simple observation that if a sum of positive quantities goes to zero, then all summands go to zero.

If the weight functions v^k consider only violated constraints, $k \geq 0$, then it follows from Proposition 4.9.11 that recurrence (5.109) defines an extrapolated simultaneous projection method. In this case one can apply Theorem 5.10.2 in order to prove the convergence of a sequence generated by (5.109). In the next subsection we prove that the convergence follows without assumptions that v^k considers only violated constraints and that the corresponding sequence of weight functions $\{w^k\}_{k=0}^{\infty}$ (see (4.112)) is approximately semi-regular. It suffices to suppose instead that $\{v^k\}_{k=0}^{\infty}$ is proper.

5.12.2 Convergence Theorem

Theorem 5.12.2. Let $x^0 \in \mathbb{R}^n$ and a sequence $\{x^k\}_{k=0}^{\infty}$ be generated by the surrogate constraints method (5.109), where $\liminf_k \lambda_k (2 - \lambda_k) > 0$ and the sequence of weight functions $\{v^k\}_{k=0}^{\infty}$ is proper. Then x^k converges to a solution x^* of the system $Ax \leq b$.

Proof. The operator S_{v^k} is a cutter and $\text{Fix } S_{v^k} = C$, $k \geq 0$ (see Theorem 4.9.6). Therefore, equality (4.109) and Theorem 2.1.39 yield

$$\left\| x^{k+1} - z \right\|^2 \leq \left\| x^k - z \right\|^2 - \lambda_k (2 - \lambda_k) \left[\frac{v^k(x^k)^\top (Ax^k - b)}{\| A^\top v^k(x^k) \|} \right]^2,$$

for any $z \in C$. Consequently, $\| x^k - z \|$ converges as a decreasing sequence, x^k is bounded and

$$\lim_k \frac{v^k(x^k)^\top (Ax^k - b)}{\| A^\top v^k(x^k) \|} = 0.$$

We can suppose without loss of generality that $v^k(x^k) \in \Delta_m$. Since x^k is bounded, $A^\top v^k(x^k)$ is also bounded. Therefore, $\lim_k v^k(x^k)^\top (Ax^k - b) = 0$. By (5.110), we have

$$\liminf_k \sum_{i=1}^{m} [(a_i^\top x^k - \beta_i)_+]^2 = \liminf_k \left\| (Ax^k - b)_+ \right\|^2 = 0.$$

This implies

$$\liminf_k \| P_{C_i} x^k - x^k \| = \liminf_k (a_i^\top x^k - \beta_i)_+ = 0$$

for all $i \in I$. Corollary 3.7.1 (iii) for $X = \mathbb{R}^n$, $T_k := S_{v^k}$ and $S := P_{C_i}$ yields now the convergence of x^k to $x^* \in \text{Fix } P_{C_i} = C_i$ for any $i \in I$, i.e., $x^* \in C$. □

5.12.3 Examples of Proper Control

Now we give examples of proper sequences of weight functions $v^k : \mathbb{R}^n \to \mathbb{R}^m_+ \setminus \{0\}$. We suppose for simplicity that A has normalized rows, i.e., $\|a_i\| = 1$, $i \in I$.

Example 5.12.3. Let $\{x^k\}_{k=0}^{\infty}$ be generated by the surrogate constraints method (5.109), where $\{v^k\}_{k=0}^{\infty}$ is a sequence of weight functions $v^k : \mathbb{R}^n \to \mathbb{R}^m_+ \setminus \{0\}$ defined by $v^k(x) := W^k(x)^\top (Ax - b)_+$, $x \in \mathbb{R}^n$, $k \geq 0$, with $W^k := \operatorname{diag} w^k$ and $\{w^k\}_{k=0}^{\infty}$ being a sequence of weight functions $w^k : \mathbb{R}^n \to \Delta_m$ which are approximately semi-regular with respect to $\{P_{C_i}\}_{i \in I}$ with $C_i := H_-(a_i, \beta_i)$, $i \in I$. It is clear that v^k considers only violated constraints, consequently, v^k is essential (see Lemma 4.9.9). We show that the sequence $\{v^k\}_{k=0}^{\infty}$ is proper. We have

$$v^k(x^k)^\top (Ax^k - b) = (Ax^k - b)_+^\top W^k(x^k)(Ax^k - b)$$
$$= \sum_{i \in I} \omega_i^k(x^k)[(a_i^\top x^k - \beta_i)_+]^2$$

If $\lim_k v^k(x^k)^\top (Ax^k - b) = 0$, then, for any sequence $\{i_k\}_{k=0}^{\infty} \subseteq I$, we have

$$\lim_k \omega_{i_k}^k(x^k)[(a_{i_k}^\top x^k - \beta_{i_k})_+]^2 = 0.$$

Since $\{v^k\}_{k=0}^{\infty}$ is approximately semi-regular, $\liminf_k \|P_{C_i} x^k - x^k\| = 0$, i.e., $\liminf_k (a_i^\top x^k - \beta_i)_+ = 0$ for all $i \in I$, or, equivalently, $\liminf_k \|(Ax^k - b)_+\| = 0$. In particular, a constant sequence $v^k := v$ is proper, where $v : \mathbb{R}^n \to \mathbb{R}^m_+ \setminus \{0\}$ is defined by

$$v(x) := W(x)(Ax - b)_+,$$

for $x \notin C$, $W(x) := \operatorname{diag} w(x)$ and w has one of the following forms:

(i) w is constant with positive weights, i.e., $w(x) =: w \in \operatorname{ri} \Delta_m$ for all $x \in \mathbb{R}^n$. Note that w is regular (see Example 5.8.3 (i)), consequently, it is approximately regular (see Lemma 5.8.7 (i)). If we take $w = (\frac{1}{m}, \frac{1}{m}, \ldots, \frac{1}{m})$, then we obtain $v_i(x) = \frac{1}{m}(a_i^\top x - \beta_i)_+$, i.e., the weights $v_i(x)$ are proportional to the residua of violated constraints. Since the rescaling of weights does not change the operator S_v defined by (4.109), we can apply, equivalently, $v_i(x) := (a_i^\top x - \beta_i)_+$ or $v_i(x) := \frac{(a_i^\top x - \beta_i)_+}{\sum_{j \in I}(a_j^\top x - \beta_j)_+}$. The latter weights were proposed by Yang and Murty [351, page 167].

(ii) w considers only almost violated constraints, i.e., $\omega_j(x) = 0$ for all $j \in I$ such that

$$(a_j^\top x - \beta_j)_+ < \gamma \max_{i \in I}(a_i^\top x - \beta_i)_+,$$

where $\gamma \in (0, 1]$. Note that w is approximately regular (see Example 5.8.3 (v) and Lemma 5.8.7 (i)). If we take $\gamma = 1$, then the surrogate constraints method reduces to a successive projection method with the remotest set control.

Example 5.12.4. Let $\{x^k\}_{k=0}^{\infty}$ be generated by the surrogate constraints method (5.109), where $\{v^k\}_{k=0}^{\infty}$ is a sequence containing a subsequence $\{v^{n_k}\}_{k=0}^{\infty}$ of weight functions which consider only almost remotest constraints, i.e., $v_i^{n_k}(x) = 0$ for all $i \notin J_\gamma(x)$, where

$$J_\gamma(x) := \{j \in I : (a_j^\top x - \beta_j)_+ \geq \gamma \max_{i \in I}(a_i^\top x - \beta_i)_+\}$$

$x \in \mathbb{R}^n$, with $\gamma \in (0, 1]$. We show that $\{v^k\}_{k=0}^{\infty}$ is proper. Suppose for simplicity that $v^k(x) \in \Delta_m$, $x \in \mathbb{R}^n$. Let $j_{n_k} = j_{n_k}(x^{n_k}) \in J_\gamma(x^{n_k})$ be such that $v_{j_{n_k}}^{n_k}(x^{n_k}) \geq \frac{1}{m}$, $k \geq 0$. The existence of such j follows from the fact that $v^{n_k}(x^{n_k}) \in \Delta_m$. We have

$$v^{n_k}(x^{n_k})^\top (Ax^{n_k} - b) = \sum_{i \in I} v_i^{n_k}(x^{n_k})(a_i^\top x^{n_k} - \beta_i)$$

$$= \sum_{i \in J_\gamma(x^{n_k})} v_i^{n_k}(x^{n_k})(a_i^\top x^{n_k} - \beta_i)_+$$

$$\geq v_{j_{n_k}}^{n_k}(x^{n_k})(a_{j_{n_k}}^\top x^{n_k} - \beta_{j_{n_k}})_+$$

$$\geq \frac{1}{m}\gamma \max_{i \in I}(a_i^\top x^{n_k} - \beta_i)_+.$$

Therefore, $\lim_k v^k(x^k)^\top (Ax^k - b) = 0$ yields $\liminf_k \max_{i \in I}(a_i^\top x^k - \beta_i)_+ = 0$ which is equivalent to $\liminf_k \|(Ax^k - b)_+\| = 0$. Note that the successive projection method with the remotest set control is a special case of the method with $\gamma = 1$.

Example 5.12.5. Let $\{x^k\}_{k=0}^{\infty}$ be generated by the surrogate constraints method (5.109), where $\{v^k\}_{k=0}^{\infty}$ is a sequence of essential weight functions containing a subsequence $\{v^{n_k}\}_{k=0}^{\infty}$ satisfying the following condition

$$v_i^{n_k}(x^{n_k}) \begin{cases} \geq \delta \text{ for some } i \in \text{Argmax}_{i \in I}(a_i^\top x^{n_k} - \beta_i) \\ = 0 \text{ for } i \notin I(x^{n_k}), \end{cases} \tag{5.113}$$

for all $x^{n_k} \notin C$, where $I(x) := \{i \in I : a_i^\top x > \beta_i\}$, $k \geq 0$, and $\delta > 0$ is a small predefined constant. We show that $\{v^k\}_{k=0}^{\infty}$ is proper. Let $x^{n_k} \notin C$ and $j_{n_k} \in I(x^{n_k})$ be such that $v_i^{n_k}(x^{n_k}) \geq \delta$ and

$$(a_{j_{n_k}}^\top x^{n_k} - \beta_{j_{n_k}})_+ = \max_{i \in I}(a_i^\top x^{n_k} - \beta_i)_+.$$

Then we have

$$v^{n_k}(x^{n_k})^\top (Ax^{n_k} - b) = \sum_{i \in I} v_i^{n_k}(x^{n_k})(a_i^\top x^{n_k} - \beta_i)$$

$$= \sum_{i \in I(x^{n_k})} v_i^{n_k}(x^{n_k})(a_i^\top x^{n_k} - \beta_i)_+$$

$$\geq v_{j_{n_k}}^{n_k}(x^{n_k})(a_{j_{n_k}}^\top x^{n_k} - \beta_{j_{n_k}})_+$$

$$\geq \delta \max_{i \in I}(a_i^\top x^{n_k} - \beta_i)_+.$$

Suppose that $\lim_k v^k(x^k)^\top(Ax^k - b) = 0$. Then $\liminf_k \max_{i \in I}(a_i^\top x^k - \beta_i)_+ = 0$ which is equivalent to $\liminf_k \|(Ax^k - b)_+\| = 0$, i.e., $\{v^k\}_{k=0}^\infty$ is proper.

Remark 5.12.6. Yang and Murty proved the finite convergence of sequences $\{x^k\}_{k=0}^\infty$ generated by the surrogate constraints method (5.109) with weight functions v^k satisfying the condition

$$v_i^k(x^k) \begin{cases} \geq \delta & \text{for } i \in I_\varepsilon(x^k) \\ = 0 & \text{for } i \notin I_\varepsilon(x^k), \end{cases} \tag{5.114}$$

for predefined small constants $\varepsilon, \delta > 0$, where $I_\varepsilon(x) := \{i \in I : a_i^\top x^k - \beta_i > \varepsilon\}$ and $x^k \notin C_\varepsilon := \bigcap_{i \in I}\{x \in \mathbb{R}^n : a_i^\top x - \beta_i > \varepsilon\}$ (see [351, Theorem 3.3]) Note that if a sequence $\{v^k\}_{k=0}^\infty$ satisfies condition (5.114), then any subsequence $\{v^{n_k}\}_{k=0}^\infty$ satisfies condition (5.113) for all $x^{n_k} \notin C_\varepsilon$. We leave it to the reader to check that the result of [351, Theorem 3.3] can be derived from Example 5.12.5.

5.13 SCM with Residual Selection

In this section we present surrogate constraints methods for solving a consistent system of linear inequalities $Au \leq b$, which employ a special kind of surrogate projection operators. For any $x \notin C$ the surrogate projection $S_v x$ given by (4.109) is determined by an essential weight $v = v(x)$ (see Definition 4.9.4) related to a constraints selection $L := L(x) \subseteq I := \{1, 2, \ldots, m\}$. We use the same notation as in Sect. 5.12.

5.13.1 General Properties

We start with some relationships between surrogate projections and metric projections onto solution sets of equality and inequality systems. Let $x \notin C$ be a current approximation of the solution, $L := L(x) \subseteq I$ be a nonempty subset of indices (a constraints selection) and let and $r := |L|$. Let A_L denote the submatrix of A with rows a_i, $i \in L$, and b_L—the subvector of b with coordinates β_i, $i \in L$. Further, denote

$$C_L := \bigcap_{i \in L} C_i = \{u \in \mathbb{R}^n : A_L u \leq b_L\}$$

and

$$H_L := \{u \in \mathbb{R}^n : A_L u = b_L\}.$$

Suppose, for simplicity, that $L = \{1, 2, \ldots, r\}$, i.e.,

$$A = \begin{bmatrix} A_L \\ A_{I \setminus L} \end{bmatrix}, b = \begin{bmatrix} b_L \\ b_{I \setminus L} \end{bmatrix} \text{ and } v(x) = \begin{bmatrix} v_L(x) \\ v_{I \setminus L}(x) \end{bmatrix}.$$

Denote

$$S_{v_L} x := x - \frac{v_L(x)^\top (A_L x - b_L)}{\|A_L^\top v_L(x)\|^2} A_L^\top v_L(x)$$

(cf. (4.109)).

Lemma 5.13.1. *Let $x \notin C$ and $L := L(x) \subseteq I$ be such that A_L has full row rank. Then $P_{C_L} x = P_{H_L} x$ if and only if*

$$(A_L A_L^\top)^{-1}(A_L x - b_L) \geq 0.$$

Proof. Consider the following convex minimization problem

$$
\begin{aligned}
&\text{minimize } \tfrac{1}{2}\|u - x\|^2 \\
&\text{subject to } A_L u = b_L \\
&\quad u \in \mathbb{R}^n
\end{aligned}
\tag{5.115}
$$

whose solution is $\bar{u} = P_{H_L} x$. The corresponding KKT-point $(\bar{u}, \bar{y}) \in \mathbb{R}^n \times \mathbb{R}^r$ has the form

$$\bar{u} = x - A_L^\top (A_L A_L^\top)^{-1}(A_L x - b_L)$$
$$\bar{y} = (A_L A_L^\top)^{-1}(A_L x - b_L)$$

(i) Suppose that $(A_L A_L^\top)^{-1}(A_L x - b_L) \geq 0$. Then $(\bar{u}, \bar{y})$ is also a KKT-point related to the convex minimization problem

$$
\begin{aligned}
&\text{minimize } \tfrac{1}{2}\|u - x\|^2 \\
&\text{subject to } A_L u \leq b_L \\
&\quad u \in \mathbb{R}^n
\end{aligned}
\tag{5.116}
$$

whose solution is $u^* = P_{C_L}$. Therefore, $\bar{u} = u^*$.

(ii) Suppose that $\bar{u} = P_{H_L} x = P_{C_L} x$, i.e.,

$$\bar{u} = x - A_L^\top (A_L A_L^\top)^{-1}(A_L x - b_L).$$
$$\tag{5.117}$$

Let y^* be a vector of Lagrange multipliers for problem (5.116). Then $(\bar{u}, y^*)$ satisfies the related KKT-system, consequently,

$$\bar{u} = x - A_L^\top y^* = 0.$$

This equality together with (5.117) give

$$A_L^\top y^* = A_L^\top (A_L A_L^\top)^{-1}(A_L x - b_L).$$

Since A_L has full row rank, we have

$$y^* = (A_L A_L^\top)^{-1}(A_L x - b_L)$$

which is nonnegative, because y^* is a vector of Lagrange multipliers for problem (5.116).   □

Lemma 5.13.2. *Let $x \notin C$, $L := L(x) \subseteq I$ be such that A_L has full row rank and $v_L := v_L(x) \in \mathbb{R}^r_+$ be essential for the system $A_L u \le b_L$ i.e., $v_L^\top(A_L x - b_L) > 0$. Then $S_{v_L} x = P_{H_L} x$ if and only if*

$$v_L = \alpha(A_L A_L^\top)^{-1}(A_L x - b_L) \tag{5.118}$$

for some $\alpha > 0$. Consequently, if any of both conditions is satisfied, then $S_{v_L} x = P_{C_L} x$.

Proof. ($\Longleftarrow$) Let v_L be defined by (5.118). Applying (4.109) and (4.5) with $A := A_L$ and $v(x) := v_L(x)$, we obtain

$$S_{v_L} x = x - \frac{v_L^\top(A_L x - b_L)}{\left\| A_L^\top v_L \right\|^2} A_L^\top v_L$$

$$= x - \frac{v_L^\top(A_L x - b_L)}{v_L^\top A_L A_L^\top v_L} A_L^\top v_L$$

$$= x - A_L^\top(A_L A_L^\top)^{-1}(A_L x - b_L) = P_{H_L} x.$$

($\Longrightarrow$) Suppose that $S_{v_L} x = P_{H_L} x$. By (4.5) and (4.109) with $A := A_L$, we have

$$\frac{v_L^\top(A_L x - b_L)}{\left\| A_L^\top v_L \right\|^2} A_L^\top v_L = A_L^\top(A_L A_L^\top)^{-1}(A_L x - b_L).$$

Since A_L has full row rank,

$$v_L = \alpha(A_L A_L^\top)^{-1}(A_L x - b_L),$$

where

$$\alpha = \frac{\|A_L^\top v_L\|^2}{v_L^\top (A_L x - b_L)}$$

which is positive, because v_L is essential for the system $A_L u \le b_L$ and $A_L^\top v_L \ne 0$ (see Lemma 4.9.5).   □

5.13.2   Description of the Method

Now we describe one iteration of the *surrogate constraints method with residual selection*. Suppose that x^k is obtained in the kth iteration. The next iteration consists of two steps:

Step 1 (*Constraints selection*). Determine a subset $L_k := L(x^k)$ such that

(a) A_L has full row rank and
(b) The vector

$$v_{L_k} := v_{L_k}(x^k) = (A_{L_k} A_{L_k}^\top)^{-1}(A_{L_k} x^k - b_{L_k}) \qquad (5.119)$$

is nonnegative and essential for the system $A_{L_k} u \le b_{L_k}$.

Step 2 (*Actualization*). Evaluate

$$x^{k+1} = x^k + \lambda_k (S_{v_{L_k}} x^k - x^k), \qquad (5.120)$$

where $\lambda_k \in (0, 2)$.

By Lemmas 5.13.1 and 5.13.2, we have $S_{v_{L_k}} x^k = P_{C_{L_k}} x^k = P_{H_{L_k}} x^k$.
A simple way to obtain a constraints selection satisfying conditions (a) and (b) above is to take $L_k := L(x^k) = \{i_k\} = \{i(x^k)\}$, where

$$i_k := \text{argmax}\{\frac{a_i^\top x^k - \beta_i}{\|a_i\|} : i \in I\}, \qquad (5.121)$$

consequently, $v^k = v(x^k) = \delta_{i_k}$. It is clear that v^k is essential for the system $Au \le b$. The surrogate constraints method with L_k and v^k defined by (5.121) reduces to the successive projection method with the remotest set control.

It turns out that the surrogate constraints method with residual selection L_k satisfying conditions (a) and (b) above provides steps which are no shorter than in the latter method if $i_k \in L_k$. This property leads to the convergence of sequences generated by (5.120).

Theorem 5.13.3. *Let $x^0 \in \mathbb{R}^n$ and a sequence $\{x^k\}_{k=0}^\infty$ be generated by the surrogate constraints method (5.120), where $\liminf_k \lambda_k (2 - \lambda_k) > 0$, v_{L_k} is given*

*by (5.119), L_k is obtained by a residual selection and L_k contains i_k defined by
(5.121). Then x^k converges to a solution x^* of the system $Ax \le b$.*

Proof. By Lemma 5.13.2 and by the inclusion $C_{L_k} \subseteq C_{i_k}$, we have

$$\left\| S_{v_{L_k}} x^k - x^k \right\| = \left\| P_{C_{L_k}} x^k - x^k \right\| \ge \left\| P_{C_{i_k}} x^k - x^k \right\| \ge \left\| P_{C_i} x^k - x^k \right\|$$

for all $i \in I$. Consequently, condition (3.11) with $T_k := S_{v_{L_k}}$ and $S := P_{C_i}$,
$i \in I$, is satisfied. Because A_{L_k} has full row rank, $A_{L_k} A_{L_k}^\top$ is positive definite
and v_{L_k} is essential. Since for any v the surrogate projection S_v is a cutter with
Fix $S_v = C$ (see Theorem 4.9.6), Corollary 3.7.1 yields the convergence of x^k to a
point $x^* \in C := \{x \in \mathbb{R}^n : Ax \le b\}$.   □

In the next subsection we present several constructions of constraints selection
$L(x)$ for $x \notin C$ leading to full row rank matrix A_L for which the vector $v_L :=
(A_L A_L^\top)^{-1}(A_L x - b_L)$ is nonnegative and essential for the system $A_L u \le b_L$. More
general constructions of this type and their applications to the convex minimization
problems can be found in [62–65, 72, 73, 154, 155, 230–232].

5.13.3 Obtuse Cone Selection

Let $x \notin C$. In an *obtuse cone selection* a subset $L := L(x)$ is constructed recursively
in such a way that $A_L x > b_L$, A_L has full row rank and $(A_L A_L^\top)^{-1} \ge 0$. Therefore,
the vector $v_L \in \mathbb{R}^r$ given by

$$v_L := (A_L A_L^\top)^{-1}(A_L x - b)$$

is nonnegative and essential for the system $A_L u \le b_L$. Consequently, the vector
$v := (v_L, v_{I \setminus L})$ with $v_{I \setminus L} = 0 \in \mathbb{R}^{m-r}$ is essential for the system $Au \le b$. The
name "obtuse cone selection" can be explained in the following way. Denote $C_L :=
\text{cone}\{a_i : i \in L\}$ and suppose that A_L has full row rank. Then C_L is obtuse (in
$\text{Lin}\{a_i : i \in L\}$) if and only if $(A_L A_L^\top)^{-1} \ge 0$ (see [232, Lemma 3.1] or [65,
Lemma 1.6]).

   Denote $K = K(x) := \{i \in I : a_i^\top x > \beta_i\}$. For any $z \in C$ and for $w := x - z$ we
have $A_{K} w > 0$. Let $L \subseteq K$ be a current selection such that the matrix A_L has full
row rank and $(A_L A_L^\top)^{-1} \ge 0$. The following lemmas give conditions under which
the update $L' := L \cup \{l\} \subseteq K$ maintains the described above properties of A_L, i.e.,
$A_{L'}$ has full row rank and $(A_{L'} A_{L'}^\top)^{-1} \ge 0$. Denote

$$A_{L'} := \begin{bmatrix} A_L \\ a_i^\top \end{bmatrix}.$$

Recall that A_L^+ denotes the Moore–Penrose pseudoinverse of A_L which for a full
row matrix A_L has the form $A_L^+ = A_L^\top (A_L A_L^\top)^{-1}$. The proof of the following
Lemma can be found in [65, Corollary 2.7].

Lemma 5.13.4. *Let $A_L^\top w > 0$ for some $w \in \mathbb{R}^n$. If A_L has full row rank, $(A_L A_L^\top)^{-1} \geq 0$ and $a_i^\top A_L^+ \leq 0$, then $A_{L'}$ has full row rank and $(A_{L'} A_{L'}^\top)^{-1} \geq 0$.*

Let $x = x^k \notin C$. If $L = \{l\}$, where $l \in K$, then, of course, A_L has full row rank (as a nonzero vector a_l) and $(A_L A_L^\top)^{-1} = \|a_l\|^{-2} \geq 0$. If there is $i \in K$ such that $a_i^\top a_l \leq 0$, then

$$a_i^\top A_L^+ = a_i^\top a_l \|a_l\|^{-2} \leq 0.$$

Applying Lemma 5.13.4 we see that L' has the same properties as L: the matrix $A_{L'}$ has full row rank and $(A_{L'} A_{L'}^\top)^{-1} \geq 0$. By repeated application of Lemma 5.13.4 with $L := L'$ we obtain a new subset $L' \subseteq K$ such that $A_{L'}$ has full row rank and $(A_{L'} A_{L'}^\top)^{-1} \geq 0$.

Let $L := L(x)$ be constructed by an obtuse cone selection. Since $A_L x > b_L$, the vector $v_L := (A_L A_L^\top)^{-1}(A_L x - b)$ has nonnegative coordinates. Therefore, all assumptions of Lemma 5.13.2 are satisfied, consequently, $S_{v_L} x = P_{H_L} x = P_{C_L} x$. If in each iteration of (5.120) we start the above defined construction of the subset $L_k := L(x^k)$ with $l_k := \operatorname{argmax}\{\frac{a_i^\top x^k - \beta_i}{\|a_i\|} : i \in I\}$, then Theorem 5.13.3 yields the convergence of sequences generated by (5.120) to a solution of the system $Ax \leq b$.

5.13.4 Regular Obtuse Cone Selection

Now we present a special case of the obtuse cone selection. The corollary below follows from Lemma 5.13.4. We use the same notation as in Sect. 5.13.3.

Corollary 5.13.5. *Let $A_{L'}^\top w > 0$ for some $w \in \mathbb{R}^n$. If A_L has full row rank, $(A_L A_L^\top)^{-1} \geq 0$ and $A_L a_i \leq 0$, then $A_{L'}$ has full row rank and $(A_{L'} A_{L'}^\top)^{-1} \geq 0$.*

Proof. Suppose that A_L has full row rank, $(A_L A_L^\top)^{-1} \geq 0$ and $A_L a_i \leq 0$. Then, by Lemma 5.13.4, we have

$$a_i^\top A_L^+ = a_i^\top A_L^\top (A_L A_L^\top)^{-1} \leq 0.$$

Consequently, $A_{L'}$ has full row rank and $(A_{L'} A_{L'}^\top)^{-1} \geq 0$. $\qquad\square$

Let $x \notin C$. Applying Corollary 5.13.5 and starting with $L = \{l\}$, where $l \in K$, one can construct recursively a maximal subset $L \subseteq K$ such that $a_i^\top a_j \leq 0$ for all $i, j \in L, i \neq j$. Then, of course, A_L has full row rank and $(A_L A_L^\top)^{-1} \geq 0$. The subset L constructed in this way is called a *regular obtuse cone selection* (ROCS).

The regular obtuse cone selection is easier to perform than the general obtuse cone selection presented in Sect. 5.13.3 and the residual selection presented in Sect. 5.13.2. Application of all these methods leads in practice to an essential acceleration of the convergence, in particular, if the solution set is "flat" (see the numerical results presented in [72, 73, 155]).

5.14   Exercises

Exercise 5.14.1.   Show that Theorem 5.1.5 remains true if A and B are closed affine subspaces with a common point.

Exercise 5.14.2.   Show that the result of [351, Theorem 3.3] can be derived from Example 5.12.5.

Exercise 5.14.3.   Prove equalities (5.75) and (5.76).

References

1. S. Agmon, The relaxation method for linear inequalities. Can. J. Math. **6**, 382–392 (1954)
2. R. Aharoni, A. Berman, Y. Censor, An interior point algorithm for the convex feasibility problem. Adv. Appl. Math. **4**, 479–489 (1983)
3. R. Aharoni, Y. Censor, Block-iterative projection methods for parallel computation of solutions to convex feasibility problems. Lin. Algebra Appl. **120**, 165–175 (1989)
4. D. Alevras, M.W. Padberg, *Linear Optimization and Extensions. Problems and Solutions* (Springer, Berlin, 2001)
5. A. Aleyner, S. Reich, Block-iterative algorithms for solving convex feasibility problems in Hilbert and in Banach spaces. J. Math. Anal. Appl. **343**, 427–435 (2008)
6. A. Aleyner, S. Reich, Random products of quasi-nonexpansive mappings in Hilbert space. J. Convex Anal. **16**, 633–640 (2009)
7. M. Altman, On the approximate solution of linear algebraic equations. Bulletin de l'Académie Polonaise des Sciences Cl. III **3** , 365–370 (1957)
8. I. Amemiya, T. Ando, Convergence of random products of contractions in Hilbert space. Acta Sci. Math. (Szeged) **26**, 239–244 (1965)
9. R. Ansorge, Connections between the Cimmino-method and the Kaczmarz-method for solution of singular and regular systems of equations. Computing **33**, 367–375 (1984)
10. G. Appleby, D.C. Smolarski, A linear acceleration row action method for projecting onto subspaces. Electron. Trans. Numer. Anal. **20**, 253–275 (2005)
11. N. Aronszajn, Theory of reproducing kernels. Trans. Am. Math. Soc. **68**, 337–404 (1950)
12. A. Auslender, *Optimisation, méthodes numériques* (Masson, Paris, 1976)
13. V.N. Babenko, Convergence of the Kaczmarz projection algorithm. Zh. Vychisl. Mat. Mat. Fiz. **24**, 1571–1573 (1984) (in Russian)
14. J.B. Baillon, R.E. Bruck, S. Reich, On the asymptotic behavior of nonexpansive mappings and semigroups in Banach spaces. Houston J. Math. **4**, 1–9 (1978)
15. S. Banach, Sur les opérations dans les ensembles abstraits et leur application aux équations intégrals. Fundamenta Mathematicae **3**, 133–181 (1922)
16. H.H. Bauschke, A norm convergence result on random products of relaxed projections in Hilbert space. Trans. Am. Math. Soc. **347**, 1365–1373 (1995)
17. H.H. Bauschke, Projection Algorithms and Monotone Operators, Ph.D. Thesis, Department of Mathematics, Simon Fraser, University, Burnaby, BC, Canada, 1996
18. H.H. Bauschke, The approximation of fixed points of compositions of nonexpansive mapping in Hilbert space. J. Math. Anal. Appl. **202**, 150–159 (1996)
19. H.H. Bauschke, The composition of the projections onto closed convex sets in Hilbert space is asymptotically regular. Proc. Am. Math. Soc. **131**, 141–146 (2002)

A. Cegielski, *Iterative Methods for Fixed Point Problems in Hilbert Spaces*,
Lecture Notes in Mathematics 2057, DOI 10.1007/978-3-642-30901-4,
© Springer-Verlag Berlin Heidelberg 2012

20. H.H. Bauschke, J. Borwein, On the convergence of von Neumann's alternating projection algorithm for two sets. Set-Valued Anal. **1**, 185–212 (1993)
21. H.H. Bauschke, J. Borwein, Dykstra's alternating projection algorithm for two sets. J. Approx. Theor. **79**, 418–443 (1994)
22. H.H. Bauschke, J. Borwein, On projection algorithms for solving convex feasibility problems. SIAM Rev. **38**, 367–426 (1996)
23. H.H. Bauschke, J.M. Borwein, A.S. Lewis, The method of cyclic projections for closed convex sets in Hilbert space. Contemp. Math. **204**, 1–38 (1997)
24. H.H. Bauschke, P.L. Combettes, A weak-to-strong convergence principle for Fejér-monotone methods in Hilbert spaces. Math. Oper. Res. **26**, 248–264 (2001)
25. H.H. Bauschke, P.L. Combettes, S.G. Kruk, Extrapolation algorithm for affine-convex feasibility problems. Numer. Algorithms **41**, 239–274 (2006)
26. H.H. Bauschke, P.L. Combettes, D.R. Luke, Phase retrieval, error reduction algorithm, and Fienup variants: A view from convex optimization. J. Opt. Soc. Am. A **19**, 1334–1345 (2002)
27. H.H. Bauschke, P.L. Combettes, D.R. Luke, Hybrid projection-reflection method for phase retrieval. J. Opt. Soc. Am. A **20**, 1025–1034 (2003)
28. H.H. Bauschke, P.L. Combettes, D.R. Luke, Finding best approximation pairs relative to two closed convex sets in Hilbert spaces. J. Approx. Theor. **127**, 178–192 (2004)
29. H.H. Bauschke, P.L. Combettes, D.R. Luke, A strongly convergent reflection method for finding the projection onto the intersection of two closed convex sets in a Hilbert space. J. Approx. Theor. **141**, 63–69 (2006)
30. H.H. Bauschke, F. Deutsch, H. Hundal, S-H. Park, Accelerating the convergence of the method of alternating projections. Trans. Am. Math. Soc. **355**, 3433–3461 (2003)
31. H.H. Bauschke, S.G. Kruk, Reflection-projection method for convex feasibility problems with an obtuse cone. J. Optim. Theor. Appl. **120**, 503–531 (2004)
32. H.H. Bauschke, E. Matoušková, S. Reich, Projection and proximal point methods: convergence results and counterexamples. Nonlinear Anal. **56**, 715–738 (2004)
33. M.H. Bazaraa, H.D. Sherali, C.M. Shetty, *Nonlinear Programming, Theory and Algorithms*, 3rd edn. (Wiley, Hoboken, 2006)
34. M. Benzi, C.D. Meyer, A direct projection method for sparse linear systems. SIAM J. Sci. Comput. **16**, 1159–1176 (1995)
35. A. Berman, R.J. Plemmons, *Nonnegative Matrices in the Mathematical Sciences* (Academic, New York, 1979)
36. V. Berinde, *Iterative Approximation of Fixed Points* (Springer, Berlin, 2007)
37. M. Bertero, P. Boccacci, *Introduction to Inverse Problems in Imaging* (Institute of Physics Publishing, Bristol, 1998)
38. D.P. Bertsekas, *Nonlinear Programming* (Athena Scientific, Belmont, 1995)
39. D. Blatt, A.O. Hero, Energy-based sensor network source localization via projection onto convex sets (POCS). IEEE Trans. Signal Process. **54**, 3614–3619 (2006)
40. J.M. Borwein, A. Lewis, *Convex Analysis and Nonlinear Optimization, Theory and Examples* (Springer, New York, 2000)
41. R. Bramley, A. Sameh, Row projection methods for large nonsymmetric linear systems. SIAM J. Sci. Stat. Comput. **13**, 168–193 (1992)
42. L.M. Bregman, Finding the common point of convex sets by the method of successive projection (in Russian). Dokl. Akad. Nauk SSSR **162**, 487–490 (1965); English translation in: Soviet Math. Dokl. **6**, 688–692 (1965)
43. L.E.J. Brouwer, Über Abbildung von Mannigfaltigkeiten. Math. Ann. **71**, 97–115 (1912)
44. F.E. Browder, Fixed-point theorems for noncompact mappings in Hilbert space. Proc. Nat. Acad. Sci. USA **53**, 1272–1276 (1965)
45. F.E. Browder, Nonexpansive nonlinear operators in a Banach space. Proc. Nat. Acad. Sci. USA **54**, 1041–1044 (1965)
46. F.E. Browder, Convergence of approximants to fixed points of nonexpansive nonlinear mappings in Banach spaces. Arch. Rational Mech. Anal. **24**, 82–90 (1967)

47. F.E. Browder, Convergence theorems for sequences of nonlinear operators in Banach spaces. Math. Zeitschr. **100**, 201–225 (1967)
48. F.E. Browder, W.V. Petryshyn, The solution by iteration of nonlinear functional equations in Banach spaces. Bull. Am. Math. Soc. **72**, 571–575 (1966)
49. R.E. Bruck, Nonexpansive projections on subsets of Banach spaces. Pac. J. Math. **47**, 341–355 (1973)
50. R.E. Bruck, Random products of contractions in metric and Banach spaces. J. Math. Anal. Appl. **88**, 319–332 (1982)
51. R.E. Bruck, S. Reich, Nonexpansive projections and resolvents of accretive operators in Banach spaces. Houston J. Math. **3**, 459–470 (1977)
52. R.S. Burachik, J.O. Lopes, B.F. Svaiter, An outer approximation method for the variational inequality problem. SIAM J. Contr. Optim. **43**, 2071–2088 (2005)
53. D. Butnariu, Y. Censor, On the behavior of a block-iterative projection method for solving convex feasibility problems. Int. J. Comp. Math. **34**, 79–94 (1990)
54. D. Butnariu, Y. Censor, P. Gurfil, E. Hadar, On the behavior of subgradient projections methods for convex feasibility problems in Euclidean spaces. SIAM J. Opt. **19**, 786–807 (2008)
55. C. Byrne, Iterative oblique projection onto convex sets and the split feasibility problem. Inverse Probl. **18**, 441–453 (2002)
56. C. Byrne, A unified treatment of some iterative algorithms in signal processing and image reconstruction. Inverse Probl. **20**, 103–120 (2004)
57. C.L. Byrne, *Applied Iterative Methods* (AK Peters, Wellesley, 2008)
58. C.L. Byrne, Bounds on the largest singular value of a matrix and the convergence of simultaneous and block-iterative algorithms for sparse linear systems. Int. Trans. Oper. Res. **16**, 465–479 (2009)
59. T.D. Capricelli, P.L. Combettes, Parallel block-iterative reconstruction algorithms for binary tomography. Electron. Notes Discr. Math. **20**, 263–280 (2005)
60. G. Casssiani, G. Böhm, A. Vesnaver, R. Nicolich, A geostatistical framework for incorporating seismic tomography auxiliary data into hydraulic conductivity estimation. J. Hydrol. **206**, 58–74 (1998)
61. J. Cea, *Optimisation: théorie et algorithmes* (Dunod, Paris, 1971); Polish translation: Optymalizacja: Teoria i algorytmy (PWN, Warszawa, 1976)
62. A. Cegielski, Relaxation Methods in Convex Optimization Problems (in Polish). Monographs, vol. 67, Institute of Mathematics, Higher College of Engineering, Zielona Góra, 1993
63. A. Cegielski, in *Projection Onto an Acute Cone and Convex Feasibility Problems*, ed. by J. Henry i J.-P. Yvon. Lecture Notes in Control and Inform. Sci., vol. 197 (Springer, London, 1994), pp. 187–194
64. A. Cegielski, A method of projection onto an acute cone with level control in convex minimization. Math. Program. **85**, 469–490 (1999)
65. A. Cegielski, Obtuse cones and Gram matrices with nonnegative inverse. Lin. Algebra Appl. **335**, 167–181 (2001)
66. A. Cegielski, A generalization of the Opial's theorem. Contr. Cybern. **36**, 601–610 (2007)
67. A. Cegielski, Convergence of the projected surrogate constraints method for the linear split feasibility problems. J. Convex Anal. **14**, 169–183 (2007)
68. A. Cegielski, Projection methods for the linear split feasibility problems. Optimization **57**, 491–504 (2008)
69. A. Cegielski, Generalized relaxations of nonexpansive operators and convex feasibility problems. Contemp. Math. **513**, 111–123 (2010)
70. A. Cegielski, Y. Censor, in *Opial-Type Theorems and the Common Fixed Point Problem*, ed. by H.H. Bauschke, R.S. Burachik, P.L. Combettes, V. Elser, D.R. Luke, H. Wolkowicz. Fixed-Point Algorithms for Inverse Problems in Science and Engineering, Springer Optimization and Its Applications, vol. 49 (Springer, New York, 2011), pp. 155–183
71. A. Cegielski, Y. Censor, Extrapolation and local acceleration of an iterative process for common fixed point problems. J. Math. Anal. Appl. **394**, 809–818 (2012)

72. A. Cegielski, R. Dylewski, Selection strategies in projection methods for convex minimization problems. Discuss. Math. Differ. Incl. Contr. Optim. **22**, 97–123 (2002)

73. A. Cegielski, R. Dylewski, Residual selection in a projection method for convex minimization problems. Optimization **52**, 211–220 (2003)

74. A. Cegielski, R. Dylewski, Variable target value relaxed alternating projection method. Comput. Optim. Appl. **47**, 455–476 (2010)

75. A. Cegielski, A. Suchocka, Incomplete alternating projection method for large inconsistent linear systems. Lin. Algebra Appl. **428**, 1313–1324 (2008)

76. A. Cegielski, A. Suchocka, Relaxed alternating projection methods. SIAM J. Optim. **19**, 1093–1106 (2008)

77. A. Cegielski, R. Zalas, Methods for variational inequality problem over the intersection of fixed point sets of quasi-nonexpansive operators (2012) Numer. Funct. Anal. Optim. (in print)

78. Y. Censor, Row-action methods for huge and sparse systems and their applications. SIAM Rev. **23**, 444–466 (1981)

79. Y. Censor, Iterative methods for convex feasibility problems. Ann. Discrete Math. **20**, 83–91 (1984)

80. Y. Censor, An automatic relaxation method for solving interval linear inequalities. J. Math. Anal. Appl. **106**, 19–25 (1985)

81. Y. Censor, Parallel application of block-iterative methods in medical imaging and radiation therapy. Math. Program. **42**, 307–325 (1988)

82. Y. Censor, Binary steering in discrete tomography reconstruction with sequential and simultaneous iterative algorithms. Lin. Algebra Appl. **339**, 111–124 (2001)

83. Y. Censor, M.D. Altschuler, W.D. Powlis, On the use of Cimmino's simultaneous projections method for computing a solution of the inverse problem in radiation therapy treatment planning. Inverse Probl. **4**, 607–623 (1988)

84. Y. Censor, A. Ben-Israel, Y. Xiao, J.M. Galvin, On linear infeasibility arising in intensity-modulated radiation therapy inverse planning. Lin. Algebra Appl. **428**, 1406–1420 (2008)

85. Y. Censor, T. Bortfeld, B. Martin, A. Trofimov, A unified approach for inversion problems in intensity-modulated radiation therapy. Phys. Med. Biol. **51**, 2353–2365 (2006)

86. Y. Censor, P.P.B. Eggermont, D. Gordon, Strong underrelaxation in Kaczmarz's method for inconsistent systems. Numer. Math. **41**, 83–92 (1983)

87. Y. Censor, T. Elfving, New methods for linear inequalities. Lin. Algebra Appl. **42**, 199–211 (1982)

88. Y. Censor, T. Elfving, A multiprojection algorithm using Bregman projections in a product space. Numer. Algorithm **8**, 221–239 (1994)

89. Y. Censor, T. Elfving, Block-iterative algorithms with diagonal scaled oblique projections for the linear feasibility problems. SIAM J. Matrix Anal. Appl. **24**, 40–58 (2002)

90. Y. Censor, T. Elfving, Iterative algorithms with seminorm-induced oblique projections. Abstr. Appl. Anal. **8**, 387–406 (2003)

91. Y. Censor, T. Elfving, G.T. Herman, in *Averaging Strings of Sequential Iterations for Convex Feasibility Problems*, ed. by D. Butnariu, Y. Censor, S. Reich. Inherently Parallel Algorithms in Feasibility and Optimization and their Applications (Elsevier, Amsterdam, 2001), pp. 101–113

92. Y. Censor, T. Elfving, G.T. Herman, T. Nikazad, On diagonally relaxed orthogonal projection methods. SIAM J. Sci. Comput. **30**, 473–504 (2008)

93. Y. Censor, T. Elfving, N. Kopf, T. Bortfeld, The multiple-sets split feasibility problem and its applications for inverse problems. Inverse Probl. **21**, 2071–2084 (2005)

94. Y. Censor, A. Gibali, Projections onto super-half-spaces for monotone variational inequality problems in finite-dimensional spaces. J. Nonlinear Convex Anal. **9**, 461–475 (2008)

95. Y. Censor, A. Gibali, S. Reich, The subgradient extragradient method for solving variational inequalities in Hilbert space. J. Optim. Theory Appl. **148**, 318–335 (2011)

96. Y. Censor, A. Gibali, S. Reich, Extensions of Korpelevich's extragradient method for the variational inequality problem in Euclidean space. Optimization **61**, 1119–1132 (2012)

97. Y. Censor, A. Gibali, S. Reich, Strong convergence of subgradient extragradient methods for the variational inequality problem in Hilbert space. Optim. Meth. Software **26**, 827–845 (2011)

98. Y. Censor, D. Gordon, R. Gordon, Component averaging: An efficient iterative parallel algorithm for large and sparse unstructured problems. Parallel Comput. **27**, 777–808 (2001)

99. Y. Censor, D. Gordon, R. Gordon, BICAV: A block-iterative, parallel algorithm for sparse systems with pixel-related weighting. IEEE Trans. Med. Imag. **20**, 1050–1060 (2001)

100. Y. Censor, G.T. Herman, On some optimization techniques in image reconstruction from projections. Appl. Numer. Math. **3**, 365–391 (1987)

101. Y. Censor, A.N. Iusem, S.A. Zenios, An interior point method with Bregman functions for the variational inequality problem with paramonotone operators. Math. Program. **81**, 373–400 (1998)

102. Y. Censor, A. Lent, Cyclic subgradient projections. Math. Program. **24**, 233–235 (1982)

103. Y. Censor, A. Motova, A. Segal, Perturbed projections and subgradient projections for the multiple-sets split feasibility problem. J. Math. Anal. Appl. **327**, 1244–1256 (2007)

104. Y. Censor, A. Segal, The split common fixed point problem for directed operators. J. Convex Anal. **16**, 587–600 (2009)

105. Y. Censor, A. Segal, On the string averaging method for sparse common fixed point problems. Int. Trans. Oper. Res. **16**, 481–494 (2009)

106. Y. Censor, A. Segal, Sparse string-averaging and split common fixed points. Contemp. Math. **513**, 125–142 (2010)

107. Y. Censor, E. Tom, Convergence of string-averaging projection schemes for inconsistent convex feasibility problems. Optim. Meth. Software **18**, 543–554 (2003)

108. Y. Censor, S.A. Zenios, *Parallel Optimization, Theory, Algorithms and Applications* (Oxford University Press, New York, 1997)

109. A.E. Çetin, H. Özaktaş, H.M. Ozaktas, Resolution enhancement of low resolution wavefields with POCS algorithm. Electron. Lett. **9**, 1808–1810 (2003)

110. W. Chen, D. Craft, T.M. Madden, K. Zhang, H.M. Kooy, G.T. Herman, A fast optimization algorithm for multicriteria intensity modulated proton therapy planning. Med. Phys. **7**, 4938–4945 (2010)

111. W. Chen, G.T. Herman, Efficient controls for finitely convergent sequential algorithms. ACM Trans. Math. Software **37**, 1–23 (2010)

112. W. Cheney, A.A. Goldstein, Proximity maps for convex sets. Proc. Am. Math Soc. **10**, 448–450 (1959)

113. C.E. Chidume, Quasi-nonexpansive mappings and uniform asymptotic regularity. Kobe J. Math. **3**, 29–35 (1986)

114. Ch. Chidume, *Geometric Properties of Banach Spaces and Nonlinear Iterations* (Springer, London, 2009)

115. H. Choi, R.G. Baraniuk, Multiple wavelet basis image denoising using Besov ball projections. IEEE Signal Process. Lett. **11**, 717–720 (2004)

116. G. Cimmino, Calcolo approssimato per le soluzioni dei sistemi di equazioni lineari. La Ricerca Scientifica, II **9**, 326–333 (1938)

117. P.L. Combettes, Inconsistent signal feasibility problems: Least-square solutions in a product space. IEEE Trans. Signal Process. **42**, 2955–2966 (1994)

118. P.L. Combettes, in *The Convex Feasibility Problem in Image Recovery*, ed. by P. Hawkes. Advances in Imaging and Electron Physics, vol. 95 (Academic, New York, 1996), pp. 155–270

119. P.L. Combettes, Hilbertian convex feasibility problem: Convergence of projection methods. Appl. Math. Optim. **35**, 311–330 (1997)

120. P.L. Combettes, Convex set theoretic image recovery by extrapolated iterations of parallel subgradient projections. IEEE Trans. Image Process. **6**, 493–506 (1997)

121. P.L. Combettes, in *Quasi-Fejérian Analysis of Some Optimization Algorithm*, ed. by D. Butnariu, Y. Censor, S. Reich. Inherently Parallel Algorithms in Feasibility and Optimization and their Applications (Elsevier, Amsterdam, 2001), pp. 115–152

122. P.L. Combettes, Solving monotone inclusions via compositions of nonexpansive averaged operators. Optimization **53**, 475–504 (2004)
123. P.L. Combettes, P. Bondon, Hard-constrained inconsistent signal feasibility problems. IEEE Trans. Signal Process. **47**, 2460–2468 (1999)
124. P.L. Combettes, S.A. Hirstoaga, Equilibrium programming in Hilbert spaces. J. Nonlinear Convex Anal. **6**, 117–136 (2005)
125. P.L. Combettes, H. Puh, Iterations of parallel convex projections in Hilbert spaces. Numer. Funct. Anal. Optim. **15**, 225–243 (1994)
126. G. Crombez, A geometrical look at iterative methods for operators with fixed points. Numer. Funct. Anal. Optim. **26**, 157–175 (2005)
127. G. Crombez, A hierarchical presentation of operators with fixed points on Hilbert spaces. Numer. Funct. Anal. Optim. **27**, 259–277 (2006)
128. Y.-H. Dai, Fast algorithms for projection on an ellipsoid. SIAM J. Optim. **16**, 986–1006 (2006)
129. A. Dax, A note of the convergence of linear stationary iterative process. Lin. Algebra Appl. **129**, 131–142 (1990)
130. A. Dax, Linear search acceleration of iterative methods. Lin. Algebra Appl. **130**, 43–63 (1990)
131. A. Dax, On hybrid acceleration of a linear stationary iterative process. Lin. Algebra Appl. **130**, 99–110 (1990)
132. A. Dax, The convergence of linear stationary iterative processes for solving singular unstructured systems of linear equations. SIAM Rev. **32**, 611–635 (1990)
133. L. Debnath, P. Mikusiński, *Hilbert Spaces with Applications*, 2nd edn. (Academic, San Diego, 1999)
134. A.R. De Pierro, A.N. Iusem, A simultaneous projections method for linear inequalities. Lin. Algebra Appl. **64**, 243–253 (1985)
135. A.R. De Pierro, A.N. Iusem, A parallel projection method of finding a common point of a family of convex sets. Pesquisa Operacional **5**, 1–20 (1985)
136. A.R. De Pierro, A.N. Iusem, A finitely convergent "row-action" method for the convex feasibility problem. Appl. Math. Optim. **17**, 225–235 (1988)
137. A.R. De Pierro, A.N. Iusem, On the asymptotic behavior of some alternate smoothing series expansion iterative methods. Lin. Algebra Appl. **130**, 3–24 (1990)
138. F. Deutsch, in *Applications of von Neumann's Alternating Projections Algorithm*, ed. by P. Kenderov. Mathematical Methods in Operations Research (Sophia, Bulgaria, 1983), pp. 44–51
139. F. Deutsch, in *The Method of Alternating Orthogonal Projections*, ed. by S.P. Singh. Approximation Theory, Spline Functions and Applications (Kluwer Academic, The Netherlands, 1992), pp. 105–121
140. F. Deutsch, *Best Approximation in Inner Product Spaces* (Springer, New York, 2001)
141. F. Deutsch, in *Accelerating the Convergence of the Method of Alternating Projections via a Line Search: A Brief Survey*, ed. by D. Butnariu, Y. Censor, S. Reich. Inherently Parallel Algorithms in Feasibility and Optimization and their Application, Studies in Computational Mathematics, vol. 8 (Elsevier Science, Amsterdam, 2001), pp. 203–217
142. F. Deutsch, H. Hundal, The rate of convergence for the cyclic projections algorithm, I. Angles between convex sets. J. Approx. Theor. **142**, 36–55 (2006)
143. F. Deutsch, H. Hundal, The rate of convergence for the cyclic projections algorithm, II. Norms of nonlinear operators. J. Approx. Theor. **142**, 56–82 (2006)
144. F. Deutsch, I. Yamada, Minimizing certain convex functions over the intersection of the fixed point sets of nonexpansive mappings. Numer. Funct. Anal. Optim. **19**, 33–56 (1998)
145. J.B. Diaz, F.T. Metcalf, On the set of subsequential limit points of successive approximations. Trans. Am. Math. Soc. **135**, 459–485 (1969)
146. L.T. Dos Santos, A parallel subgradient projections method for the convex feasibility problem. J. Comput. Appl. Math. **18**, 307–320 (1987)
147. W.G. Dotson Jr., On the Mann iterative process. Trans. Am. Math. Soc. **149**, 65–73 (1970)

148. W.G. Dotson, Fixed points of quasi-nonexpansive mappings. J. Austral. Math. Soc. **13**, 167–170 (1972)
149. J. Douglas, H.H. Rachford, On the numerical solution of heat conduction problems in two or three space variables. Trans. Am. Math. Soc. **82**, 421–439 (1956)
150. R. Dudek, Iterative method for solving the linear feasibility problem. J. Optim Theor. Appl. **132**, 401–410 (2007)
151. J. Dye, M.A. Khamsi, S. Reich, Random products of contractions in Banach spaces. Trans. Am. Math. Soc. **325**, 87–99 (1991)
152. J.M. Dye, S. Reich, On the unrestricted iteration of projections in Hilbert space. J. Math. Anal. Appl. **156**, 101–119 (1991)
153. J. Dye, S. Reich, Unrestricted iterations of nonexpansive mappings in Hilbert space. Nonlinear Anal. **18**, 199–207 (1992)
154. R. Dylewski, Selection of Linearizations in Projection Methods for Convex Optimization Problems (in Polish), Ph.D. thesis, University of Zielona Góra, Institute of Mathematics, 2003
155. R. Dylewski, Projection method with residual selection for linear feasibility problems. Discuss. Math. Differ. Incl. Contr. Optim. **27**, 43–50 (2007)
156. M.G. Eberle, M.C. Maciel, Finding the closest Toeplitz matrix. Computat. Appl. Math. **22**, 1–18 (2003)
157. J. Eckstein, D.P. Bertsekas, On the Douglas–Rachford splitting method and the proximal point algorithm for maximal monotone operators. Math. Program. **55**, 293–318 (1992)
158. I. Ekeland, R. Témam, *Convex Analysis and Variational Problems* (North-Holland, Amsterdam, 1976)
159. T. Elfving, A projection method for semidefinite linear systems and its applications. Lin. Algebra Appl. **391**, 57–73 (2004)
160. T. Elfving, T. Nikazad, Stopping rules for Landweber-type iteration. Inverse Probl. **23**, 1417–1432 (2007)
161. V. Elser, I. Rankenburg, P. Thibault, Searching with iterated maps. Proc. Natl. Acad. Sci. USA **104**, 418–423 (2007)
162. L. Elsner, I. Koltracht, P. Lancaster, Convergence properties of ART and SOR algorithms. Numer. Math. **59**, 91–106 (1991)
163. L. Elsner, I. Koltracht, M. Neumann, On the convergence of asynchronous paracontractions with application to tomographic reconstruction from incomplete data. Lin. Algebra Appl. **130**, 65–82 (1990)
164. L. Elsner, I. Koltracht, M. Neumann, Convergence of sequential and asynchronous nonlinear paracontractions. Numer. Math. **62**, 305–319 (1992)
165. F. Facchinei, J.-S. Pang, *Finite-Dimensional Variational Inequalities and Complementarity Problems, Volume I, Volume II* (Springer, New York, 2003)
166. M. Fiedler, V. Pták, On matrices with non-positive off-diagonal elements and positive principal minors. Czech. Math. J. **12**, 382–400 (1962)
167. S. Fitzpatrick, R.R. Phelps, Differentiability of the metric projection in Hilbert space. Trans. Am. Math. Soc. **270**, 483–501 (1982)
168. S.D. Flåm, J. Zowe, Relaxed outer projections, weighted averages and convex feasibility. BIT **30**, 289–300 (1990)
169. R. Fletcher, *Practical Methods of Optimization* (Wiley, Chichester, 1987)
170. K. Friedricks, On certain inequalities and characteristic value problems for analytic functions and for functions of two variables. Trans. Am. Math. Soc. **41**, 321–364 (1937)
171. M. Fukushima, A relaxed projection method for variational inequalities. Math. Program. **35**, 58–70 (1986)
172. E.M. Gafni, D.P. Bertsekas, Two metric projection methods for constrained optimization. SIAM J. Contr. Optim. **22**, 936–964 (1984)
173. A. Galántai, *Projectors and Projection Methods* (Kluwer Academic, Boston, 2004)
174. A. Galántai, On the rate of convergence of the alternating projection method in finite dimensional spaces. J. Math. Anal. Appl. **310**, 30–44 (2005)

175. U. García-Palomares, Parallel projected aggregation methods for solving the convex feasibility problem. SIAM J. Optim. **3**, 882–900 (1993)

176. U. García-Palomares, A superlinearly convergent projection algorithm for solving the convex inequality problem. Oper. Res. Lett. **22**, 97–103 (1998)

177. W.B. Gearhart, M. Koshy, Acceleration schemes for the method of alternating projections. J. Comput. Appl. Math. **26**, 235–249 (1989)

178. C. Geiger, Ch. Kanzow, *Numerische Verfahren zur Lösung unrestingierter Optimierungsaufgaben* (Springer, Berlin, 1999)

179. C. Geiger, Ch. Kanzow, *Theorie und Numerik restringierter Optimierungsaufgaben* (Springer, Berlin, 2002)

180. J.R. Giles, *Convex Analysis with Application in Differentiation of Convex Functions* (Pitman Advanced Publishing Program, Boston, 1982)

181. P.E. Gill, W. Murray, M.H. Wright, *Numerical Linear Algebra and Optimization* (Addison-Wesley, Redwood City, 1991)

182. W. Glunt, T.L. Hayden, R. Reams, The nearest 'doubly stochastic' matrix to a real matrix with the same first moment. Numer. Lin. Algebra Appl. **5**, 475–482 (1998)

183. K. Goebel, *Concise Course on Fixed Points Theorems* (Yokohama Publishing, Yokohama, 2002); Polish translation: *Twierdzenia o punktach stałych* (Wydawnictwo UMCS, Lublin, 2005)

184. K. Goebel, W.A. Kirk, *Topics in Metric Fixed Point Theory* (Cambridge University Press, Cambridge, 1990); Polish translation: *Zagadnienia metrycznej teorii punktów stałych* (Wydawnictwo UMCS, Lublin, 1999)

185. K. Goebel, S. Reich, *Uniform Convexity, Hyperbolic Geometry, and Nonexpansive Mappings* (Marcel Dekker, New York, 1984)

186. J.L. Goffin, The relaxation method for solving systems of linear inequalities. Math. Oper. Res. **5**, 388–414 (1980)

187. J.L. Goffin, On the finite convergence of the relaxation method for solving systems of inequalities. Operations Research Center, Report ORC 71–36, University of California, Berkeley, 1971

188. D. Göhde, Zum Prinzip der kontraktiven Abbildung. Math. Nachr. **30**, 251–258 (1965)

189. R. Gordon, R. Bender, G.T. Herman, Algebraic reconstruction techniques (ART) for three-dimensional electron microscopy and X-ray photography. J. Theoret. Biol. **29**, 471–481 (1970)

190. D. Gordon, R. Gordon, Component-averaged row projections: a robust, block-parallel scheme for sparse linear systems. SIAM J. Sci. Comput. **27**, 1092–1117 (2005)

191. A. Granas, J. Dugundji, *Fixed Point Theory* (Springer, New York, 2003)

192. K.M. Grigoriadis, A.E. Frazho, R.E. Skelton, Application of alternating convex projection methods for computation of positive Toeplitz matrices. IEEE Trans. Signal Process. **42**, 1873–1875 (1994)

193. K.M. Grigoriadis, R.E. Skelton, Low-order control design for LMI problems using alternating projection methods. Automatica **32**, 1117–1125 (1996)

194. K.M. Grigoriadis, R.E. Skelton, Alternating convex projection methods for discrete-time covariance control design. J. Optim. Theor. Appl. **88**, 399–432 (1996)

195. J. Gu, H. Stark, Y. Yang, Wide-band smart antenna design using vector space projection methods. IEEE Trans. Antenn. Propag. **52**, 3228–3236 (2004)

196. L.G. Gurin, B.T. Polyak, E.V. Raik, The method of projection for finding the common point in convex sets. Zh. Vychisl. Mat. Mat. Fiz. **7**, 1211–1228 (1967) (in Russian); English translation in: USSR Comput. Math. Phys. **7**, 1–24 (1967)

197. R. Haller, R. Szwarc, Kaczmarz algorithm in Hilbert space. Studia Math. **169**, 123–132 (2005)

198. I. Halperin, The product of projection operators. Acta Sci. Math. (Szeged) **23**, 96–99 (1962)

199. H.W. Hamacher, K.-H. Küfer, Inverse radiation therapy planning – a multiple objective optimization approach. Discrete Appl. Math. **118**, 145–161 (2002)

200. S.-P. Han, A successive projection method. Math. Program. (Ser. A) **40**, 1–14 (1988)

201. M. Hanke, W. Niethammer, On the acceleration of Kaczmarz's method for inconsistent linear systems. Lin. Algebra Appl. **130**, 83–98 (1990)

202. Y. Haugazeau, *Sur les inéquations variationnelles et la minimisation de fonctionnelles convexes* (Thèse, Université de Paris, Paris, 1968)
203. H. He, S. Liu, H. Zhou, An explicit method for finding common solutions of variational inequalities and systems of equilibrium problems and fixed points of an infinite family of nonexpansive mappings. Nonlinear Anal. **72**, 3124–3135 (2010)
204. G.T. Herman, A relaxation method for reconstructing objects from noisy X-rays. Math. Program. **8**, 1–19 (1975)
205. G.T. Herman, *Fundamentals of Computerized Tomography: Image Reconstruction from Projections*, 2nd edn. (Springer, London, 2009)
206. G.T. Herman, W. Chen, A fast algorithm for solving a linear feasibility problem with application to intensity-modulated radiation therapy. Lin. Algebra Appl. **428**, 1207–1217 (2008)
207. N.J. Higham, Computing a nearest symmetric positive semidefinite matrix. Lin. Algebra Appl. **103**, 103–118 (1988)
208. N.J. Higham, Computing the nearest correlation matrix – a problem from finance. IMA J. Numer. Anal. **22**, 329–343 (2002)
209. J.-B. Hiriart-Urruty, C. Lemaréchal, *Convex Analysis and Minimization Algorithms, Vol I, Vol II* (Springer, Berlin, 1993)
210. J.-B. Hiriart-Urruty, C. Lemaréchal, *Fundamentals of Convex Analysis* (Springer, Berlin, 2001)
211. S.A. Hirstoaga, Iterative selection methods for common fixed point problems. J. Math. Anal. Appl. **324**, 1020–1035 (2006)
212. J. Höffner, P. Decker, E.L. Schmidt, W. Herbig, J. Rittler, P. Weiß, Development of a fast optimization preview in radiation treatment planning. Strahlentherapie und Onkologie **172**, 384–394 (1996)
213. H.S. Hundal, An alternating projection that does not converge in norm. Nonlinear Anal. **57**, 35–61 (2004)
214. J.K. Hunter, B. Nachtergaele, *Applied Analysis* (World Scientific, Singapore, 2000)
215. A.N. Iusem, A.R. De Pierro, Convergence results for an accelerated nonlinear Cimmino algorithm. Numer. Math. **49**, 367–378 (1986)
216. A.N. Iusem, A.R. De Pierro, A simultaneous iterative method for computing projections on polyhedra. SIAM J. Contr. Optim. **25**, 231–243 (1987)
217. A.N. Iusem, A.R. De Pierro, On the convergence properties of Hildreth's quadratic programming algorithm. Math. Program. (Ser. A) **47**, 37–51 (1990)
218. A.N. Iusem, B.F. Svaiter, A row-action method for convex programming. Math. Program. **64**, 149–171 (1994)
219. B.K. Jennison, J.P. Allebach, D.W. Sweeney, Iterative approaches to computer-generated holography. Opt. Eng. **28**, 629–637 (1989)
220. M. Jiang, G. Wang, Development of iterative algorithms for image reconstruction. J. X-Ray Sci. Tech. **10**, 77–86 (2002)
221. M. Jiang, G. Wang, Convergence studies on iterative algorithms for image reconstruction. IEEE Trans. Med. Imag. **22**, 569–579 (2003)
222. B. Johansson, T. Elfving, V. Kozlov, Y. Censor, P.-E. Forssén, G. Granlund, The application of an oblique-projected Landweber method to a model of supervised learning. Math. Comput. Model. **43**, 892–909 (2006)
223. S. Kaczmarz, Angenäherte Auflösung von Systemen linearer Gleichungen. Bulletin International de l'Académie Polonaise des Sciences et des Lettres **A35**, 355–357 (1937); English translation: S. Kaczmarz, Approximate solution of systems of linear equations. Int. J. Contr. **57**, 1269–1271 (1993)
224. A.C. Kak, M. Slaney, *Principles of Computerized Tomographic Imaging* (IEEE, New York, 1988)
225. I.G. Kazantsev, S. Schmidt, H.F. Poulsen, A discrete spherical x-ray transform of orientation distribution functions using bounding cubes. Inverse Probl. **25**, 105009 (2009)

226. D. Kinderlehrer, G. Stampacchia, *An Introduction to Variational Inequalities and Their Applications* (Academic, New York, 1980)

227. W.A. Kirk, A fixed point theorem for mappings which do not increase distances. Am. Math. Mon. **72**, 1004–1006 (1965)

228. Yu.N. Kiseliov, Algorithms of projection of a point onto an ellipsoid. Lithuanian Math. J. **34**, 141–159 (1994)

229. K.C. Kiwiel, Block-iterative surrogate projection methods for convex feasibility problems. Lin. Algebra Appl. **215**, 225–259 (1995)

230. K.C. Kiwiel, The efficiency of subgradient projection methods for convex optimization. I. General level methods. SIAM J. Contr. Optim. **34**, 660–676 (1996)

231. K.C. Kiwiel, The efficiency of subgradient projection methods for convex optimization, II. Implementations and extensions. SIAM J. Contr. Optim. **34**, 677–697 (1996)

232. K.C. Kiwiel, Monotone Gram matrices and deepest surrogate inequalities in accelerated relaxation methods for convex feasibility problems. Lin. Algebra Appl. **252**, 27–33 (1997)

233. K.C. Kiwiel, B. Łopuch, Surrogate projection methods for finding fixed points of firmly nonexpansive mappings. SIAM J. Opt. **7**, 1084–1102 (1997)

234. A. Kiełbasiński, H. Schwetlick, *Numerical Linear Algebra* (in German) (Verlag Harri Deutsch, Thun, 1988); Polish translation: *Numeryczna algebra liniowa* (WNT, Warszawa, 1992)

235. E. Kopecká, S. Reich, A note on the von Neumann alternating projections algorithm. J. Nonlinear Convex Anal. **5**, 379–386 (2004)

236. E. Kopecká, S. Reich, Another note on the von Neumann alternating projections algorithm. J. Nonlinear Convex Anal. **11**, 455–460 (2010)

237. G.M. Korpelevich, The extragradient method for finding saddle points and other problems. Ekonomika i Matematicheskie Metody **12**, 747–756 (1976)

238. M.A. Krasnosel'skiĭ, Two remarks on the method of successive approximations (in Russian). Uspehi Mat. Nauk **10**, 123–127 (1955)

239. S. Kwapień, J. Mycielski, On the Kaczmarz algorithm of approximation in infinite-dimensional spaces. Studia Math. **148**, 5–86 (2001)

240. L. Landweber, An iteration formula for Fredholm integral equations of the first kind. Am. J. Math. **73**, 615–624 (1951)

241. S. Lee, P.S. Cho, R.J. Marks, S.Oh, Conformal radiotherapy computation by the method of alternating projections onto convex sets. Phys. Med. Biol. **42**, 1065–1086 (1997)

242. S.-H. Lee, K.-R. Kwon, Mesh watermarking based projection onto two convex sets. Multimedia Syst. **13**, 323–330 (2008)

243. A. Lent, in *A Convergent Algorithm for Maximum Entropy Image Restoration with a Medical X-ray Application*, ed. by R. Shaw. Image Analysis and Evaluation (SPSE, Washington DC), pp. 249–257

244. A. Lent, Y. Censor, Extensions of Hildreth's row-action method for quadratic programming. SIAM J. Contr. Optim. **18**, 444–454 (1980)

245. A.W.-C. Liew, H. Yan, N.-F. Law, POCS-based blocking artifacts suppression using a smoothness constraint set with explicit region modeling. IEEE Trans. Circ. Syst. Video Tech. **15**, 795–800 (2005)

246. P.L. Lions, B. Mercier, Splitting algorithms for the sum of two nonlinear operators. SIAM J. Numer. Anal. **16**, 964–979 (1979)

247. C. Liu, An acceleration scheme for row projection methods. J. Comput. Appl. Math. **57**, 363–391 (1995)

248. Y.M. Lu, M. Karzand, M. Vetterli, Demosaicking by alternating projections: theory and fast one-step implementation. IEEE Trans. Image Process. **19**, 2085–2098 (2010)

249. P.-E. Maingé, Inertial iterative process for fixed points of certain quasi-nonexpansive mappings. Set-Valued Anal. **15**, 67–79 (2007)

250. P.-E. Maingé, Extension of the hybrid steepest descent method to a class of variational inequalities and fixed point problems with nonself-mappings. Numer. Funct. Anal. Optim. **29**, 820–834 (2008)

251. P.-E. Maingé, New approach to solving a system of variational inequalities and hierarchical problems. J. Optim. Theor. Appl. **138**, 459–477 (2008)

252. W.R. Mann, Mean value methods in iteration. Proc. Am. Math. Soc. **4**, 506–510 (1953)

253. Şt. Măruşter, The solution by iteration of nonlinear equations in Hilbert spaces. Proc. Am. Math. Soc. **63**, 69–73 (1977)

254. Şt. Măruşter, Quasi-nonexpansivity and the convex feasibility problem. An. Ştiinţ. Univ. Al. I. Cuza Iaşi Inform. (N.S.) **15**, 47–56 (2005)

255. Şt. Măruşter, C. Popîrlan, On the Mann-type iteration and the convex feasibility problem. J. Comput. Appl. Math. **212**, 390–396 (2008)

256. Şt. Măruşter, C. Popîrlan, On the regularity condition in a convex feasibility problem. Nonlinear Anal. **70**, 1923–1928 (2009)

257. E. Masad, S. Reich, A note on the multiple-set split convex feasibility problem in Hilbert space. J. Nonlinear Convex Anal. **8**, 367–371 (2007)

258. E. Matoušková, S. Reich, The Hundal example revisited. J. Nonlinear Convex Anal. **4**, 411–427 (2003)

259. S.F. McCormick, The methods of Kaczmarz and row orthogonalization for solving linear equations and least squares problems in Hilbert space. Indiana Univ. Math. J. **26**, 1137–1150 (1977)

260. Yu.I. Merzlyakov, On a relaxation method of solving systems of linear inequalities (in Russian). Zh. Vychisl. Mat. Mat. Fiz. **2**, 482–487 (1962)

261. D. Michalski, Y. Xiao, Y. Censor, J.M. Galvin, The dose-volume constraint satisfaction problem for inverse treatment planning with field segments. Phys. Med. Biol. **49**, 601–616 (2004)

262. W. Mlak, *Introduction to Hilbert Spaces* (in Polish) (PWN, Warsaw, 1982)

263. W. Mlak, *Hilbert Spaces and Operator Theory* (Kluwer Academic, Boston, 1991)

264. J.-J. Moreau, Décomposition orthogonale d'un espace hilbertien selon deux cônes mutuellement polaires. C. R. Acad. Sci. Paris **255**, 238–240 (1962)

265. J. Moreno, B. Datta, M. Raydan, A symmetry preserving alternating projection method for matrix model updating. Mech. Syst. Signal Process. **23**, 1784–1791 (2009)

266. T.S. Motzkin, I.J. Schoenberg, The relaxation method for linear inequalities. Can. J. Math. **6**, 393–404 (1954)

267. J. Musielak, *Introduction to Functional Analysis* (in Polish) (PWN, Warszawa, 1989)

268. J. Mycielski, S. Świerczkowski, Uniform approximation with linear combinations of reproducing kernels. Studia Math. **121**, 105–114 (1996)

269. N. Nadezhkina, W. Takahashi, Strong convergence theorem by a hybrid method for nonexpansive mappings and Lipschitz-continuous monotone mappings. SIAM J. Optim. **16**, 1230–1241 (2006)

270. F. Natterer, *The Mathematics of Computerized Tomography* (Wiley, Chichester, 1986)

271. J. von Neumann, in *Functional Operators – Vol. II. The Geometry of Orthogonal Spaces.* Annals of Mathematics Studies, vol. 22 (Princeton University Press, Princeton, 1950) (Reprint of mimeographed lecture notes first distributed in 1933)

272. O. Nevanlinna, S. Reich, Strong convergence of contraction semigroups and of iterative methods for accretive operators in Banach spaces. Israel J. Math. **32**, 44–58 (1979)

273. N. Ogura, I. Yamada, Non-strictly convex minimization over the fixed point set of an asymptotically shrinking nonexpansive mapping. Numer. Funct. Anal. Optim. **23**, 113–137 (2002)

274. N. Ogura, I. Yamada, Nonstrictly convex minimization over the bounded fixed point set of a nonexpansive mapping. Numer. Funct. Anal. Optim. **24**, 129–135 (2003)

275. S. Oh, R.J. Marks, L.E. Atlas, Kernel synthesis for generalized time-frequency distributions using the method of alternating projections onto convex sets. IEEE Trans. Signal Process. **42**, 1653–1661 (1994)

276. J.G. O'Hara, P. Pillay, H.-K. Xu, Iterative approaches to finding nearest common fixed points of nonexpansive mappings in Hilbert spaces. Nonlinear Anal. **54**, 1417–1426 (2003)

277. S.O. Oko, Surrogate methods for linear inequalities. J. Optim. Theor. Appl. **72**, 247–268 (1992)

278. Z. Opial, Nonexpansive and Monotone Mappings in Banach Spaces. Lecture Notes 67-1, Center for Dynamical Systems, Brown University, Providence, RI, 1967

279. Z. Opial, Weak convergence of the sequence of successive approximations for nonexpansive mappings. Bull. Am. Math. Soc. **73**, 591–597 (1967)

280. S.C. Park, M.K. Park, M.G. Kang, Super-resolution image reconstruction: A technical overview. IEEE Signal Process. Mag. **20**, 21–36 (2003)

281. J. Park, D.C. Park, R.J. Marks, M. El-Sharkawi, Recovery of image blocks using the method of alternating projections. IEEE Trans. Image Process. **14**, 461–474 (2005)

282. S.N. Penfold, R.W. Schulte, Y. Censor, A.B. Rosenfeld, Total variation superiorization schemes in proton computed tomography image reconstruction. Med. Phys. **37**, 5887–5895 (2010)

283. W.V. Petryshyn, T.E. Williamson Jr., Strong and weak convergence of the sequence of successive approximations for quasi-nonexpansive mappings. J. Math. Anal. Appl. **43**, 459–497 (1973)

284. G. Pierra, Decomposition through formalization in a product space. Math. Program. **28**, 96–115 (1984)

285. C. Popa, Least-squared solution of overdetermined inconsistent linear systems using Kaczmarz's relaxation. Int. J. Comp. Math. **55**, 79–89 (1995)

286. C. Popa, Extensions of block-projections methods with relaxation parameters to inconsistent and rank-deficient least-squares problems. BIT **38**, 151–176 (1998)

287. C. Popa, R. Zdunek, Kaczmarz extended algorithm for tomographic image reconstruction from limited data. Math. Comput. Simulat. **65**, 579–598 (2004)

288. S. Prasad, Generalized array pattern synthesis by the method of alternating orthogonal projections. IEEE Trans. Antenn. Propag. **28**, 328–332 (1980)

289. J.L. Prince, A.S. Willsky, A geometric projection-space reconstruction algorithm. Lin. Algebra Appl. **130**, 151–191 (1990)

290. E. Pustylnik, S. Reich, A.J. Zaslavski, Convergence of infinite products of nonexpansive operators in Hilbert space. J. Nonlinear Convex Anal. **11**, 461–474 (2010)

291. E. Pustylnik, S. Reich, A.J. Zaslavski, Convergence of non-cyclic infinite products of operators. J. Math. Anal. Appl. **380**, 759–767 (2011)

292. B. Qu, N. Xiu, A note on the CQ algorithm for the split feasibility problem. Inverse Probl. **21**, 1655–1665 (2005)

293. B. Qu, N. Xiu, A new halfspace-relaxation projection method for the split feasibility problem. Lin. Algebra Appl. **428**, 1218–1229 (2008)

294. S. Reich, Weak convergence theorems for nonexpansive mappings in Banach spaces. J. Math. Anal. Appl. **67**, 274–276 (1979)

295. S. Reich, A limit theorem for projections. Lin. Multilinear Algebra **13**, 281–290 (1983)

296. S. Reich and I. Shafrir, The asymptotic behavior of firmly nonexpansive mappings. Proc. Am. Math. Soc. **101**, 246–250 (1987)

297. S. Reich, A.J. Zaslavski, Attracting mappings in Banach and hyperbolic spaces. J. Math. Anal. Appl. **253**, 250–268 (2001)

298. R.T. Rockafellar, *Convex Analysis* (Princeton University Press, Princeton, 1970)

299. R.T. Rockafellar, Monotone operators and the proximal point algorithm. SIAM J. Contr. Optim. **14**, 877–898 (1976)

300. W. Rudin, *Functional Analysis*, 2nd edn. (McGraw-Hill, New York, 1991); Polish translation: *Analiza funkcjonalna* (PWN, Warszawa, 2002)

301. A.A. Samsonov, E.G. Kholmovski, D.L. Parker, C.R. Johnson, POCSENSE: POCS-based reconstruction for sensitivity encoded magnetic resonance imaging. Magn. Reson. Med. **52**, 1397–1406 (2004)

302. J. Schauder, Der Fixpunktsatz in Funktionalräumen. Studia Math. **2**, 171–180 (1930)

303. D. Schott, A general iterative scheme with applications to convex optimization and related fields. Optimization **22**, 885–902 (1991)

304. H.D. Scolnik, N. Echebest, M.T. Guardarucci, M.C. Vacchino, A class of optimized row projection methods for solving large nonsymmetric linear systems. Appl. Numer. Math. **41**, 499–513 (2002)

305. H.D. Scolnik, N. Echebest, M.T. Guardarucci, M.C. Vacchino, Acceleration scheme for parallel projected aggregation methods for solving large linear systems. Ann. Oper. Res. **117**, 95–115 (2002)
306. H.D. Scolnik, N. Echebest, M.T. Guardarucci, M.C. Vacchino, Incomplete oblique projections for solving large inconsistent linear systems. Math. Program. **111**, 273–300 (2008)
307. A. Segal, Directed Operators for Common Fixed Point Problems and Convex Programming Problems, Ph.D. Thesis, University of Haifa, Haifa, Israel, 2008
308. H.F. Senter, W.G. Dotson Jr., Approximating fixed points of nonexpansive mappings. Proc. Am. Math. Soc. **44**, 375–380 (1974)
309. A. Serbes, L. Durak, Optimum signal and image recovery by the method of alternating projections in fractional Fourier domains. Comm. Nonlinear Sci. Numer. Simulat. **15**, 675–689 (2010)
310. N.T. Shaked, J. Rosen, Multiple-viewpoint projection holograms synthesized by spatially incoherent correlation with broadband functions. J. Opt. Soc. Am. A **25**, 2129–2138 (2008)
311. G. Sharma, Set theoretic estimation for problems in subtractive color. Color Res. Appl. **25**, 333–348 (2000)
312. K.K. Sharma, S.D. Joshi, Extrapolation of signals using the method of alternating projections in fractional Fourier domains. Signal Image Video Process. **2**, 177–182 (2008)
313. K.T. Smith, D.C. Solman, S.L. Wagner, Practical and mathematical aspects of the problem of reconstructing objects from radiographs. Bull. Am. Math. Soc. **83**, 1227–1270 (1977)
314. R.A. Soni, K.A. Gallivan, W.K. Jenkins, Low-complexity data reusing methods in adaptive filtering. IEEE Trans. Signal Process. **52**, 394–405 (2004)
315. H. Stark, P. Oskoui, High resolution image recovery from image-plane arrays, using convex projections. J. Opt. Soc. Am. A **6**, 1715–1726 (1989)
316. H. Stark, Y. Yang, *Vector Space Projections. A Numerical Approach to Signal and Image Processing, Neural Nets and Optics* (Wiley, New York, 1998)
317. J. Stoer, R. Bulirsch, *Introduction to Numerical Analysis*, 3rd edn. (Springer, New York, 2002)
318. C. Sudsukh, Strong convergence theorems for fixed point problems, equilibrium problems and applications. Int. J. Math. Anal. (Ruse) **3**, 1867–1880 (2009)
319. S. Świerczkowski, A model of following. J. Math. Anal. Appl. **222**, 547–561 (1998)
320. W. Takahashi, Y. Takeuchi, R. Kubota, Strong convergence theorems by hybrid methods for families of nonexpansive mappings in Hilbert spaces. J. Math. Anal. Appl. **341**, 276–286 (2008)
321. K. Tanabe, Projection method for solving a singular system of linear equations and its applications. Numer. Math. **17**, 203–214 (1971)
322. K. Tanabe, Characterization of linear stationary iterative processes for solving a singular system of linear equations. Numer. Math. **22**, 349–359 (1974)
323. G. Tetzlaff, K. Arnold, A. Raabe, A. Ziemann, Observations of area averaged near-surface wind- and temperature-fields in real terrain using acoustic travel time tomography. Meteorologische Zeitschrift **11**, 273–283 (2002)
324. Ch. Thieke, T. Bortfeld, A. Niemierko, S. Nill, From physical dose constraints to equivalent uniform dose constraints in inverse radiotherapy planning. Med. Phys. **30**, 2332–2339 (2003)
325. J. van Tiel, *Convex Analysis, An Introductory Text* (Wiley, Chichester, 1984)
326. M.J. Todd, *Some Remarks on the Relaxation Method for Linear Inequalities*, Technical Report, vol. 419, Cornell University, Cornell, Ithaca, 1979
327. Ph.L. Toint, Global convergence of a class of trust region methods for nonconvex minimization in Hilbert space. IMA J. Numer. Anal. **8**, 231–252 (1988)
328. W. Treimer, U. Feye-Treimer, Two dimensional reconstruction of small angle scattering patterns from rocking curves. Physica B **241–243**, 1228–1230 (1998)
329. J.A. Tropp, I.S. Dhillon, R.W. Heath, T. Strohmer, Designing structured tight frames via an alternating projection method. IEEE Trans. Inform. Theor. **51**, 188–209 (2005)
330. M.R. Trummer, SMART – an algorithm for reconstructing pictures from projections. J. Appl. Math. Phys. **34**, 746–753 (1983)

331. P. Tseng, On the convergence of the products of firmly nonexpansive mappings. SIAM J. Optim. **2**, 425–434 (1992)
332. A. Van der Sluis, H.A. Van der Vorst, in *Numerical Solution of Large Sparse Linear Algebraic Systems Arising from Tomographic Problems*, ed. by G. Nolet. Seismic Tomography (Reidel, Dordrecht, 1987)
333. V.V. Vasin, A.L. Ageev, *Ill-Posed Problems with A Priori Information* (VSP, Utrecht, 1995)
334. S. Webb, *Intensity Modulated Radiation Therapy* (Institute of Physics Publishing, Bristol, 2001)
335. S. Webb, *The Physics of Conformal Radiotherapy* (Institute of Physics Publishing, Bristol, 2001)
336. R. Webster, *Convexity* (Oxford University Press, Oxford, 1994)
337. R. Wegmann, Conformal mapping by the method of alternating projections. Numer. Math. **56**, 291–307 (1989)
338. R. Wittmann, Approximation of fixed points of nonexpansive mappings. Arch. Math. **58**, 486–491 (1992)
339. P. Wolfe, Finding the nearest point in a polytope. Math. Program. **11**, 128–149 (1976)
340. B.J. van Wyk, M.A. van Wyk, A POCS-based graph matching algorithm. IEEE Trans. Pattern Anal. Mach. Intell. **26**, 1526–1530 (2004)
341. Y. Xiao, Y. Censor, D. Michalski, J.M. Galvin, The least-intensity feasible solution for aperture-based inverse planning in radiation therapy. Ann. Oper. Res. **119**, 183–203 (2003)
342. H.-K. Xu, Iterative algorithms for nonlinear operators. J. Lond. Math. Soc. **66**, 240–256 (2002)
343. H.-K. Xu, An iterative approach to quadratic optimization. J. Optim. Theor. Appl. **116**, 659–678 (2003)
344. H.-K. Xu, A variable Krasnosel'skiĭ-Mann algorithm and the multiple-set split feasibility problem. Inverse Probl. **22**, 2021–2034 (2006)
345. I. Yamada, in *The Hybrid Steepest Descent Method for the Variational Inequality Problem Over the Intersection of Fixed Point Sets of Nonexpansive Mappings*, ed. by D. Butnariu, Y. Censor, S. Reich. Inherently Parallel Algorithms in Feasibility and Optimization and their Applications, Studies in Computational Mathematics, vol. 8 (Elsevier Science, Amsterdam, 2001), pp. 473–504
346. I. Yamada, N. Ogura, Hybrid steepest descent method for variational inequality problem over the fixed point set of certain quasi-nonexpansive mappings. Numer. Funct. Anal. Optim. **25**, 619–655 (2004)
347. I. Yamada, N. Ogura, Adaptive projected subgradient method for asymptotic minimization of sequence of nonnegative convex functions. Numer. Funct. Anal. Optim. **25**, 593–617 (2004)
348. I. Yamada, N. Ogura, N. Shirakawa, in *A Numerically Robust Hybrid Steepest Descent Method for the Convexly Constrained Generalized Inverse Problems*, ed. by Z. Nashed, O. Scherzer. Inverse Problems, Image Analysis and Medical Imaging, American Mathematical Society, Contemp. Math., vol. 313 (2002), pp. 269–305
349. I. Yamada, N. Ogura, Y. Yamashita, K. Sakaniwa, Quadratic optimization of fixed points of nonexpansive mappings in Hilbert space. Numer. Funct. Anal. Optim. **19**, 165–190 (1998)
350. Q.Z. Yang, *The relaxed CQ algorithm solving the split feasibility problem*. Inverse Probl. **20**, 1261–1266 (2004)
351. K. Yang, K.G. Murty, New iterative methods for linear inequalities. JOTA **72**, 163–185 (1992)
352. Q. Yang, J. Zhao, Generalized KM theorems and their applications. Inverse Probl. **22**, 833–844 (2006)
353. Y. Yao, Y.-C. Liou, Weak and strong convergence of Krasnoselski–Mann iteration for hierarchical fixed point problems. Inverse Probl. **24**, 015015, 8 (2008)
354. D. Youla, Generalized image restoration by the method of alternating orthogonal projections. IEEE Trans. Circ. Syst. **25**, 694–702 (1978)
355. M. Yukawa, I. Yamada, Pairwise optimal weight realization – Acceleration technique for set-theoretic adaptive parallel subgradient projection algorithm. IEEE Trans. Signal Process. **54**, 4557–4571 (2006)

356. M. Zaknoon, *Algorithmic Developments for the Convex Feasibility Problem*, Ph.D. Thesis, University of Haifa, Haifa, Israel, 2003
357. E.H. Zarantonello, in *Projections on Convex Sets in Hilbert Space and Spectral Theory*, ed. by E.H. Zarantonello. Contributions to Nonlinear Functional Analysis (Academic, New York, 1971), pp. 237–424
358. E. Zeidler, *Nonlinear Functional Analysis and Its Applications, III – Variational Methods and Optimization* (Springer, New York, 1985)
359. J. Zhang, A.K. Katsaggelos, in *Image Recovery Using the EM Algorithm*, ed. by V.K. Madisetti, D.B. Williams. Digital Signal Processing Handbook (CRC Press LLC, Boca Raton, 1999)
360. D.F. Zhao, The principles and practice of iterative alternating projection algorithm: Solution for non-LTE stellar atmospheric model with the method of linearized separation. Chin. Astron. Astrophys. **25**, 305–316 (2001)
361. J. Zhao, Q. Yang, Several solution methods for the split feasibility problem. Inverse Probl. **21**, 1791–1799 (2005)

Glossary of Symbols

$\mathcal{H}$	Hilbert space
$\langle x, y \rangle$	Inner product of $x, y \in \mathcal{H}$
$\|x\|$	Norm of $x \in \mathcal{H}$ induced by $\langle \cdot, \cdot \rangle$
$\langle x, y \rangle_G$	Inner product of $x, y \in \mathbb{R}^n$ induced by a positive definite matrix G
$\|x\|_G := \sqrt{\langle x, x \rangle_G}$	The norm of $x \in \mathbb{R}^n$ induced by $\langle \cdot, \cdot \rangle_G$
$\sphericalangle(x, y)$	Angle between nonzero vectors $x, y \in \mathcal{H}$
X	A nonempty closed convex subset of $\mathcal{H}$
$I := \{1, 2, ..., m\}$	Finite subset of indices
x_+, x_-	Positive and the negative part of $x \in \mathbb{R}^n$
$\mathbb{R}^n_+, \mathbb{R}{-}^n$	Nonnegative and the nonpositive orthant
Δ_m	Standard simplex in $\mathbb{R}^m$
$|J|$	The number of elements of a finite subset J
$V^\perp$	Subspace orthogonal to a subspace $V \subseteq \mathcal{H}$
$B(x, \rho)$	Ball with a centre x and radius $\rho > 0$
C'	Complement of a subset $C \subseteq \mathcal{H}$
bd C	Boundary of a subset $C \subseteq \mathcal{H}$
int C	Interior of a subset $C \subseteq \mathcal{H}$
cl C	Closure of a subset $C \subseteq \mathcal{H}$
Lin S	Linear subspace generated by $S \subseteq \mathcal{H}$
aff S	Affine subspace generated by $S \subseteq \mathcal{H}$
$H(a, \beta)$	Hyperplane $\{x \in \mathcal{H} : \langle a, x \rangle = \beta\}$
$H_-(a, \beta)$	Half-space $\{x \in \mathcal{H} : \langle a, x \rangle \leq \beta\}$
$H_+(a, \beta)$	Half-space $\{x \in \mathcal{H} : \langle a, x \rangle \geq \beta\}$
f_+, f_-	Positive and the negative part of a function f
$\text{Argmin}_{x \in X} f(x)$	Subset of minimizers of $f : X \to \mathbb{R}$
$\text{argmin}_{x \in X} f(x)$	Minimizer of $f : X \to \mathbb{R}$

A. Cegielski, *Iterative Methods for Fixed Point Problems in Hilbert Spaces*,
Lecture Notes in Mathematics 2057, DOI 10.1007/978-3-642-30901-4,
© Springer-Verlag Berlin Heidelberg 2012

$S(f, \alpha)$	Sublevel set of a function f at a level $\alpha \in \mathbb{R}$
epi f	Epigraph of a function f
$f'(x, s)$	Directional derivative of a function f at x in a direction s
Df, f', DT	Derivative of a function f or of an operator T
$\nabla f(x)$	Gradient of a function f at x
$\nabla^2 f(x)$	Hessian of a function f at x
diag v	Diagonal matrix with a vector v at the main diagonal
$A^\top$	Matrix transposed to a matrix A
A^+	Moore–Penrose pseudoinverse of a matrix A
cone S	Conical hull of a subset $S \subseteq \mathcal{H}$
conv S	Convex hull of a subset $S \subseteq \mathcal{H}$
ri C	Relative interior of a subset $C \subseteq \mathcal{H}$
C^*	Polar cone to $C \subseteq \mathcal{H}$
$N_C(x)$	Normal cone to a convex subset $C \subseteq \mathcal{H}$ at $x \in \mathcal{H}$
$T_C(x)$	Tangent cone to a convex subset $C \subseteq \mathcal{H}$ at $x \in \mathcal{H}$
$\partial f(x)$	Subdifferential of a function f at $x \in \mathcal{H}$
$g_f(x)$	Subgradient of a function f at $x \in \mathcal{H}$
P_C	Metric projection onto a subset $C \subseteq \mathcal{H}$
P_a	Metric projection onto Lin$\{a\}$, where $a \in \mathcal{H}$
T_λ	Relaxation of an operator T
Fix T	Subset of fixed points of an operator T
$d(\cdot, C)$	Distance function to a subset $C \subseteq \mathcal{H}$
$d(A, B)$	Distance between subsets $A, B \subseteq \mathcal{H}$
$L(\mathcal{H}_1, \mathcal{H}_2)$	Space of all bounded linear operators $A : \mathcal{H}_1 \to \mathcal{H}_2$
$\lambda_{\max}(A)$	Largest eigenvalue of a nonnegative operator $A : \mathcal{H} \to \mathcal{H}$
Fej T	See page 46
Sep T	See page 55

Glossary of Acronyms

A. Cegielski, *Iterative Methods for Fixed Point Problems in Hilbert Spaces*,
Lecture Notes in Mathematics 2057, DOI 10.1007/978-3-642-30901-4,
© Springer-Verlag Berlin Heidelberg 2012

Index

LECTURE NOTES IN MATHEMATICS

Ⓓ Springer

Edited by J.-M. Morel, B. Teissier; P.K. Maini

Editorial Policy (for the publication of monographs)

1. Lecture Notes aim to report new developments in all areas of mathematics and their applications - quickly, informally and at a high level. Mathematical texts analysing new developments in modelling and numerical simulation are welcome.

   Monograph manuscripts should be reasonably self-contained and rounded off. Thus they may, and often will, present not only results of the author but also related work by other people. They may be based on specialised lecture courses. Furthermore, the manuscripts should provide sufficient motivation, examples and applications. This clearly distinguishes Lecture Notes from journal articles or technical reports which normally are very concise. Articles intended for a journal but too long to be accepted by most journals, usually do not have this "lecture notes" character. For similar reasons it is unusual for doctoral theses to be accepted for the Lecture Notes series, though habilitation theses may be appropriate.

2. Manuscripts should be submitted either online at www.editorialmanager.com/lnm to Springer's mathematics editorial in Heidelberg, or to one of the series editors. In general, manuscripts will be sent out to 2 external referees for evaluation. If a decision cannot yet be reached on the basis of the first 2 reports, further referees may be contacted: The author will be informed of this. A final decision to publish can be made only on the basis of the complete manuscript, however a refereeing process leading to a preliminary decision can be based on a pre-final or incomplete manuscript. The strict minimum amount of material that will be considered should include a detailed outline describing the planned contents of each chapter, a bibliography and several sample chapters.

   Authors should be aware that incomplete or insufficiently close to final manuscripts almost always result in longer refereeing times and nevertheless unclear referees' recommendations, making further refereeing of a final draft necessary.

   Authors should also be aware that parallel submission of their manuscript to another publisher while under consideration for LNM will in general lead to immediate rejection.

3. Manuscripts should in general be submitted in English. Final manuscripts should contain at least 100 pages of mathematical text and should always include

   – a table of contents;
   – an informative introduction, with adequate motivation and perhaps some historical remarks: it should be accessible to a reader not intimately familiar with the topic treated;
   – a subject index: as a rule this is genuinely helpful for the reader.

   For evaluation purposes, manuscripts may be submitted in print or electronic form (print form is still preferred by most referees), in the latter case preferably as pdf- or zipped psfiles. Lecture Notes volumes are, as a rule, printed digitally from the authors' files. To ensure best results, authors are asked to use the LaTeX2e style files available from Springer's web-server at:

   ftp://ftp.springer.de/pub/tex/latex/svmonot1/ (for monographs) and
   ftp://ftp.springer.de/pub/tex/latex/svmultt1/ (for summer schools/tutorials).

Additional technical instructions, if necessary, are available on request from lnm@springer.com.

4. Careful preparation of the manuscripts will help keep production time short besides ensuring satisfactory appearance of the finished book in print and online. After acceptance of the manuscript authors will be asked to prepare the final LaTeX source files and also the corresponding dvi-, pdf- or zipped ps-file. The LaTeX source files are essential for producing the full-text online version of the book (see http://www.springerlink.com/openurl.asp?genre=journal&issn=0075-8434 for the existing online volumes of LNM). The actual production of a Lecture Notes volume takes approximately 12 weeks.

5. Authors receive a total of 50 free copies of their volume, but no royalties. They are entitled to a discount of 33.3 % on the price of Springer books purchased for their personal use, if ordering directly from Springer.

6. Commitment to publish is made by letter of intent rather than by signing a formal contract. Springer-Verlag secures the copyright for each volume. Authors are free to reuse material contained in their LNM volumes in later publications: a brief written (or e-mail) request for formal permission is sufficient.

Addresses:
Professor J.-M. Morel, CMLA,
École Normale Supérieure de Cachan,
61 Avenue du Président Wilson, 94235 Cachan Cedex, France
E-mail: morel@cmla.ens-cachan.fr

Professor B. Teissier, Institut Mathématique de Jussieu,
UMR 7586 du CNRS, Équipe "Géométrie et Dynamique",
175 rue du Chevaleret
75013 Paris, France
E-mail: teissier@math.jussieu.fr

For the "Mathematical Biosciences Subseries" of LNM:

Professor P. K. Maini, Center for Mathematical Biology,
Mathematical Institute, 24-29 St Giles,
Oxford OX1 3LP, UK
E-mail : maini@maths.ox.ac.uk

Springer, Mathematics Editorial, Tiergartenstr. 17,
69121 Heidelberg, Germany,
Tel.: +49 (6221) 4876-8259

Fax: +49 (6221) 4876-8259
E-mail: lnm@springer.com